Navigation, Control and Mission Planning Advances for Safe, Efficient and Autonomous Drones

Navigation, Control and Mission Planning Advances for Safe, Efficient and Autonomous Drones

Guest Editors

Nadjim Horri
Samir Khan
Vaios Lappas

MDPI

Basel • Beijing • Wuhan • Barcelona • Belgrade • Novi Sad • Cluj • Manchester

Guest Editors

Nadjim Horri
School of Future Transport Engineering
Coventry University
Coventry
UK

Samir Khan
Graduate School of Engineering
The University of Tokyo
Tokyo
Japan

Vaios Lappas
Department of Aerospace Science & Technology
National & Kapodistrian University of Athens
Athens
Greece

Editorial Office
MDPI AG
Grosspeteranlage 5
4052 Basel, Switzerland

This is a reprint of the Special Issue, published open access by the journal *Drones* (ISSN 2504-446X), freely accessible at: https://www.mdpi.com/journal/religions/special_issues/286348L728.

For citation purposes, cite each article independently as indicated on the article page online and as indicated below:

Lastname, A.A.; Lastname, B.B. Article Title. *Journal Name* **Year**, *Volume Number*, Page Range.

ISBN 978-3-7258-6171-2 (Hbk)
ISBN 978-3-7258-6172-9 (PDF)
https://doi.org/10.3390/books978-3-7258-6172-9

Contents

About the Editors

Nadjim Horri

Nadjim Horri is a lecturer (Assistant Professor) in aerospace control at University of Leicester where he is a member of the aerospace research group, an associate member of the green energy and transportation group and works closely with Space Park Leicester. His research interests include the guidance, navigation and control of autonomous vehicles from aircraft to spacecraft and automotive applications. This includes onboard vehicle parameter estimation, aircraft fault estimation and recovery, adaptive and robust aircraft control, and mission planning for unmanned aircraft. Before joining Leicester, Nadjim was at Coventry University where he led the path following control part of the InnovateUK Assured CAV Parking project with Horiba Mira and delivered consultancies on hybrid UAV avionics design for AIE-UK and on flight simulation for a tailplane stall flight incident investigation for AIBN Norway. During his time at the Surrey Space Centre, he led the control for the AAReST reconfigurable space telescope mission and for STRaND-2, a formation of two nanosatellites.

Samir Khan

Samir Khan is an associate professor at the University of Tokyo. He has developed machine learning solutions for aeronautical systems. He has developed research on fault analysis including no fault found scenarios, condition monitoring from track-side feedback sensors, IoT technologies development, analytics platforms for maintenance decision-making and intelligent monitoring of intermittent failures and false alarms in electronic systems. He is a chartered engineer, and a member of IEEE and IET.

Vaios Lappas

Vaios Lappas is a Professor in the Department of Aerospace Science and Technology (NKUA) at the University of Athens. He has a Bachelor in Aeronautical Engineering from Ryerson University (Toronto, Canada) and MSc from International Space University (Strasbourg, France). He was previously a professor at University of Surrey. He participated as Principal Investigator and Co-Investigator in research programs and space missions with funding by Airbus Defense and Space, European Space Agency and NASA. He participates as a national representative in the field of control and navigation systems to the European Defense Agency. His research interests are focused on Aerospace systems and technology.

Preface

Drones have considerably evolved in the last two decades across the air, ground and marine domains, with an increased emphasis on their safety, autonomy and performance. In the UK, advancing drone autonomy for commercial aerial vehicles was the main focus of a policy paper jointly developed by the UK government and the sector. More generally, recent research is increasingly focused on the challenges associated with vehicle motion planning, navigation and control to perform increasingly demanding tasks in challenging environments. Recent advances in artificial intelligence, including deep and reinforcement learning present new opportunities in this field but concerns remain due to the need to ensure safety with trustworthy and explainable solutions. This Reprint presents some of the latest advances in path planning, navigation and control for autonomous ground, aerial and marine vehicles in challenging or uncertain environments.

Nadjim Horri, Samir Khan, and Vaios Lappas
Guest Editors

drones

MDPI

Article

An Improved Probabilistic Roadmap Planning Method for Safe Indoor Flights of Unmanned Aerial Vehicles

Qingeng Jin, Qingwu Hu *, Pengcheng Zhao, Shaohua Wang * and Mingyao Ai

School of Remote Sensing and Information Engineering, Wuhan University, Wuhan 430079, China
* Correspondence: huqw@whu.edu.cn (Q.H.); shwang@whu.edu.cn (S.W.)

Abstract: Unmanned aerial vehicles (UAVs) have been widely used in industry and daily life, where safety is the primary consideration, resulting in their use in open outdoor environments, which are wider than complex indoor environments. However, the demand is growing for deploying UAVs indoors for specific tasks such as inspection, supervision, transportation, and management. To broaden indoor applications while ensuring safety, the quadrotor is notable for its motion flexibility, particularly in the vertical direction. In this study, we developed an improved probabilistic roadmap (PRM) planning method for safe indoor flights based on the assumption of a quadrotor model UAV. First, to represent and model a 3D environment, we generated a reduced-dimensional map using a point cloud projection method. Second, to deploy UAV indoor missions and ensure safety, we improved the PRM planning method and obtained a collision-free flight path for the UAV. Lastly, to optimize the overall mission, we performed postprocessing optimization on the path, avoiding redundant flights. We conducted experiments to validate the effectiveness and efficiency of the proposed method on both desktop and onboard PC, in terms of path-finding success rate, planning time, and path length. The results showed that our method ensures safe indoor UAV flights while significantly improving computational efficiency.

Keywords: indoor environment; point cloud; unmanned aerial vehicle; path planning and optimization; probabilistic roadmap

1. Introduction

Small unmanned aerial vehicles (UAVs) have considerably evolved and are increasingly applied in many fields, such as agriculture [1,2], monitoring [3,4], transportation [5,6], delivery [7,8], and rescue [9,10], necessitating additional research on the use of UAVs for mobile robotics, photogrammetry, and monitoring, to name but a few. The core advantage of using UAVs is that they can operate and execute missions in hazardous and dangerous situations. However, certain safety challenges must be considered when integrating UAVs, including attention cost, psychological impact, and physical risks [11]. As such, safety is the most-debated topic of designing and using UAVs [12]. The reason why UAVs are not yet safe, especially indoors, is that they still have flaws such as poor environmental perception and low strain capacity, which means that during autonomous UAV flight in complex environments, safety hazards cannot be completely avoided, posing potential threats to life and property.

Different UAV models, each with their own unique traits, are appropriate for different application scenarios. Among the different models of the UAV, the quadrotor has been extensively developed, researched, and applied over time. The notable quadrotor advantages are its flexibility, adaptivity, and ease of construction [13]. The quadrotor is an aircraft with four rotors with associated propellers, which is capable of hovering, jerking, vertical takeoff and landing, and horizontal flight. Numerous quadrotors are being fabricated for academic research or commercial use, e.g., Pixhawk [14], DraganFlyer X4 [15], and DJI M300 [16], and the boundaries of modeling theories [17–20] and control methods [21–24] are being advanced. Given that we mainly focused on UAV path planning rather than UAV control in this study, we did not consider the aspect of UAV control. However, several autopilot products or software [25–27] cover the underlying control of UAVs, allowing researchers to perform and evaluate the designed planning methods on supported platforms.

Despite most UAVs being used in open and spacious outdoor situations rather than cramped and unpredictable indoor situations, demand has been growing for deploying UAVs indoors for specific missions, including manufacturing, inspection, and service [28]. Unlike executing a flight outdoors, UAVs cannot use the global navigation satellite system (GNSS) as a location mechanism when indoors. As a result, indoor UAV localization systems typically use a visual camera [29], laser [30], wireless communication [31], or motion capture systems [32] as alternative techniques to assist with indoor tasks. Recently, systems based on new technology such as the Internet of Things (IoT) [33], deep-learning-based monocular cameras [34], and optimal mass flow sensors [35] have been proposed. Additionally, some advanced systems have been developed using more conventional sensors providing solutions for indoor UAV applications, including camera-based tracking systems [36], light detection and ranging (LiDAR)-based systems [37], and multisensor fusion systems [38].

To ensure the safety of UAV mission planning and autonomous flight, especially in indoor environments, UAV systems need to combine local environment awareness with global cognition. By creating a map of the indoor environment, the UAV can realize global cognition of the environment in which it is located, also providing the basis for a series of subsequent operations such as UAV path planning, mission configuration, and autonomous flight. Conversely, when relying solely on the UAV forming local perception of the surrounding environment, only emergency operations can be executed, such as obstacle avoidance, thus passively rather than actively ensuring safe UAV flights at the planning level. To allow UAVs to better understand indoor environments, a commonly used option is the application of payload sensors, where LiDAR provides benefits in illumination tolerance and object structure description. With a point cloud of the scenario as the only output, subsequent processes, such as environment modeling and path planning, can be performed. Based on this, numerous indoor UAV tasks can be accomplished, e.g., non-GPS navigation, interior inspection, and autonomous delivery.

The environmental awareness and cognition of a UAV enable UAV control at the planning level. To complete a UAV flight mission, the first step is deciding where to travel and stop, followed by controls to maintain proper movement during these tasks. Additionally, the UAV needs to be resilient to unforeseen circumstances, especially dynamic obstacles. The latter tasks were outside our scope in this study: our focus was on the process of creating a global traversal path for a UAV in an environment, i.e., path or route planning.

Environment modeling and path search are two major components of path planning. Many methods have been proposed by scholars in these fields, the most important of which are geometric, topological, and cell decomposition methods. The geometric method abstracts the elements of the environment using geometric shapes, such as a convex region combined with polynomial forms [39]. Although these methods can intuitively describe the surroundings, their computational volume noticeably increases when the obstacles are dense and the layout is complicated; additionally, dealing with irregularly shaped obstacles is challenging. With the topological methods, the environment is abstracted as a graph structure, which considerably simplifies the environment and increases path search efficiency, but creating a topological graph is complicated and time-consuming. The core idea behind the cell decomposition method is to divide a large environment into small cells using a simple environment discretization method, which is easy to compute and applies well to situations with many obstacles and complex morphology, reducing the complexity of the environment model and increasing planning efficiency. The grid map [40] is a typical cell decomposition method used in environment modeling.

The purpose of path search is to use specific algorithms to find a path in the environment so that the predetermined performance function returns an optimal value. According to the basic principle, path search algorithms can be divided into three types: traditional, sample-based, and simulation bionic methods.

Traditional methods were first proposed, but have since been enhanced. Dijkstra's algorithm [41], Floyd–Warshall algorithm [42], A* [43], D* [44], and artificial potential field [45] are a few examples. They are typically simple to calculate and easy to implement, which laid the foundation for resolving the path planning problem. Li et al. [46] proposed a universal path planning method for UAVs that can solve the shortest safe path and a safe least-cost path. González et al. [47] proposed an indoor path-planning algorithm for UAV-based contact inspection that can calculate a path in a few milliseconds. Xu et al. [48] designed a resilient, enhanced algorithm for UAV 3D route planning based on the A* and artificial potential fields algorithms, overcoming the insufficient anti-disturbance ability of the traditional route planning algorithm.

Simulation bionic methods are inspired by natural entities or events, and they achieve their goal by simulating and grouping individual behaviors and interactions, providing a new possibility for solving complex problems. Typical methods include the genetic algorithm [49], ant colony algorithm [50], memetic algorithm [51], etc. The simulation bionic methods are practical for UAV path planning. For example, Liu et al. [52] designed a modified sparrow search algorithm and increased the efficiency in solving the UAV route planning problem. Wang et al. [53] introduced a modified mayfly algorithm, which performed well in the UAV route planning problem. Li et al. [54] combined cellular automata with the spanning tree algorithm to construct a route network in low-altitude airspace, providing a solution for the distribution of logistics UAV. However, although the concept of the simulation bionic methods can be intuitively understood, it has disadvantages in terms of the complexity of description and difficulty in modeling, which do not meet the time requirement and resource limitations of the onboard system.

Sample-based methods involve sampling subgoals in the configuration space to extend the search until the final goal is reached. Rapidly exploring random tree (RRT) [55] and probabilistic roadmap (PRM) [56] are two typical sample-based methods. RRT employs spatial sampling to determine the expansion direction of the search tree. This method can perform fast searches and is adaptable in high-dimensional space; however, it is not well-suited for environments with narrow spaces, and it does not guarantee an optimal path result. PRM, on the contrary, samples available spatial positions serving as nodes in a route network, from which a path search algorithm is used to solve an optimal path from beginning to end. This method converts the path search problem from an entire continuous space to a discrete graph with nodes and edges, making it applicable to path planning problems with high-dimensional and complex constraints. However, because this method is only probabilistically complete, the path result may not be optimal among all possible sample networks.

RRT is more suitable for exploring environments without prior information, whereas PRM is more suitable for selecting a better path in an environment with prior information, so is more suitable for applications that consider the safety of UAV flight. In 2004, Pettersson et al. [57] used the PRM algorithm for an autonomous UAV. Methods based on PRM have been continuously proposed and tend to be combined with other methods. Chen et al. [58] proposed an improved probabilistic roadmap with a potential field function for a quadrotor UAV. Mohanta et al. [59] proposed a knowledge-based fuzzy-probabilistic roadmap for mobile robot navigation. Combining PRM with the A* algorithm is feasible [60], and it works well with the ant colony algorithm [61].

With the development of path planning methods, we have been constantly striving to increase operational effectiveness, reduce computing time, and maintain an optimal path result, to better meet the requirements of safe UAV flight. The sample-based method for path planning achieves an appropriate balance between effectiveness and efficiency. First, this method is well-suited for cell decomposition modeling, where the grid map is a common, mature, and easily used form to describe the environment. Second, as long as any interrelationship exists between cells in the map, it is capable of further simplifying the map by abstracting it into a graph composed of nodes and edges. Lastly, its path search shows high-quality performance in terms of computational efficiency and path result.

In this study, we focused on improving the computational efficiency of path planning, because it determines whether a UAV can complete the planning and execute an autonomous flight in real-time, i.e., in a few seconds. Due to the fact that indoor environments may be compact and complex, as well as sometimes dynamic and unpredictable, a UAV should be able to finish planning as quickly as possible to handle emergency situations. Based on a typical quadrotor model, we designed an indoor environment reduced-dimensional modeling method that employs point cloud projection to create a downscaled raster map of an indoor environment, reducing the indoor space from 3D to 2D while retaining necessary environmental information such as boundaries and obstacles. We used an adjacency relationship of the grids in a raster map to represent the spatial location relationship in 3D indoor space, thus markedly simplifying the environment. We combined several 2D maps into a multilayer map to produce an improved path-solving result in a complex environment where a single 2D map is not enough to effectively describe the actual situation. Furthermore, we developed an improved PRM planning method, which is an exploratory path search method that converts the path search in indoor environments into a graph search based on sampled nodes. Although the obtained paths may not be the shortest in length due to the sampling randomness, the search capability of the algorithm is remarkably improved, and solving for feasible paths in complex indoor environments is easier. The results of experiments showed that the proposed method substantially reduces the planning time compared with that of the basic PRM algorithm, and it performs well even on a resource-limited computing platform, whereas the postprocessing optimization of the generated paths further improves path quality to meet real-world requirements regarding the timely generation of autonomous UAV flight paths, thereby ensuring UAV flight safety.

2. Generation of Reduced-Dimensional Raster Map Based on Point Cloud Projection

A quadrotor UAV has flexible 3D mobility, i.e., loose constraints on vertical and horizonal motions, which considerably facilitates describing, representing, and modeling an environment. Therefore, we modeled an indoor environment by a point cloud projection method; then, we generated reduced-dimensional raster maps to represent various altitude ranges of the environment, based on which we designed and implemented an improved probabilistic roadmap planning method to obtain mission paths for the UAV. Additionally, we optimized the path by post-process to account for the efficiency and UAV flight safety. An overview of the method workflow is shown in Figure 1.

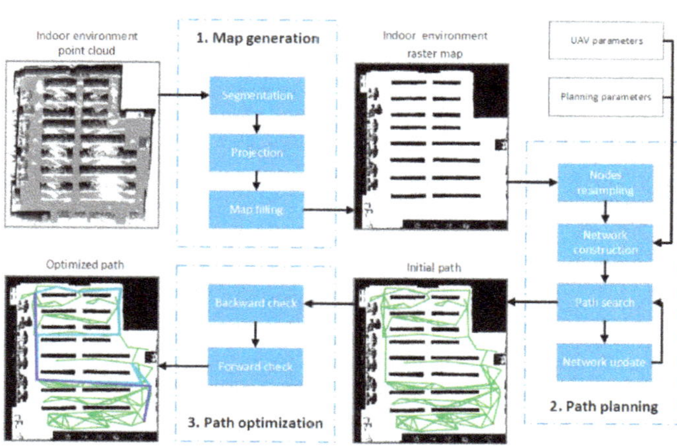

Figure 1. Overview of method workflow.

2.1. Kinematic and Dynamic UAV Model Assumptions

A quadrotor UAV is composed of four rotors attached to the ends of four arms by a symmetric frame. As the direct power source of flight, the rotors can adjust each spinning speed to change the lift force generated by the attached propellers, allowing for flexible horizontal and vertical movements, constant motion, or relative stillness. The control system of a quadrotor is an underactuated system, with six degrees of freedom outputs (three translational motions and three rotational motions) controlled by only four inputs (the spinning speed of four rotors).

In this study, we selected the quadrotor as the assumed type of UAV in the modeling and planning. However, the actual type of the UAV was not our main concern as the current autopilot products and software provide good encapsulation and integration of underlying executions of the UAV, which do not require complex user control. The main reason why we used the quadrotor as the kinematic and dynamic model is that it is capable of flexible mobility in both the vertical and horizontal directions, especially hovering and jerking, which is highly automated by the autopilot.

Some assumptions are required to properly introduce the kinematic model of a quadrotor UAV. We assume that it has a symmetric and rigid structure, with propellers of equal height on the rotors, and the mass center is the same as the space center of the UAV.

We first define two coordinate systems: body inertial frame $O_b X_b Y_b Z_b$ and fix inertial frame $O_f X_f Y_f Z_f$. The center O_b is defined the same as the mass center of the UAV, with the forward and upward directions of the UAV being the x-direction and z-direction, respectively; the y-direction is determined by the right-hand rule. For simplicity, the fixed inertial frame has the same definition as the first body inertial frame of the UAV, and it does not change once determined.

As such, the UAV state q^T in the environment can be described as

$$q^T = \{x, y, z, \psi, \theta, \phi\}, \tag{1}$$

where (x, y, z) denotes the position of the center of the quadrotor in a fixed frame, and (ψ, θ, ϕ) denotes the orientation of the quadrotor in the body frame represented in Euler angles (yaw, pitch, and roll), which can be further transformed to the fixed frame by

$$R_b^f = \begin{bmatrix} c(\psi)c(\theta) & c(\psi)s(\theta)s(\phi) - s(\psi)c(\phi) & c(\psi)s(\theta)c(\phi) + s(\psi)s(\phi) \\ s(\psi)c(\theta) & s(\psi)s(\theta)s(\phi) + c(\psi)c(\phi) & s(\psi)s(\theta)c(\phi) - c(\psi)s(\phi) \\ -s(\theta) & c(\theta)s(\phi) & c(\theta)c(\phi) \end{bmatrix}, \tag{2}$$

where c() and s() denote cos and sin operators, respectively.

To further simplify the dynamic model for ease of implementation, we overlook pitch and roll controls for now, because they strongly impact the flight stability of the UAV, and delegating these controls to the autopilot would be preferable. Translational and yaw controls are necessary, as the former is used to change the spatial position of the UAV, while the latter is used to adjust the heading direction. Thus, the following control states remain:

$$\dot{q}^T = \left\{\dot{x}, \dot{y}, \dot{z}, \dot{\psi}\right\}, \tag{3}$$

where $\{\dot{x}, \dot{y}, \dot{z}\}$ denotes the respective speeds with reference to the fixed frame, and $\dot{\psi}$ denotes the change rate of the yaw angle in the body frame.

These speed control parameters are inputs to the UAV kinematic model to keep the UAV on the resulting path solved at the planning level; they are also outputs of the UAV dynamic model, where traction and torque are inputs. However, because the implementation of dynamic modeling can be delegated to the autopilot and we focused more on planning methods than control methods for the UAV, we do not provide further discussion on this topic.

2.2. Indoor Environment Rasterization

In simple indoor environments, describing the inter-relationships of boundary surfaces and obstacle shapes is relatively simple, facilitating the vectorization of environmental elements. However, for complex indoor environments, the traditional vectorization method has limitations. For example, when a room is irregularly shaped, an increase in the number of walls causes the constraints of the boundary to become more complex, and more parameters must be added to the model to completely describe the entire environment, which also substantially affects the efficiency of modeling.

We implemented a downscaling modeling method for indoor environments based on point cloud projection that avoids the vectorization of environment elements and generates a reduced-dimensional raster map based on point cloud coordinate values. The method converts the original three-dimensional space to two-dimensional space and transforms the spatial location relationship between environment elements into the adjacency relationship between elements in the raster map, which considerably reduces the complexity of modeling, improves efficiency, and is more compatible.

The reduced-dimensional raster map consists of small elements called grids, each of which represents a specific size in space. They can be classified based on their values to distinguish boundaries, obstacles, and free space in the environment. Due to the fact that both the boundaries and obstacles are impassable on a map, they can be represented and grouped together as obstacle grids. In addition to obstacle grids, free grids exist in the map, which composes the entire set of map grids.

If we add an attribute and set a value of the free grid $value_{free} = class0$ and the obstacle grid $value_{obstacle} = class1$, we obtain a simple environment reduced-dimensional raster map, as shown in Figure 2, which is essentially a binary image with a size of $height * width$.

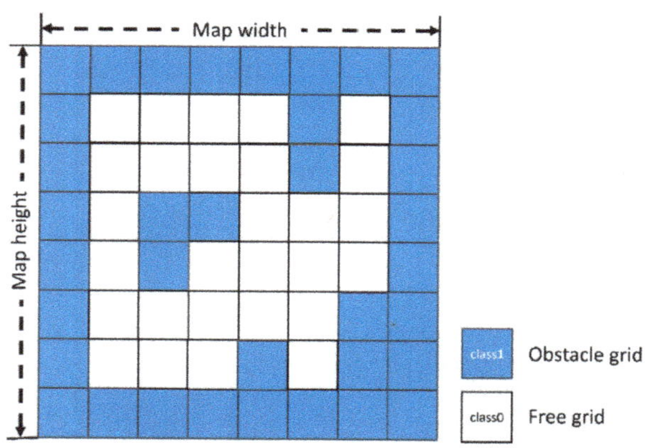

Figure 2. Environment reduced-dimensional raster map.

2.3. Indoor Environment Point Cloud Projection

To preserve the relationship between obstacles and free space as much as possible throughout the conversion of a 3D point cloud to a 2D map, we adopt the indoor environment point cloud projection method, which projects the target 3D point cloud onto a parametric model. In this study, we projected the 3D point cloud of the indoor environment onto a horizontal plane along the vertical direction to provide a vertical view that serves as a reduced-dimensional raster map model of this indoor environment.

In addition, because the UAV has a limited field of vision and concentrates on a specific region rather than a global one, only the points within a range close to the UAV are projected. Thus, we further determine the range for projection according to the detection distance of the UAV:

$$\begin{aligned} x_{min} &= x_{UAV} - d_{UAV},\ x_{max} = x_{UAV} + d_{UAV}, \\ y_{min} &= y_{UAV} - d_{UAV},\ y_{max} = y_{UAV} + d_{UAV}, \\ z_{min} &= z_{UAV} - h/2,\ z_{max} = z_{UAV} + h/2, \end{aligned} \tag{4}$$

where x_{min}, x_{max}, y_{min}, y_{max}, z_{min}, and z_{max} denote the projection range; $(x_{UAV}, y_{UAV}, z_{UAV})$ denotes the UAV position (or another specified center position of the focused point cloud); d_{UAV} denotes the detection range of the UAV and h denotes the specified range for altitude. All the variables are in the same local coordinates as the point cloud.

Any 3D point $p(x, y, z)$ in the point cloud satisfies

$$\begin{aligned} x_{min} &< x < x_{max}, \\ y_{min} &< y < y_{max}, \\ z_{min} &< z < z_{max}. \end{aligned} \tag{5}$$

The conversion from 3D point p to 2D map grid p' is as follows:

$$p' = \begin{bmatrix} x' \\ y' \end{bmatrix} = \begin{bmatrix} round(x - x_{min}) \\ round(y - y_{min}) \end{bmatrix} * s, \tag{6}$$

where $round()$ denotes the rounding sign to resample the 3D point to a 2D point and s denotes the map resolution scaling factor.

The reason why we chose a vertical projection is that indoor objects, in most cases, are vertically placed on the floor, and the free space also extends in the vertical direction. The vertical view is the most widely used map form in robot mapping and navigation applications, reflecting its usefulness, effectiveness, and representativeness.

However, the projection method may disregard the vertical structure of obstacles, particularly in complex environments. To overcome this issue, we further vary the z_{min} and z_{max} in the projection and construct grid maps representing the free space and obstacles at various altitudes. For example, a multilayer grid map $M\{m^1, m^2, \ldots, m^n\}$ consists of grid maps at various altitudes, and the ith map m^i is formed by the projection of point cloud P^i:

$$\begin{aligned} &\forall p^i(x^i, y^i, z^i) \in P^i, \\ &x_{min} < x^i < x_{max}, \\ &y_{min} < y^i < y_{max}, \\ &z^i_{min} < z^i < z^i_{max}, \end{aligned} \tag{7}$$

where z^i_{min} varies from $(z_{UAV} - nh/2)$ to $(z_{UAV} + nh/2)$ with a step size of h.

As a result, the UAV can search for a path not only in one single map but also by merging maps at various altitudes if necessary. In simple cases, a grid map of the near range of the UAV altitude is sufficient to solve a feasible path; in more complex cases, we first search for a path at the current altitudes, and if it is not feasible, we continue searching for subsequent paths in adjacent maps from where the initial search ended. As such, we achieve an appropriate balance between effectiveness and efficiency.

2.4. Indoor Environment Reduced-Dimensional Raster Map Generation

In the reduced-dimensional raster map of indoor environments, the map grid is divided into two categories: obstacle and free grids. The obstacle grids represent two types of environmental elements: boundaries and obstacles. Before constructing a map, the original point cloud of the indoor environment must be preprocessed to generate a usable map of the indoor environment.

Preprocessing commonly comprises the segmentation of floor and ceiling points, denoising, and other processes. Removal of the floor and ceiling points is necessary; otherwise, they obscure the location of the free space and cover the entire projection surface. We used random sample consensus (RANSAC) [62] to segment and extract the floor and ceiling points. RANSAC not only satisfied our segmentation requirements, but also provided us with the parameters of the extracted planes that could be used to determine the UAV altitude in the environment. Based on this, we divide the space into varying altitudes and generate a map at each altitude.

The process of generating the reduced-dimensional raster map is shown in Figure 3. More specifically, the procedure entails the following steps:

1. Extract the floor and ceiling points using point cloud segmentation, and remove them from the original point cloud. Segmentation range and height can be specified.
2. Calculate the maximum and minimum values of the remaining point clouds on the X and Y axes for the height and width of the map image, respectively. Scaling up of the image is optional to increase model accuracy.
3. Iteratively read the 3D position values of each point, and convert them into map grid coordinates by projection.
4. Repeat step 3 until all points have been traversed and obtain a binary image of the raster map.
5. Vary the specified height of projection to generate raster maps at various altitudes.

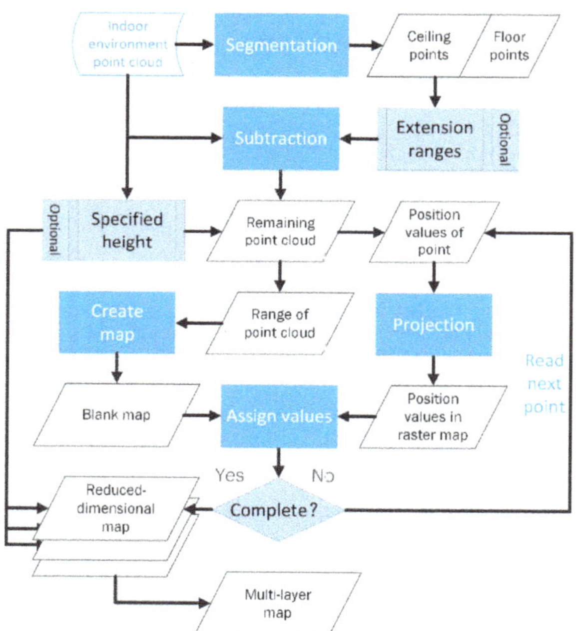

Figure 3. Workflow of the method used to generate reduced-dimensional raster map of indoor environment based on point cloud projection.

3. Improved Probabilistic Roadmap Planning for Safe UAV Flight

3.1. Basic PRM Algorithm

The PRM algorithm, which is essentially a graph-based path search method, is based on the fundamental concept of randomly generating sampling points in free space that serve as graph nodes. After verifying the connectivity of nodes and constructing a connection network, the PRM algorithm conducts a search and then solves a path from the source to the goal.

The PRM algorithm considerably simplifies the environment by discretizing the space into a graph, and is applicable to high-dimensional spaces with complex constraints. However, it is time-consuming and inefficient in network initialization. Additionally, its stability is restricted by the number of sampling nodes and their random locations. The algorithm is therefore probabilistically complete.

The workflow of the basic PRM algorithm mainly includes three parts: spatial sampling, edge generation, and path search. The pseudo-code for the basic PRM is shown in Figure 4. More specifically, the procedure entails the following steps:

(1) Define a node set N; add the source node n_{src} and the goal node n_{goal}.
(2) Generate a node n_{rand} by random sampling in the entire map.
(3) Perform a collision check on n_{rand}. If it passes, add n_{rand} to N; otherwise, return to step 2.
(4) Repeat steps 2 and 3 until M nodes in total have been generated, completing spatial sampling.
(5) Define an edge set E.
(6) Traverse n_m in N and select other nodes n_k to generate edge $e_{m,k}$; perform a collision check on it. If it passes, add it to E.
(7) Repeat step 6 until all M nodes have been traversed, completing edge generation.
(8) Define a graph $G(N, E)$ and deploy a path search algorithm to solve the shortest path P from n_{src} to n_{goal}, completing the path search.

Algorithm 1 Basic Probabilistic Roadmap. It takes two set of node coordinates determining source n_{src} and goal n_{goal} positions and number of sampling nodes M. The algorithm process includes three processes, SAMPLE(), CREATE_EDGES() and FIND_PATH().

SAMPLE(n_{src}, n_{goal}, M):

Input: Node coordinates of source n_{src} and goal n_{goal}, number of sampling node M.

N.add(n_{src}, n_{goal});
N.add(n_{goal});
while N.size() < M **do**
n_{rand} ← RANDOM_STATE();
if CHECK_COLLISION(n_{rand}) == COLLISION_FREE
N.add_node(n_{rand});
return N

CREATE_EDGES(N):

Input: A set of sampling nodes N.

for m = 1 **to** N.size() **do**
for k = 1 **to** N.size() **do**
if n_m != n_k
$e_{m,k}$ ← EDGE(n_m, n_k);
if CHECK_COLLISION ($e_{m,k}$) == COLLISION_FREE
E.add_edge($e_{m,k}$);
return E

FIND_PATH(N, E, n_{src}, n_{goal}):

Input: A set of sampling nodes N and a set of connection edges E, Node coordinates of source n_{src} and goal n_{goal}.

G.init(N, E);
P ← PATH_SOLVE(G, n_{src}, n_{goal});
return P

Figure 4. Pseudocode of basic PRM.

3.2. Improvement Strategies for PRM Algorithm

The basic PRM algorithm has disadvantages in terms of stability and efficiency.

Its insufficient stability is caused by its reliance on the number of sampling nodes and their random locations. When the number of randomly generated nodes in the space is small, or the distribution is unfavorably located, as shown in Figure 5a,c, it may fail to form a network connecting the source and goal, instead generating several disconnected local networks. Nevertheless, the PRM algorithm is probabilistically complete, which means that as long as the random nodes are distributed throughout the space, a feasible path must be found. Therefore, the stability issue can be mitigated by appropriately increasing the number of nodes according to the complexity of the actual indoor environment.

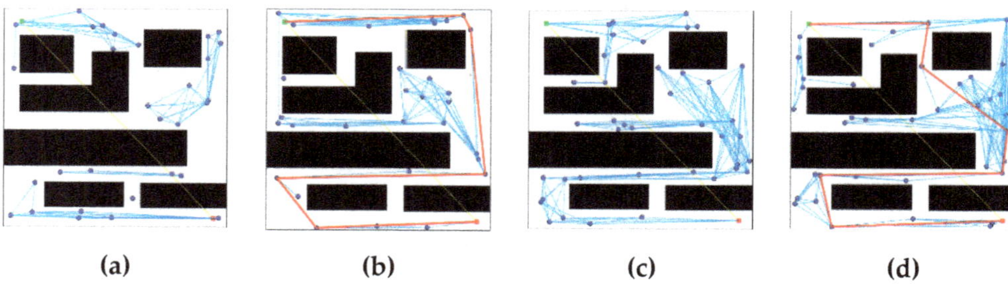

Figure 5. Node distribution under different node numbers (n): (**a**) $n = 30$, undesirable distribution; (**b**) $n = 30$, desirable distribution; (**c**) $n = 40$, undesirable distribution; (**d**) $n = 40$, desirable distribution.

The inefficiency is that some steps in the algorithm, particularly edge generation, are time-consuming. Each edge necessitates a collision check during generation to ensure a collision-free network. Furthermore, as the distance between nodes increases, the likelihood of obstacles between them increases, and generating a collision-free edge becomes more difficult. An effective solution is to reduce the number of collision checks and edge generations between distant nodes to improve the efficiency of the algorithm while having less impact on network connectivity.

To further reduce edge collision checks, we can adopt a strategy of constructing first and checking later, i.e., we do not perform the collision check on every pair of nodes in the process of edge generation after spatial sampling, but perform the collision check after solving a candidate path. Moreover, we eliminate the infeasible edges in the candidate path and find a new path that can reconnect the remaining edges. This strategy restricts the collision check of all edges to only the candidate path and its neighboring nodes and edges, therefore substantially lowering the number of collision checks and improving the efficiency of the algorithm.

3.3. Network Construction Based on Connection Distance

As the distance between nodes increases, the likelihood of obstacles between them increases, resulting in their invisibility and the impossibility of constructing collision-free edges. On the basis of this insight, we developed a method of network construction based on connection distance.

First, we set a "connection distance" parameter to determine whether to generate a connection edge between two nodes. During the edge generation process, the distance between each node is calculated when traversing each node to the other nodes. If it is above the threshold, the edge is not connected, and the subsequent collision check is skipped; otherwise, the collision check of the edge is performed again, and if it passes, a connected edge is generated between them.

The moderate connection distance is important. If the connection distance is too large, many colliding edges are still unnecessarily checked; however, if the connection distance is too small, network connectivity may be reduced or the network may become disconnected, which will affect the subsequent path search results, as shown in Figure 6.

This method improves the efficiency of the algorithm with little impact on the network connectivity by reducing the connection of nodes whose distance exceeds an acceptable threshold. Comparative networks with different connection distances *cdis* are shown in Figure 6, where the number of collision-free edges e_{free} and the number of colliding edges $e_{collided}$ are counted. As the connection distance increases, the number of colliding edges rapidly increases, while the number of collision-free edges slowly increases. This demonstrates that a proper connection distance can effectively reduce invalid checks for the colliding edges while assuring minimal disruption of network connectivity.

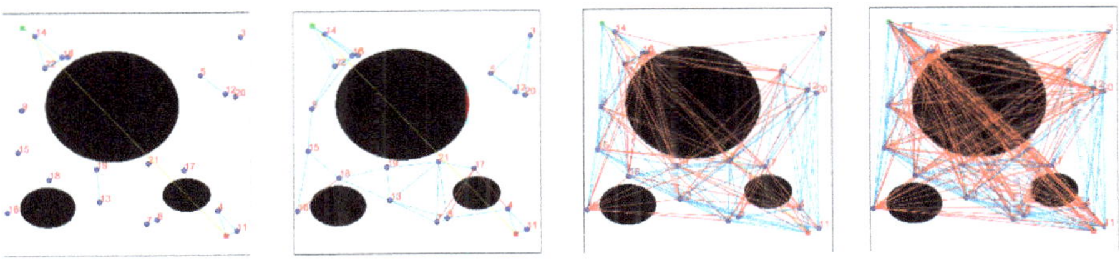

(a) $e_{free} = 13$, $e_{collided} = 0$. **(b)** $e_{free} = 36$, $e_{collided} = 4$. **(c)** $e_{free} = 82$, $e_{collided} = 69$. **(d)** $e_{free} = 92$, $e_{collided} = 139$.

Figure 6. Networks with connection distances of 0.1 (**a**), 0.2 (**b**), 0.5 (**c**), and 1 (**d**), consisting of collision-free edges (e_{free}) indicated by blue lines and colliding edges ($e_{collided}$) by red lines. With increasing connection distance, the number of colliding edges rapidly increases, while the number of collision-free edges slowly increases.

3.4. Path Local Check and Incremental Update

Based on the strategy of constructing first and checking later, we propose a method for path local checking and incremental updating.

After spatial sampling of the nodes, a network is constructed based on the connection distance, but no edge check is conducted at this stage. Hence, colliding edges may exist between the invisible nodes in the network, for which we then conduct a path search from the source to the goal for an initial path. On this path, we execute a minimum number of collision checks by incremental update. If an edge is collided, preventing direct passage between two nodes on the edge, we remove it from the network and search for a new path connecting the two nodes. These steps are repeated until all edges on the path pass the collision check, i.e., the entire path satisfies the no-collision requirement.

The workflow of the improved PRM algorithm mainly includes two parts: network initialization and path update. The pseudocode of the improved PRM is shown in Figure 7. More specifically, the procedure entails the following steps:

(1) Define a node set N; add the source node n_{src} and the goal node n_{goal}.
(2) Generate a node n_{rand} by random sampling in the entire map.
(3) Perform a collision check on n_{rand}. If it passes, add n_{rand} to N; otherwise, return to step 2.
(4) Repeat steps 2 and 3 until M nodes in total have been generated.
(5) Define an edge set E.
(6) Traverse n_m in N and select other nodes n_k to generate edge $e_{m,k}$; add it to E without performing collision checking.
(7) Repeat step 6 until all M nodes have been traversed, completing network initialization.
(8) Define a graph $G(N, E)$, and traverse a current node n_{cur} starting from n_{src}.
(9) Find the nearest neighbor n of n_{cur} and perform a collision check on $e_{n_{cur},n}$. If it passes, add n to result path P and move n_{cur} backward to n; otherwise, remove n from G and update the network.
(10) Repeat step 9 until reaching n_{goal}, completing path update.

The methods of local path check and incremental update essentially comprise a path search strategy that reduces ineffective collision checks on edges, decreasing the time required and increasing algorithm efficiency. The method has strong applicability in environments that are not extremely complex, and a feasible path can be quickly solved with minimal redundancy in simple environments. The procedure of local check and incremental update is shown in Figure 8. For colliding edges (red line) in the initial path, a new path (yellow lines) connecting two segments of the initial path is searched, which serves as a newly available local path for a final collision-free path (green lines).

Algorithm 2 Improved Probabilistic Roadmap. Similarly to the Basic Probabilistic Roadmap, it takes two sets of node coordinates determining source n_{src} and goal n_{goal} positions and number of sampling nodes M. The algorithm process includes **INITIALIZE_NETWORK()** and **UPDATE_PATH()**.

INITIALIZE_NETWORK(n_{src}, n_{goal}, M):

Input: Node coordinates of source n_{src} and goal n_{goal}, number of sampling node M.

N.add(n_{src}, n_{goal});
N.add(n_{goal});
while N.size() < M **do**
$n_{rand} \leftarrow$ RANDOM_STATE();
if CHECK_COLLISION(n_{rand}) == COLLISION_FREE
N.add_node(n_{rand});
for m = 1 **to** N.size() **do**
for k = 1 **to** N.size() **do**
if n_m != n_k
$e_{m,k} \leftarrow$ EDGE(n_m, n_k);
E.add_edge($e_{m,k}$);
return N, E

UPDATE_PATH(N, E, n_{src}, n_{goal}):

Input: A set of sampling nodes N and a set of connection edges E, Node coordinates of source n_{src} and goal n_{goal}.

G.init(N, E);
$n_{cur} \leftarrow n_{src}$;
while n != n_{goal} **do**
$n \leftarrow$ NEAREST_NEIGHBOR(G, n_{cur});
$e_{n_{cur},n} \leftarrow$ EDGE(n_{cur}, n);
if CHECK_COLLISION ($e_{n_{cur},n}$) != COLLISION_FREE
G.remove_edge($e_{n_{cur},n}$);
else
P.add_node(n);
$n_{cur} \leftarrow n$;
return P

Figure 7. Pseudocode of improved PRM.

3.5. Path Planning in Multilayer Grid Map

In more complex cases, a single map at a certain altitude might not be appropriate for solving a feasible path if obstacles are blocking the map and dividing it into several disconnected areas. Although areas in a single map may not be connected at the same altitude, they may be connected via another area at a different altitude. Therefore, we use a multilayer grid map for path planning.

Path search and update strategies in the multilayer map are quite similar to those in a single map; however, the essential distinction lies in how the transfer areas (overlapping areas of adjacent layers) are determined for the UAV to adjust its altitude.

To detect available transfer areas, we first use an image region-growing algorithm to identify and segment the disconnected areas in each single map. Due to the fact that the total number of the areas is uncertain, we randomly sample the growth seeds on the map If the growing region contains a sufficient number of grids, we record it as a valid area and then continue to sample a new seed and search other areas until all valid areas have been segmented. Furthermore, we examine the connectivity between areas in each layer and those in the adjacent layers. If two areas have grids with the same X and Y coordinates, they are regarded as connected, and the overlapping area formed by the grids is considered to be a transfer area.

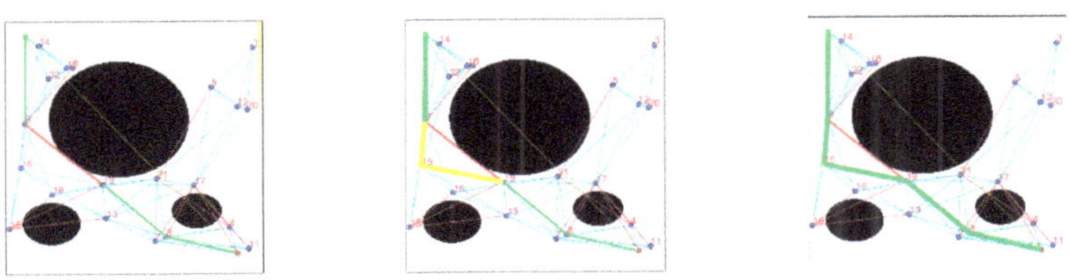

(a) Initial path **(b)** Check and update **(c)** Collision-free path

Figure 8. Path local check and incremental update: (**a**) an initial path contains collision-free edges (green lines) and a colliding edge (red line); (**b**) a new path (yellow lines) connecting two segments of the initial path is searched and updated; (**c**) a collision-free path is finally completed by adding the newly available local path.

For the path search process, we attempt to find a path from the source to the goal in a single map at the default altitude. If the path search fails, meaning that impassible obstacles may be located at this altitude, we search in its adjacent layers for any area that overlaps the current search area. If several overlapping areas exist, we use a greedy strategy to select the area with the smallest horizontal distance from the goal as the next search area. In the selected overlapping area, we additionally sample a transfer node for the UAV adjusting its altitude, which also serves as a temporary goal node of the current search area and the source node of the next search area. Thus, we accomplish the cross-layer path search for UAVs in complex indoor environments.

The pseudo-code of path planning in the multilayer map is shown in Figure 9. More specifically, the procedure entails the following steps:

Define $Map_{current}$ as the currently used map (at altitude h_{start} by default) and n_{sub_src} and n_{sub_goal} as the source and goal for the current search, respectively; they are initialized in n_{src} and n_{goal} at first.

(1) Define and initialize node set N as the network nodes in $Map_{current}$.
(2) Start the search for a path from n_{sub_src} to n_{sub_goal} on network N.
(3) If the search fails, find another map Map_{next} that has the smallest distance with n_{sub_goal} as the next map. Search the transfer area of the two maps and sample a transfer node $n_{transfer}$ in it, and set n_{sub_goal} to $n_{transfer}$. Start a new search in this configuration.
(4) If the search succeeds, record the path result in P, switch the map to Map_{next} and n_{sub_src} to $n_{transfer}$, and reset n_{sub_goal} to n_{goal}.
(5) Repeat steps 4 to 5 until P contains n_{src} and n_{goal}

Because our method is based on the pre-captured point cloud of the environment, the proposed method of map area detection considers a global perspective, i.e., with a priori knowledge of the environment. Possible enhancements for UAV discovery in an unknown environment were beyond the scope of this study.

Algorithm 3 Path Planning in Multi-Layer Grid Map. In transfer area identification, it takes each layer in the multi-layer grid map *Maps* and split it into several disconnected areas, based on which it extracts overlapping areas. In path search, it takes two set of node coordinates determining source *n_src* and goal *n_goal* positions, number of sampling nodes *M* and UAV altitude h_{start}. The algorithm process includes **SEGMENT_AREAS(), EXTRACT_OVERLAPS()** and **PATH_SEARCH_MULTI_LAYERS().**

```
SEGMENT_AREAS(Maps):
Input: Multi-layer map Maps.
for map in Maps do
    while seed != NULL
        seed ← RANDOM_FREE_GRID();
        a ← REGION_GROWING(seed);
        Areas{map, i}.add_area(a);
        i++;
return A ← Areas

EXTRACT_OVERLAPS(A):
Input: Segmented areas A in Multi-layer map.
for map1 = 1 to A.size() do
    for map2 = map1.lower_layer() to map1.upper_layer() do
        freeGrids ← AND(map1.free_grids(), map2.free_grids());
        transferAreas{map1, map2} ← INTERSECT(freeGrids, map1);
return T ← transferAreas

PATH_SEARCH_MULTI_LAYERS(Maps, T, M, n_src, n_goal, h_start):
Input: The multi-layer map Maps, a set of transfer areas T, node coordinates of
source n_src and goal n_goal, number of sampling node M and default UAV altitude
h_start.
Map_current ← Maps.get_map(h_start);
n_sub_src ← n_src;
n_sub_goal ← n_goal;
while !(P.contain(n_src) && P.contain(n_goal)) do
    N ← INITIALIZE_NETWORK(Map_current, n_sub_src, n_sub_goal, M);
    if PATH_SOLVE(Map_current, N, n_sub_src, n_sub_goal) != SUCCESS
        Map_next ← MIN_DISTANCE(T{Map_current}, n_sub_goal);
        n_trasnsfer ← SAMPLE(T{Map_current, Map_next});
        N.add_node(n_trasnsfer);
        n_sub_goal ← n_trasnsfer;
    else
        P ← PATH_SOLVE_RESULT(Map_current, N, n_sub_src, n_sub_goal);
        Map_current ← Map_next;
        n_sub_src ← n_transfer;
        n_sub_goal ← n_goal;
return P
```

Figure 9. Pseudocode of path planning in multi-layer grid map.

4. Path Postprocessing Optimization

The randomness in the sampling nodes of the PRM algorithm may result in redundant nodes in the path, which manifests as unnatural distortions in path morphology, as well as multiple visits to nodes during path checking and updating, thus increasing the final path length. In this regard, the initial path obtained from the path search should be optimized by postprocessing to remove redundant nodes and avoid unnecessary visits to the same node, thereby improving path quality and increasing the safety of UAV flight along the path.

In this study, we developed a two-step postprocessing method for path optimization consisting of a backward and a forward path connection check. The core concept of the method is to search in the initially solved path for a set of "key nodes", forming an optimal path that is collision-free with obstacles and as short a length as possible.

4.1. Backward Path Connection Check

To reduce redundant nodes, the backward connection check method starts from the source node as the first determined node and searches backward, node by node, to the farthest visible node as its direct connection node, which is also the newly determined node. To avoid multiple visits to the same node, each time the node is determined, the initial path is queried to discover if other nodes are visible. If so, it jumps to the last visible node and removes any other nodes in between.

The pseudocode of the backward path connection check is shown in Figure 10. More specifically, the procedure entails the following steps:

(1) Initialize the optimal path P' by adding n_{src}.
(2) Read the initially solved path P, starting from n_{src} as determined node n, with the neighboring node behind as the displacement node n_m.
(3) Check if P contains more than one n, i.e., multiple visits occur to n. If so, jump to the last occurrence of n on P, set it as the newly determined node n, and update the displacement node n_m; otherwise, continue to the next step.
(4) Perform collision check on the edge $e_{n,m}$. If it passes, let n_m move backward; otherwise, add n_{m-1} to P' and update n to n_{m-1} and n_m to $n+1$.
(5) Repeat steps 3 and 4 until n_{goal} is reached, then add it to P'.

Algorithm 4 Backward optimization method. It takes the solved path P after search, and results in a path P' which has less nodes than P.

Input: Solved path P' in which is a set of nodes includes source n_{src} and goal n_{goal}.

P'.add(n_{src});
$n \leftarrow n_{src}$;
$n_m \leftarrow n + 1$;
while n != n_{goal} **do**
if P.find(n) > 1
$n \leftarrow$ get_final_index(n);
$n_m \leftarrow n + 1$;
else
$e_{n,m} \leftarrow$ EDGE(n, n_m);
if CHECK_COLLISION ($e_{n,m}$) == COLLISION_FREE
$n_m \leftarrow n_{m+1}$;
else
P'.add_node(n_{m-1});
$n \leftarrow n_{m-1}$;
$n_m \leftarrow n + 1$;
return P'

Figure 10. Pseudocode of backward path connection check.

4.2. Forward Path Connection Check

The forward connection check method also starts from the source node as the first determined node but differs from the backward connection check, in that it checks the visibility starting from the goal node and searching forward node by node to the determined node. The aim of the forward connection check is to cope with the backward connection check in the case where two distant visible nodes exist with invisible nodes in between. As shown in Figure 11b, the path (cyan lines) obtained by the backward connection check returns to have the path extended around the upper, right, and lower sides of the obstacle area at the top of the map. However, in reality, the path can be directly connected down the left side of the obstacle area. At this point, the original path can be optimized by the forward connection check; as shown in Figure 11c, the resulting path (blue lines) after the forward connection check can directly connect the two nodes on the left side of the above obstacle area, avoiding the redundant path nodes.

(a) Collision-free path before optimization **(b)** Backward connection checked path **(c)** Forward connection checked path

Figure 11. Path postprocessing optimization: (**a**) an collision-free path (green lines) is obtained as an initially output of path planning; (**b**) backward connection checked path (cyan lines) firstly reduces unnecessary visits to the same nodes; (**c**) forward connection checked path (blue lines) further reduces redundant nodes.

The pseudocode of the forward path connection check is shown in Figure 12. More specifically, the procedure entails the following steps:

(1) Initialize the optimal path P'' by adding n_{src}.
(2) Input the backward optimal path P', starting from n_{src} as the determined node n and n_{goal} as the displacement node n_m.
(3) Perform collision check on edge $e_{n,m}$. If it passes, add n_m to P'', update n to n_m, and reset n_m; otherwise, let n_m move forward.
(4) Repeat steps 3 and 4 until n_{goal} is reached, then add it to P''.

The path postprocessing optimization combines the backward and forward connection checks to combine their respective benefits. First, considering the decreasing possibility of node visibility as their distance increases, the backward connection check from a fixed node is more efficient than the forward connection check from the end of the path, allowing for a faster traversal of the initial path. Second, the backward connection check, in the first step, cannot handle the situation where two distant visible nodes have invisible nodes in between, whereas the forward connection check, in the second step, can remedy this deficiency, because it checks the visibility from the other direction. Furthermore, the forward connection check, on the basis of the backward connection checked path, can further improve the path quality, avoiding the efficiency issues caused by directly performing it on the initial solved path after the path search.

The path postprocessing optimization method avoids multiple visits to the same node, minimizes unnecessary visits to redundant nodes, creates a straightening effect in the path shape, and reduces the final path length.

Algorithm 5 Forward optimization method. It takes the backward optimal path P' and results in a path P''.

Input: Backward optimal path P' including source n_{src} and goal n_{goal}.

$P''.\text{add}(n_{src})$;
$n \leftarrow n_{src}$;
$n_m \leftarrow n_{goal}$;
while n != n_{goal} **do**
while $n_m > n$ **do**
$e_{n,m} \leftarrow \text{EDGE}(n, n_m)$;
if CHECK_COLLISION($e_{n,m}$) == COLLISION_FREE
P''.add_node(n_m);
$n \leftarrow n_m$;
$n_m \leftarrow n_{goal}$;
else
$n_m \leftarrow n_{m-1}$;
return P''

Figure 12. Pseudocode of path forward connection check.

5. Experimental Results

5.1. Source Data and Environment

The source data for the experiments in this study were 3D point cloud data of an indoor environment acquired using light detection and ranging (LiDAR) scanning equipment, including two indoor scenes whose details are listed in Table 1.

Table 1. Source data of indoor environment.

Name	Description	No. Points	Data Quality	Overview
Scene 1	Wuhan University library east reading room, 2nd Floor	2,848,055	locally vacant	Figure 14a
Scene 2	Underground parking lot	6,794,787	locally vacant	Figure 14b

The map data we used in the path planning experiments included two reduced-dimensional raster maps of the aforementioned indoor environment and two virtual binary image maps [63] used for comparison. The original point cloud data were missing some scans, which necessitated manual completion of the vacant areas to ensure the integrity of the maps before the subsequent operations. The details of the four indoor environment maps are listed in Table 2, and Figure 13 provides overviews.

The proposed indoor environment modeling method involves the processing of point clouds, and we implemented the associated experiments using C/C++ programming with the Point Cloud Library (PCL) [64]. We also used the image processing library OpenCV to generate a reduced-dimensional raster map of the indoor environment.

The experimental simulation platform was MatLab, with an Intel® Core™ i7-7700HQ 2.80GHz CPU and 8GB RAM. We interpreted a reduced-dimensional raster map of the indoor environment as a two-dimensional simulation space and then implemented the path planning algorithm in the same space.

Table 2. Map data of the indoor environment.

Name	Description	Map Size	Data Quality
Map1 [1]	Binary image of simple obstacles	500 × 500 (px)	good
Map2	Binary image of a complex maze	500 × 500 (px)	good
Map_lib [2]	Reduced-dimensional raster map of Scene 1	575 × 773 (px)	locally vacant
Map_pkl	Reduced-dimensional raster map of Scene 2	682 × 625 (px)	locally vacant

[1] Binary images of Map1 and Map2 were obtained from web resources; [2] Map_lib and Map_pkl are reduced-dimensional raster maps generated using the method proposed in this study, where vacant areas in original data were filled, and external areas were set to black.

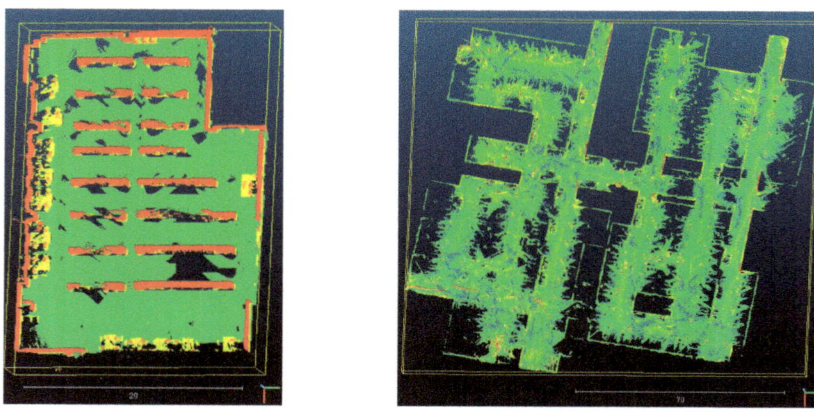

(a) Scene 1 **(b)** Scene 2

Figure 13. Indoor environment map for path planning (**a**–**d**).

To further simulate the condition of a typical UAV onboard system with limited resources, we validated and evaluated the performance of our methods using Manifold2-C, an onboard PC specially designed by DJI for their UAV products. The configuration was an Intel® Core™ i7-8550U 1.80GHz CPU with 8G RAM. We manually limited its CPU usage down to 30%; otherwise, we found that it had a faster computation speed than in the previous experimental environment, as the CPU and RAM on Manifold2-C are more up-to-date and offer better performance under the same input power.

5.2. Evaluation Metrics

We evaluated the proposed methods using three metrics: pathfinding success rate, planning time, and path length.

Pathfinding success rate is the basic metric as it indicates the practicability of the path-planning method. If the improvement in the path-planning method results in a significant decrease in the pathfinding success rate compared with the original method, even if it achieves a considerable improvement in other aspects, such changes are meaningless because the algorithm no longer satisfies the most fundamental requirement of solving a path from the source to the goal.

We focused on planning time as a metric as some UAV autonomous flight applications involve collaboration between various onboard systems. If the path planning procedure is too slow, a series of subsequent operations will need to wait and will stagnate, which not only does not meet the real-time UAV positioning and planning requirements but is also detrimental to the safety of autonomous UAV flight.

Path length reflects the quality of the path as determined by the path planning method. The shorter the path length, the shorter the flight time, which can reduce unnecessary power consumption and help the UAV avoid energy shortages when performing autonomous flight missions, thereby enhancing flight safety.

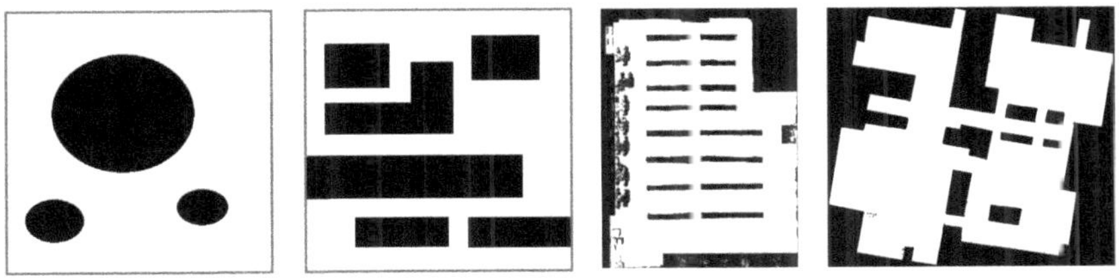

(a) Map1 (b) Map2 (c) Map_lib (d) Map_pkl

Figure 14. Point cloud overviews of indoor environment. We removed ceiling points from the point clouds for better observation (**a**,**b**).

Because our focus in this study was reducing the path planning time, it is worth noting that the algorithm tends to reach a solution faster when determining the path than when identifying the shortest path between the source and goal. On this basis, path postprocessing optimization is conducted to account for a non-shortest path length. As a result, the final path may be longer than the basic methods, but this is acceptable as long as the deviation is not excessive.

5.3. Experiments on Reduced-Dimensional Rasterization of Indoor Environment

In this study, we conducted experiments with real-world indoor scenes, including a library reading room and an underground parking lot. We performed the rasterization of the indoor environment based on the reduced-dimensional raster map generated by point cloud projection. The black grids represent impassable areas in the indoor environment, such as obstacles and boundaries, whereas the white grids represent passable and occupiable free space.

For simple environments, we projected the point cloud at the altitude midway between the floor and ceiling of the scene, within a 1 m height range. Moreover, we created a multilayer grid map to test our planning method in a more complex scenario where the distribution of obstacles varied among the different altitudes of the environment. We had no existing source data that met our needs, so we manually created a multilayer map by editing obstacles in the map that divided the maps into several areas. Different layers of the multilayer grid map are shown in Figure 15, representing the reduced-dimensional maps of the environment at different altitudes. We assumed that the altitude of Layer 1 was lower than that of Layer 2. On this particular map, if a UAV wanted to move from Area 1 to Area 2, it had to execute the following flight:

(1) Start from somewhere in Area 1 at the altitude of Layer 1.
(2) Move to the overlapping area of Areas 1 and 3.
(3) Ascend to the altitude of Layer 2.
(4) Move to the overlapping area of Areas 3 and 2.
(5) Descend to the altitude of Layer 1.
(6) Continue the flight in Area 2 at the altitude of Layer 1.

The map of the library reading room in Figure 13c shows that a portion of the obstacle areas representing the library cabinets was broken. This was due to missing scans in the original point cloud data, which we resolved by recollecting higher-quality indoor space point cloud data. The obstacles near the wall were the projections of desks and chairs. Although some space between them and the ceiling in the actual indoor space was free, this kind of space accounted for a small portion of the total space and had no substantial impact on the connectivity of the free space, so the waste of this portion of the space was still acceptable.

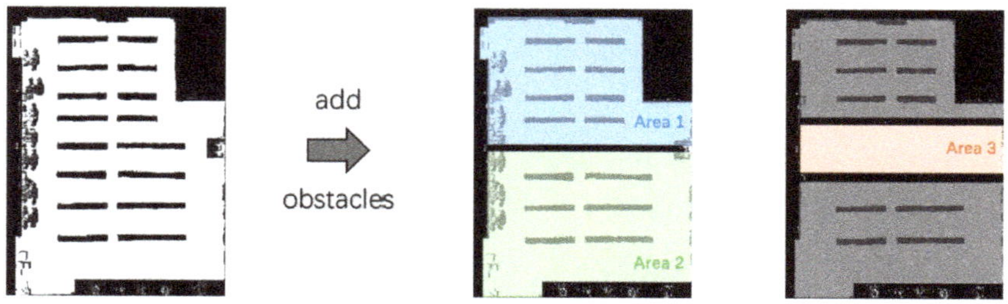

(a) Original. **(b)** Layer 1. **(c)** Layer 2.

Figure 15. Multilayer grid maps. Based on original Map_lib (**a**), we manually added an obstacle wall in the middle of Layer 1 (**b**), and divided the map into Areas 1 and 2; two obstacle walls in Layer 2 (**c**) near the aforementioned wall and enclosed Area 3.

The map of the underground parking lot in Figure 13d shows that vehicles in the parking lot did not notably interfere with the point cloud projection because we specified the projection method that extracts the height of the upper middle region of the parking lot.

Using the reduced-dimensional rasterization of the indoor environment considerably simplified the environment and met the data requirements of the subsequent path-planning experiments. Table 3 lists a comparison of the data before and after modeling of the two indoor scenes in this experiment, demonstrating that the reduced-dimensional modeling substantially reduced the data volume. The final maps generated by the method proposed in this study were essentially images, and their spatial accuracy was freely adjustable, and the corresponding data size changed with the image resolution.

Table 3. Data comparison before and after modeling.

Scene			Description	Format	Resolution	Size
Library reading room	Before		Point cloud	.pcd	2,848,055 (pts)	119 MB
	After	Small	Raster map	.bmp	575 × 773 (px)	435 KB
		Medium			1150 × 1546 (px)	1.74 MB
		Large			2300 × 3092 (px)	6.96 MB
Under-ground parking lot	Before		Point cloud	.pcd	6,794,787 (pts)	155 MB
	After	Small	Raster map	.bmp	682 × 625 (px)	418 KB
		Medium			1363 × 1250 (px)	1.62 MB
		Large			2724 × 2498 (px)	6.49 MB

5.4. Experiments on Network Construction Based on Connection Distance

The purpose of setting the connection distance parameter in network construction is to avoid collision checks between distant nodes, because connections that span a larger area are more likely to intersect with obstacles in the environment.

In the experiment, we defined the connection distance *c_dis* as follows:

$$c\_dis = w_{cd} * \sqrt{width_{map}^2 + height_{map}^2}, \quad (8)$$

where $width_{map}$ and $height_{map}$ denote the size of the grid map, and w_{cd} denotes the scale factor, i.e., connection distance weight.

Using Map_lib and Map_pkl data, we separately set w_{cd} to 0.25, 0.5, 0.75, and 1, i.e., the connection distance was 1/4, 1/2, 3/4, and 1 times the length of the map diagonal, respectively. For different numbers of nodes, we recorded the number of connected edges and network construction time, as well as the path length obtained by path search based on this network. The experimental results are listed in Table A1.

The network construction time and path length for the connection distance experiments are shown in Figure 16. With smaller connection distances (w_{cd} = 0.25), the number of constructed network edges was smaller and the construction time was shorter, but this resulted in a less successful path search and a longer path.

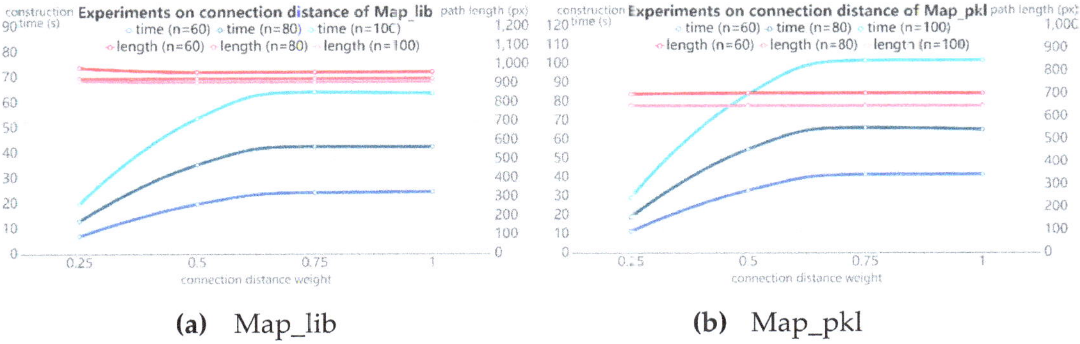

Figure 16. Construction time and path length of Map_lib (**a**) and Map_pkl (**b**).

As the connection distance weight increased, the network construction time accordingly increased, and the path length obtained from the path search tended to be shorter as a result of the increased number of possible connecting edges. However, the growth in the number of collision-free edges and the decrease in path length tended to be flat, and the number of colliding edges markedly increased. These redundant checks also resulted in a larger waste of network construction time, confirming the hypothesis that increasing distance decreases the possibility of collision-free edges. In the process of network construction, the algorithm performance could be enhanced by selecting a moderate connection distance.

5.5. Experiments on Improved Probabilistic Roadmap Planning

In this study, we used the four raster maps shown in Figure 13 to conduct probabilistic roadmap planning experiments. For each map, we positioned the source and goal near the diagonal position of the map, varied the number of nodes and connection distance, and repeated the experiment to record the pathfinding success rate, planning time, and path length of the basic and improved PRM methods.

The path planning results of the four maps are shown in Figure 17, where blue dots represent network nodes; blue and red lines represent visible and nonvisible edges, respectively; yellow lines represent updated edges during path search; cyan lines represent the forward-checked optimal path; and green lines represent the backward-checked optimal path, i.e., the final path result.

Comparing the result of the basic PRM with that of the improved PRM, we found two common cases. In the first case, the majority of nodes in both paths were the same but had slightly different path shapes; in the second case, the two paths had considerably different routes because we adopted an incremental update strategy and the path search followed the rule of greedy extension rather than that of global length shortening. Notably, as a result of our path optimization, the path length of our method was often comparable to that of the basic one, despite the route difference.

The path planning results of the multilayer grid map are shown in Figure 18. The source and the goal were the same as in Map_lib, with the source located in Area 1 and the goal in Area 2 (as shown in Figure 15). To travel from start to end, the UAV adjusted its flight altitude, i.e., ascending in the overlapping of Areas 1 and 3 and descending in the overlapping of Areas 3 and 2. The complete path from start to end was composed of three subpaths connected by two transfer nodes where the UAV performed vertical movements.

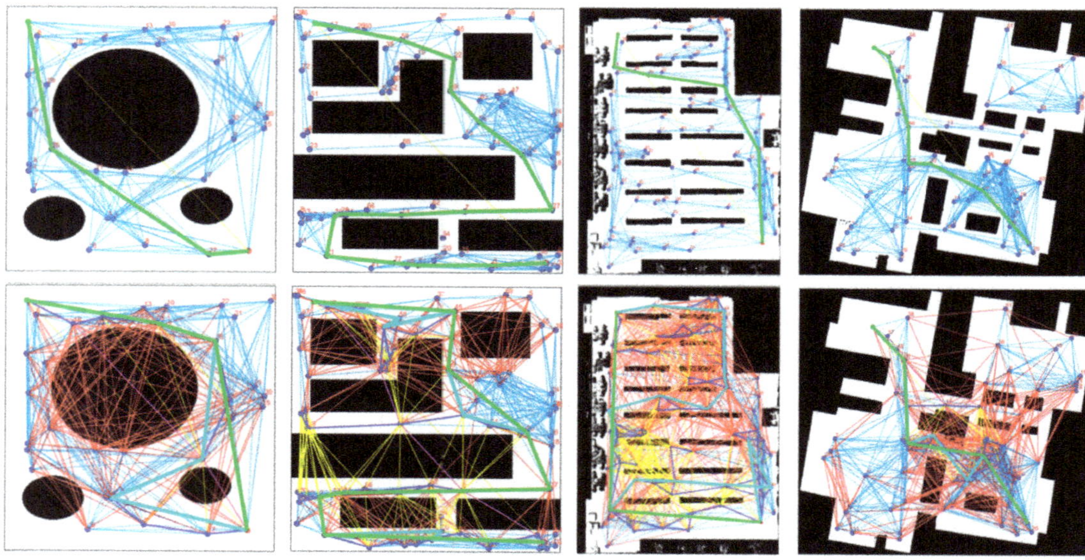

(a) $n = 30$, $cd = 0.5$. (b) $n = 75$, $cd = 0.25$. (c) $n = 80$, $cd = 0.25$. (d) $n = 60$, $cd = 0.25$.

Figure 17. Path planning results of grid maps of node number (n) and connection distance (cd). The first row is the results of basic PRM, and the second row is the results of improved PRM. The columns from left to right are Map1 (**a**), Map2 (**b**), Map_lib (**c**), and Map_pkl (**d**).

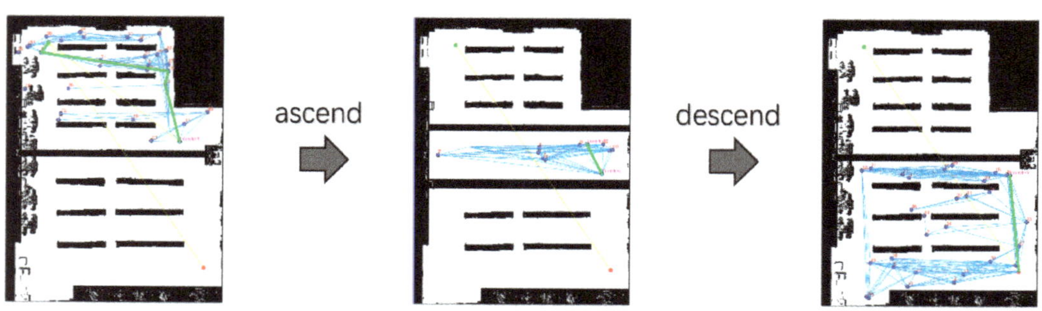

(a) Subpath in Area 1 (b) Subpath in Area 3 (c) Subpath in Area 2

Figure 18. Path planning result of multilayer grid map: three subpaths are searched in Area 1 (**a**), Area 3 (**b**), and Area 2 (**c**), respectively, and the subpaths are connected by two transfer nodes in the overlapping areas.

We repeated the experiments by configuring different connection distances for a different number of nodes. We recorded the planning time and path length of the basic and improved PRM methods. The comparative results of Map1 and Map2 are listed in Table A2. The basic PRM spent most of its time constructing the network, whereas the improved PRM spent most of its time on path search, network update, and path postprocessing optimization, as the check of network connectivity was deferred.

For Map1 and Map2, we calculated the average planning time and average path length ratio before and after the improvement, as shown in Figure 19. The planning time of the improved PRM was only approximately 30% of that of the basic PRM, resulting in a substantial reduction in path planning time. Due to some strategies adopted by the algorithm, it tended to find a path as quickly as possible rather than determine the shortest path between the source and the goal; consequently, the path length was longer than that of the basic PRM by less than 10%, which is acceptable given the considerable increase in planning time. In addition, the planning time and path length showed that as the number of nodes increased, the advantage in the planning time of the proposed method became more apparent.

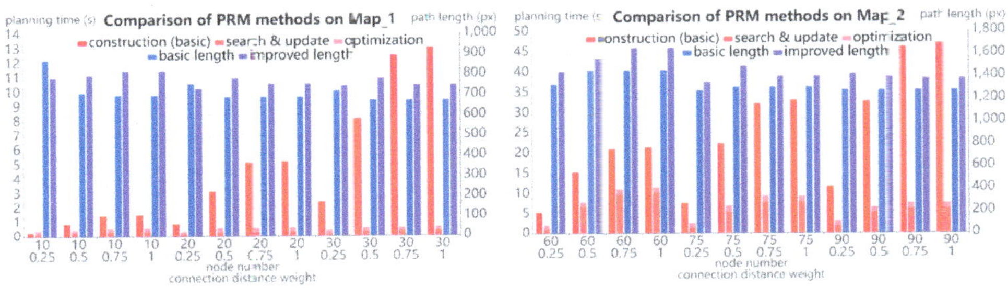

(a) Map1 **(b)** Map2

Figure 19. Planning time and path length of Map1 (**a**) and Map2 (**b**).

We also conducted the same experiments for two reduced dimensional raster maps of indoor environments, Map_lib and Map_pkl; the results are listed in Table A3. From a comparison of the results of the basic and improved PRM on Map_lib and Map_pkl, as shown in Figure 20, our conclusions were basically the same: the improved PRM provided an advantage over the basic PRM in terms of planning time, at the cost of an increase in path length, which was acceptable.

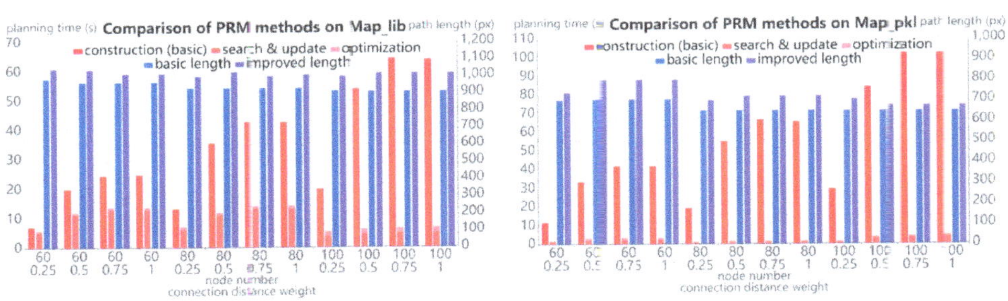

(a) Map_lib **(b)** Map_pkl

Figure 20. Planning time and path length of Map_lib (**a**) and Map_pkl (**b**).

Comparing the binary image map and the reduced-dimensional raster map of the indoor environment, we found that for the simple maps (Map1 and Map_pkl), the planning time stability in the improved PRM was high, i.e., the planning time did not notably fluctuate with changes in the number of nodes or the connection distance. For the complex maps (Map2 and Map_lib), the stability decreased, but was still considerably better than that of the basic PRM.

We also conducted experiments on an onboard PC to simulate the integration of our planning method into an autonomous UAV system. We tested our improved PRM method on the onboard PC and evaluated the planning time using the same input data as the original experiments. At first, we did not set a resource limit on the onboard PC, but due to its hardware configuration being more up-to-date than that of our desktop PC, it performed even better. Therefore, we manually limited the CPU usage down to 30% to more accurately simulate a resource-limited configuration.

The experimental results of the improved PRM on Map_lib and Map_pkl on the onboard PC (with CPU usage limited to 30%) are listed in Table A4, and they are shown in Figure 21. The planning times in both experimental environments exhibited comparable trends under the same input data. Additionally, the consumed time was fundamentally influenced by the hardware configuration, whereas our planning method maintained appropriate performance despite resource limitations. Moreover, we showed that the current onboard hardware was capable of high-level configurations (except for those dedicated to micro vehicles), so scholars can more easily deploy and test their methods on onboard computing platforms with high performance.

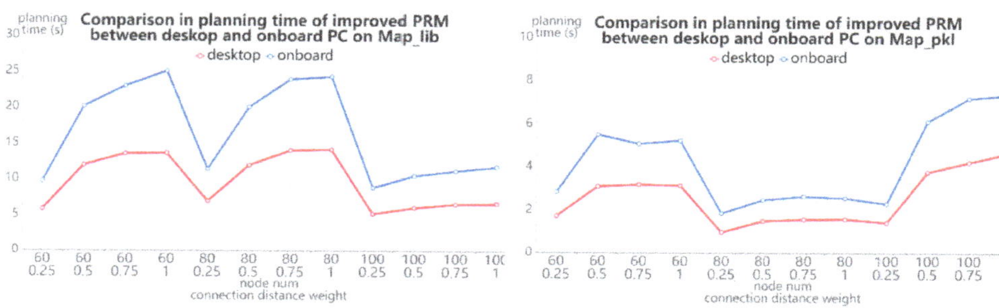

(a) Map_lib **(b)** Map_pkl

Figure 21. Planning time of Map_lib (**a**) and Map_pkl (**b**), with comparison between desktop and onboard PC configurations.

5.6. Experiments on Path Postprocessing Optimization

Using the same reduced-dimensional raster maps of the indoor environment, we conducted the experiments of path post-processing optimization. Experimental results are shown in Figure 22, where thick yellow lines represent updated edges, yellow lines represent collision-free path after path search, cyan lines represent the backward-checked optimal path, and green line represents the forward-checked optimal path, i.e., the final path result.

In Map_lib and Map_pkl, the paths were solved using the basic and improved PRM methods for comparison. The initially obtained collision-free paths were then optimized using two-step path postprocessing. The experimental results are listed in Table A5.

From the path postprocessing optimization results shown in Figure 23, we found that the proposed optimization method substantially improves the path quality for the initially obtained paths, with longer lengths and higher path repetition rates in the improved PRM, as the method minimized redundant nodes and allowed the path to attain a "straightened" route. This path postprocessing optimization method solves the problem where the initial path quality obtained by the improved PRM method is inferior to that of the basic one, completing the proposed path planning scheme. As a result, the optimization method avoids redundant motions and ensures a collision-free and direct path for the UAV, which is conducive to the safety of autonomous flight in indoor environments.

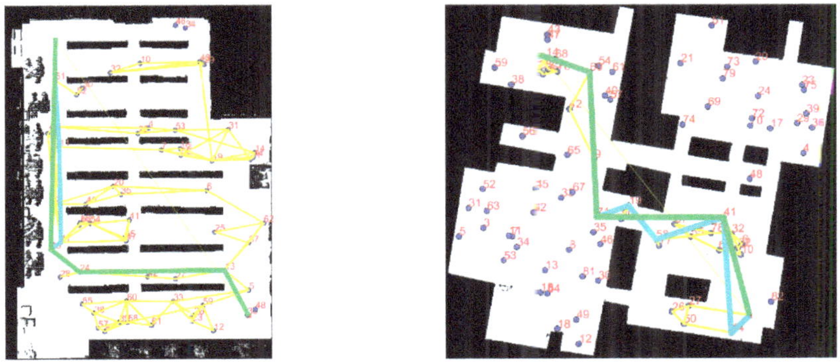

(a) Map_lib **(b)** Map_pkl

Figure 22. Path postprocessing optimization result of Map_lib (**a**) and Map_pkl (**b**).

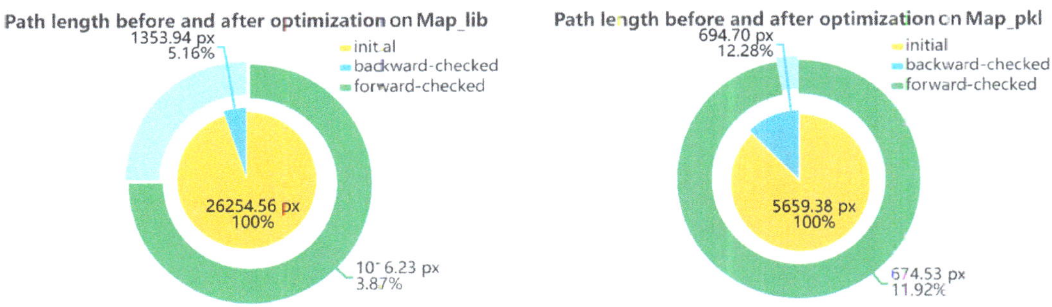

(a) Map_lib **(b)** Map_pkl

Figure 23. Path length before and after optimization on Map_lib (**a**) and Map_pkl (**b**).

6. Discussion

In this study, we primarily focused on modeling an indoor environment and improving the PRM path-planning method. We assumed the UAV was a quadrotor model and reduced the controls to 3D translation and rotation in yaw angle. This assumption is relatively simple but was sufficient for the purposes of this study, as numerous flight control products and software have already encapsulated and integrated UAV operations into simple commands. More in-depth research can be conducted on the topic of UAV control.

The data source for indoor environment modeling was a point cloud. Furthermore, we used a point cloud projection scheme to generate reduced-dimensional raster maps of indoor environments. Although this method preserves the semantic understanding of indoor environment elements and the modeling speed is faster, the problem of noise points present in the point cloud must still be solved, as determining whether these noise points represent tiny obstacles or actually are noise points is crucial to the safety of indoor UAV flights. In the actual process of map generation, we must also be problem-specific because the presence of too many noise points will impede the efficiency of solving the path, and eliminating these noise points may pose a safety risk, so we should strike a balance between these two factors.

Considering the complexity of indoor environments, we generated grid maps at various altitudes and constructed a multilayer map. This method increases the robustness of path search and the applicability of the UAV.

The proposed improved PRM planning method aims to solve the computational efficiency problem caused by the basic PRM while minimizing the impact on the path length. The results of our experiments showed that the proposed method is effective. However, the efficiency of the proposed method may decrease in overly complex indoor situations, because if collision checks on edges are not performed when constructing the original network, a large number of edge adjacency information updates are required in the subsequent search and update process; each update necessitates at least one local path search, resulting in a substantial decrease in efficiency. To avoid repeated checks, it is recommended to check the connection between all nodes at the beginning.

The key to our improvement provided by our strategy is reducing a large number of collision checks in the basic method, because collision checks performed during network construction may be meaningless for two reasons: First, as the distance between nodes increases, the possibility of visibility decreases, because more obstacles created blockages between them. Second, the farther nodes are from the source and goal nodes, the less likely they are to become candidates, so checks on them are often unnecessary in the end. As a result, we adopted the strategy of constructing first and checking later to eliminate as many unnecessary checks as possible and searched for paths based on the overall effectiveness of the candidate nodes from high to low, which can more quickly solve the paths while retaining the ability to search for all nodes because, in extreme cases after traversing through all nodes in the space, as long as a path solution exists, the path solution should be obtained.

7. Conclusions

In this study, we developed an improved probabilistic roadmap planning method for safe indoor UAV flight, under the assumption of a quadrotor UAV.

We modeled the indoor environment with point cloud projection and represented it with reduced-dimensional raster maps. The grid map model performed well in terms of environment representativity and modeling efficiency. Furthermore, our proposed multilayer gird map model, an innovation over the original single-layer model, contributes to improving path planning effectiveness, particularly in complex environments.

Based on the grid map model, we conducted path planning experiments and improved upon the basic PRM planning method in terms of network construction, search strategy, and path optimization. Our method remarkably outperformed the basic PRM in computational efficiency while maintaining a reasonable path length. It also showed high-quality performance on both desktop PCs and resource-limited onboard platforms, laying the foundation for indoor UAV applications and fulfilling the requirements for autonomous UAV flight safety.

Author Contributions: Conceptualization, Q.H. and S.W.; methodology, Q.J. and P.Z.; software, Q.J. and P.Z.; validation, Q.H., S.W. and M.A.; formal analysis, Q.H. and S.W.; investigation, Q.J., P.Z. and M.A.; resources, P.Z. and M.A.; data curation, Q.J., P.Z. and M.A.; writing—original draft preparation, Q.J.; writing—review and editing, Q.H. and S.W.; visualization, Q.J.; supervision, Q.H. and S.W.; project administration, Q.H.; funding acquisition, Q.H. All authors have read and agreed to the published version of the manuscript.

Funding: This research was funded by the National Natural Science Foundation of China (NSFC), grant number 42101441, and the Fundamental Research Funds for the Central Universities.

Data Availability Statement: Data and source codes in this study will be made available on https://github.com/Jin-qg/iPRM before 1 March 2023.

Conflicts of Interest: The authors declare no conflict of interest. The funders had no role in the design of the study; in the collection, analyses or interpretation of data; in the writing of the manuscript, or in the decision to publish the results.

Appendix A

Table A1. Path planning results.

Node Num.	1	2	Edge Num.			6	Path Length
			3	4	5		
			Map_lib				
60	0.25	80%	213.38	417.50	1260.13	7.1084	984.82
60	0.5	100%	293.30	1235.40	362.30	19.8989	963.19
60	0.75	100%	297.30	1580.00	13.70	24.4590	963.19
60	1	100%	297.30	1593.70	0.00	24.7062	963.19
80	0.25	80%	380.88	778.13	2162.00	13.0713	928.56
80	0.5	100%	495.60	2272.50	552.90	35.4468	928.21
80	0.75	100%	501.50	2807.80	11.70	42.6125	928.21
80	1	100%	501.50	2819.50	0.00	42.5122	928.21
100	0.25	100%	548.90	1246.80	3355.30	19.8673	915.75
100	0.5	100%	710.60	3534.80	905.60	53.7906	910.33
100	0.75	100%	718.50	4411.20	21.30	64.2075	910.33
100	1	100%	718.50	4432.50	0.00	63.7934	910.33
			Map_pkl				
60	0.25	90%	400.11	264.44	1226.44	11.4924	700.32
60	0.5	100%	476.50	1143.50	271.00	33.1575	704.85
60	0.75	100%	482.20	1407.90	0.90	41.4939	704.85
60	1	100%	482.20	1408.80	0.00	41.5201	704.85
80	0.25	100%	697.90	486.30	2136.80	19.0639	650.89
80	0.5	100%	866.40	2020.80	433.80	54.8623	650.74
80	0.75	100%	878.40	2440.90	1.70	66.3098	650.74
80	1	100%	878.40	2442.60	0.00	65.3203	650.74
100	0.25	100%	1050.80	747.60	3352.60	29.3165	648.71
100	0.5	100%	1264.70	3164.30	722.00	83.9546	648.69
100	0.75	100%	1280.40	3868.50	2.10	101.9095	648.69
100	1	100%	1280.40	3870.60	0.00	102.1078	648.69

1 Connection distance weight; 2 Path finding success rate; 3 Edge number of collision-free edges; 4 Edge number of collided edges; 5 Edge number of skipped; 6 Network construction time.

Table A2. PRM comparative results of Map1 and Map2.

Node Num.	1	2	Time of Improved PRM			Path Length	
		3	4	5	Total	Basic	Improved
				Map1			
10	0.25	0.2655	0.1570	0.2500	0.4070	873.69	786.16
10	0.5	0.9056	0.2551	0.1994	0.4546	711.68	797.93
10	0.75	1.4807	0.3016	0.2715	0.5731	700.87	821.82
10	1	1.5241	0.3183	0.2689	0.5872	700.87	821.82
20	0.25	0.8855	0.1327	0.2073	0.3400	757.33	732.86
20	0.5	3.1325	0.2857	0.2985	0.5842	692.63	784.42
20	0.75	5.1130	0.2876	0.3155	0.6031	691.70	756.36
20	1	5.1925	0.3107	0.3138	0.6245	691.70	756.46
30	0.25	2.3752	0.1563	0.2483	0.4046	721.99	746.97
30	0.5	8.1504	0.3155	0.2887	0.6042	675.18	783.13
30	0.75	12.5140	0.3517	0.2684	0.6201	674.90	748.98
30	1	13.0509	0.3703	0.2669	0.6372	674.90	748.98
	Average time ratio		33.0%		Average length ratio		108.9%
				Map2			
60	0.25	5.2805	1.0070	1.0493	2.0563	1330.90	1443.88
60	0.5	15.2353	6.6992	1.1010	7.8002	1454.97	1556.92
60	0.75	20.9249	9.6638	1.3777	11.042	1454.71	1651.95
60	1	21.3038	10.019	1.3912	11.410	1454.71	1651.95
75	0.25	7.5534	1.2740	1.2716	2.5456	1274.06	1348.04

Table A2. *Cont.*

Node Num.	1	2 3	Time of Improved PRM 4	5	Total	Path Length Basic	Improved
75	0.5	22.2787	5.3312	1.5318	6.8630	1306.63	1493.31
75	0.75	32.1494	7.7480	1.5358	9.2838	1306.61	1404.32
75	1	33.0011	7.7118	1.5358	9.2477	1306.61	1404.32
90	0.25	11.5591	1.7538	1.3451	3.0989	1281.55	1422.85
90	0.5	32.6026	5.1573	1.3882	6.5456	1279.15	1391.26
90	0.75	46.0452	6.1180	1.4262	7.5442	1279.09	1382.75
90	1	46.9495	6.0681	1.4318	7.4999	1279.09	1382.75
	Average time ratio		33.1%		Average length ratio		109.5%

1 Connection distance weight; 2 Time of basic PRM; 3 Network construction; 4 Update and search; 5 Path optimization.

Table A3. PRM comparative results of Map_lib and Map_pkl.

Node Num.	1	2 3	4	Time of Improved PRM 5	6	Total	Path Length 7	8
				Map1				
60	0.25	7.11	0.00	5.0644	0.8180	5.8843	984.82	1043.95
60	0.5	19.90	0.01	11.1111	0.9270	12.0476	963.19	1037.95
60	0.75	24.46	0.00	12.7196	0.9000	13.6243	963.19	1014.45
60	1	24.71	0.00	12.7823	0.9002	13.6872	963.19	1014.45
80	0.25	13.07	0.01	5.9111	1.0935	7.0123	928.56	997.52
80	0.5	35.45	0.01	11.0135	0.9830	12.0074	928.21	1024.94
80	0.75	42.61	0.01	13.0456	1.0274	14.0872	928.21	1000.56
80	1	42.51	0.01	13.1091	1.0667	14.1884	928.21	1007.73
100	0.25	19.87	0.01	3.6025	1.6199	5.2286	915.75	998.80
100	0.5	53.79	0.01	4.4192	1.6485	6.0815	910.33	1018.50
100	0.75	64.21	0.02	4.8817	1.7028	6.6048	910.33	1017.93
100	1	63.79	0.02	4.9162	1.7433	6.6801	910.33	1017.93
	Average time ratio		28.5%		Average length ratio		108.5%	
				Map2				
60	0.25	11.49	0.00	1.0242	0.7097	1.7357	700.32	738.45
60	0.5	33.16	0.01	2.0315	1.0820	3.1212	704.85	799.88
60	0.75	41.49	0.01	2.1125	1.0986	3.2173	704.85	799.88
60	1	41.52	0.00	2.0853	1.0827	3.1727	704.85	799.88
80	0.25	19.06	0.01	0.4200	0.5734	0.9996	650.89	698.75
80	0.5	54.86	0.01	0.8500	0.6561	1.5171	650.74	721.73
80	0.75	66.31	0.01	0.8879	0.6932	1.5919	650.74	721.55
80	1	65.32	0.01	0.9127	0.6798	1.6064	650.74	721.55
100	0.25	29.32	0.01	0.6826	0.7624	1.4544	648.71	705.78
100	0.5	83.95	0.02	3.0867	0.6923	3.7976	648.69	674.53
100	0.75	101.91	0.02	3.5807	0.6483	4.2493	648.69	674.53
100	1	102.11	0.02	4.0346	0.6420	4.6936	648.69	674.53
	Average time ratio		4.8%		Average length ratio		109.0%	

1 Connection distance weight; 2 Time of basic PRM; 3, 4 Network construction time; 5 Update and search; 6 Path optimization; 7 Basic.; 8 Improved.

Table A4. Improved PRM results on onboard PC (with CPU usage limited to 30%).

Node Num.	1	2	Planning Time 3	4	Total
			Map_lib		
60	0.25	0.00	8.4516	1.3514	9.8063
60	0.5	0.01	18.6095	1.6332	20.2484
60	0.75	0.01	21.7281	1.3381	23.0730
60	1	0.01	23.4918	1.6114	25.1119
80	0.25	0.01	9.7605	1.7316	11.4980

Table A4. *Cont.*

Node Num.	1	Planning Time 2	3	4	Total
80	0.5	0.01	18.7435	1.3743	20.1281
80	0.75	0.01	22.3054	1.6590	23.9764
80	1	0.01	22.6307	1.6870	24.3299
100	0.25	0.01	6.2032	2.7115	8.9240
100	0.5	0.02	7.8199	2.7303	10.5655
100	0.75	0.02	8.5419	2.7033	11.2639
100	1	0.20	8.7333	2.8929	11.8291
		Map_pkl			
60	0.25	0.00	1.8963	0.9678	2.8676
60	0.5	0.01	3.6753	1.8464	5.5278
60	0.75	0.01	3.4006	1.7047	5.1124
60	1	0.01	3.6033	1.6536	5.2637
80	0.25	0.01	0.7882	1.0823	1.8764
80	0.5	0.01	1.4127	1.0773	2.5008
80	0.75	0.01	1.5038	1.1492	2.6650
80	1	0.01	1.4970	1.0774	2.5867
100	0.25	0.07	1.0735	1.1801	2.3278
100	0.5	0.08	5.1668	0.9108	6.1574
100	0.75	0.02	6.3549	0.8481	7.2209
100	1	0.02	6.3881	0.9770	7.3810

1 Connection distance weight; 2 Network construction; 3 Update and search; 4 Path optimization.

Table A5. Path postprocessing optimization results.

Node Num.	1	Path Length of Basic PRM Initial	2	3	Path Length of Improved PRM Initial	4	5
			Map_lib				
60	0.25	984.82	981.51	981.47	19,788.58	1347.54	1043.95
60	0.5	963.19	963.19	963.19	34,983.89	1325.51	1037.95
60	0.75	963.19	963.19	963.19	38,275.40	1259.18	1014.45
60	1	963.19	963.19	963.19	38,361.68	1259.18	1014.45
80	0.25	928.56	927.74	927.74	20,592.44	1263.47	997.52
80	0.5	928.21	928.21	928.21	33,226.28	1210.95	1024.94
80	0.75	928.21	928.21	928.21	37,746.75	1258.81	1000.56
80	1	928.21	928.21	928.21	38,090.94	1330.02	1007.73
100	0.25	915.75	914.77	914.77	11,689.17	1594.60	998.80
100	0.5	910.33	910.33	910.33	13,224.70	1436.51	1018.50
100	0.75	910.33	910.33	910.33	14,401.92	1447.90	1017.93
100	1	910.33	910.33	910.33	14,672.96	1513.58	1017.93
Avg. length ratio		100%	99.95%	100.00%	100%	5.16%	75.05%
			Map_pkl				
60	0.25	700.32	700.30	700.30	3525.15	809.84	738.45
60	0.5	704.85	704.85	704.85	5544.48	894.79	799.88
60	0.75	704.85	704.85	704.85	5544.48	894.79	799.88
60	1	704.85	704.85	704.85	5544.48	894.79	799.88
80	0.25	650.89	650.74	650.74	1683.02	712.62	698.75
80	0.5	650.74	650.74	650.74	2729.33	733.67	721.73
80	0.75	650.74	650.74	650.74	2729.37	733.49	721.55
80	1	650.74	650.74	650.74	2729.37	733.49	721.55
100	0.25	648.71	648.69	648.69	2079.39	726.05	705.78
100	0.5	648.69	648.69	648.69	5480.68	694.70	674.53
100	0.75	648.69	648.69	648.69	5659.38	694.70	674.53
100	1	648.69	648.69	648.69	5659.38	694.70	674.53
Avg. length ratio		100%	100.00%	100.00%	100%	12.28%	97.10%

1 Connection distance weight; 2, 4 Backward-checked; 3, 5 Forward-checked.

References

1. Tsouros, D.C.; Bibi, S.; Sarigiannidis, P.G. A review on UAV-based applications for precision agriculture. *Information* **2019**, *10*, 349. [CrossRef]
2. Apolo-Apolo, O.; Martínez-Guanter, J.; Egea, G.; Raja, P.; Pérez-Ruiz, M. Deep learning techniques for estimation of the yield and size of citrus fruits using a UAV. *Eur. J. Agron.* **2020**, *115*, 126030. [CrossRef]
3. Carroll, S.; Satme, J.; Alkharusi, S.; Vitzilaios, N.; Downey, A.; Rizos, D. Drone-based vibration monitoring and assessment of structures. *Appl. Sci.* **2021**, *11*, 8560. [CrossRef]
4. Ren, H.; Zhao, Y.; Xiao, W.; Hu, Z. A review of UAV monitoring in mining areas: Current status and future perspectives. *Int. J. Coal Sci. Technol.* **2019**, *6*, 320–333. [CrossRef]
5. Sharma, V.; You, I.; Pau, G.; Collotta, M.; Lim, J.D.; Kim, J.N. LoRaWAN-based energy-efficient surveillance by drones for intelligent transportation systems. *Energies* **2018**, *11*, 573. [CrossRef]
6. Gupta, A.; Afrin, T.; Scully, E.; Yodo, N. Advances of UAVs toward future transportation: The State-of-the-Art, challenges, and Opportunities. *Future Transp.* **2021**, *1*, 326–350. [CrossRef]
7. Huang, H.; Savkin, A.V.; Huang, C. Scheduling of a parcel delivery system consisting of an aerial drone interacting with public transportation vehicles. *Sensors* **2020**, *20*, 2045. [CrossRef]
8. Ghelichi, Z.; Gentili, M.; Mirchandani, P.B. Logistics for a fleet of drones for medical item delivery: A case study for Louisville, KY. *Comput. Oper. Res.* **2021**, *135*, 105443. [CrossRef]
9. Pensieri, M.G.; Garau, M.; Barone, P.M. Drones as an integral part of remote sensing technologies to help missing people. *Drones* **2020**, *4*, 15. [CrossRef]
10. Liu, C.; Szirányi, T. Real-time human detection and gesture recognition for on-board uav rescue. *Sensors* **2021**, *21*, 2180. [CrossRef]
11. Jeelani, I.; Gheisari, M. Safety challenges of UAV integration in construction: Conceptual analysis and future research roadmap. *Saf. Sci.* **2021**, *144*, 105473. [CrossRef]
12. Maity, R.; Mishra, R.; Pattnaik, P.K. Flying robot path planning techniques and its trends. *Mater. Today Proc.* 2021; *in press*. [CrossRef]
13. Zhang, X.; Li, X.; Wang, K.; Lu, Y. *A Survey of Modelling and Identification of Quadrotor Robot*; Abstract and Applied Analysis; Hindawi: London, UK, 2014.
14. Meier, L.; Tanskanen, P.; Heng, L.; Lee, G.H.; Fraundorfer, F.; Pollefeys, M. PIXHAWK: A micro aerial vehicle design for autonomous flight using onboard computer vision. *Auton. Robot.* **2012**, *33*, 21–39. [CrossRef]
15. Quadcopters & Multirotor Drones—Draganfly—Driving Innovatin of Drones. Available online: https://draganfly.com/products/quadcopters-multirotors/ (accessed on 8 November 2022).
16. Matrice 300 RTK—Built Tough. Works Smart—DJI. Available online: https://www.dji.com/matrice-300 (accessed on 8 November 2022).
17. Wu, X.; Xiao, B.; Qu, Y. Modeling and sliding mode-based attitude tracking control of a quadrotor UAV with time-varying mass. *ISA Trans.* **2019**, *124*, 436–443. [CrossRef]
18. Jeon, H.; Song, J.; Lee, H.; Eun, Y. Modeling quadrotor dynamics in a wind field. *IEEE/ASME Trans. Mechatron.* **2020**, *26*, 1401–1411. [CrossRef]
19. Ji, D.; Wang, R.; Zhai, Y.; Gu, H. Dynamic modeling of quadrotor AUV using a novel CFD simulation. *Ocean. Eng.* **2021**, *237*, 109651. [CrossRef]
20. Moshayedi, A.J.; Gheibollahi, M.; Liao, L. The quadrotor dynamic modeling and study of meta-heuristic algorithms performance on optimization of PID controller index to control angles and tracking the route. *IAES Int. J. Robot. Autom.* **2020**, *9*, 256. [CrossRef]
21. Almakhles, D.J. Robust backstepping sliding mode control for a quadrotor trajectory tracking application. *IEEE Access* **2019**, *8*, 5515–5525. [CrossRef]
22. Lambert, N.O.; Drew, D.S.; Yaconelli, J.; Levine, S.; Calandra, R.; Pister, K.S. Low-level control of a quadrotor with deep model-based reinforcement learning. *IEEE Robot. Autom. Lett.* **2019**, *4*, 4224–4230. [CrossRef]
23. Yang, S.; Xian, B. Energy-based nonlinear adaptive control design for the quadrotor UAV system with a suspended payload. *IEEE Trans. Ind. Electron.* **2019**, *67*, 2054–2064. [CrossRef]
24. Zhang, X.; Wang, Y.; Zhu, G.; Chen, X.; Li, Z.; Wang, C.; Su, C.Y. Compound adaptive fuzzy quantized control for quadrotor and its experimental verification. *IEEE Trans. Cybern.* **2020**, *51*, 1121–1133. [CrossRef] [PubMed]
25. Wang, L. Review of the application of open source flight control in multi-rotor aircraft. *Int. Core J. Eng.* **2021**, *7*, 261–270.
26. García, J.; Molina, J.M. Simulation in real conditions of navigation and obstacle avoidance with PX4/Gazebo platform. *Pers. Ubiquitous Comput.* **2020**, *26*, 1171–1191. [CrossRef]
27. Kawamura, E.; Azimov, D. Integrated extremal control and explicit guidance for quadcopters. *J. Intell. Robot. Syst.* **2020**, *100*, 1583–1613. [CrossRef]
28. Khosiawan, Y.; Nielsen, I. A system of UAV application in indoor environment. *Prod. Manuf. Res.* **2016**, *4*, 2–22. [CrossRef]
29. Zhang, T.; Liu, C.; Li, J.; Pang, M.; Wang, M. A new visual inertial Simultaneous Localization and Mapping (SLAM) algorithm based on point and line features. *Drones* **2022**, *6*, 23. [CrossRef]
30. Bauersfeld, L.; Ducard, G. RTOB SLAM: Real-time onboard laser-based localization and mapping. *Vehicles* **2021**, *3*, 778–789. [CrossRef]

31. Jing, Y.; Qi, F.; Yang, F.; Cao, Y.; Zhu, M.; Li, Z.; Lei, T.; Xia, J.; Wang, J.; Lu, G. Respiration detection of ground injured human target using UWB radar mounted on a hovering UAV. *Drones* **2022**, *6*, 235. [CrossRef]
32. Kwon, J.; Hailes, S. In Scheduling UAVs to bridge communications in delay-tolerant networks using real-time scheduling analysis techniques. In Proceedings of the 2014 IEEE/SICE International Symposium on System Integration, Tokyo, Japan, 13–15 December 2014; IEEE: Piscataway, NJ, USA, 2014; pp. 363–369.
33. Potortì, F.; Palumbo, F.; Crivello, A. Sensors and sensing technologies for indoor positioning and indoor navigation. *Sensors* **2020**, *20*, 5924. [CrossRef]
34. Steenbeek, A.; Nex, F. CNN-based dense monocular visual SLAM for real-time UAV exploration in emergency conditions. *Drones* **2022**, *6*, 79. [CrossRef]
35. Moussa, M.; Zahran, S.; Mostafa, M.; Moussa, A.; El-Sheimy, N.; Elhabiby, M. Optical and mass flow sensors for aiding vehicle navigation in gnss denied environment. *Sensors* **2020**, *20*, 6567. [CrossRef] [PubMed]
36. Deng, H.; Fu, Q.; Quan, Q.; Yang, K.; Cai, K.-Y. Indoor multi-camera-based testbed for 3-D tracking and control of UAVs. *IEEE Trans. Instrum. Meas.* **2019**, *69*, 3139–3156. [CrossRef]
37. Xin, C.; Wu, G.; Zhang, C.; Chen, K.; Wang, J.; Wang, X. Research on indoor navigation system of uav based on lidar. In Proceedings of the 2020 12th International Conference on Measuring Technology and Mechatronics Automation (ICMTMA), Phuket, Thailand, 28–29 February 2020; IEEE: Piscataway, NJ, USA, 2020; pp. 763–766.
38. Youn, W.; Ko, H.; Choi, H.; Choi, I.; Baek, J.-H.; Myung, H. Collision-free autonomous navigation of a small UAV using low-cost sensors in GPS-denied environments. *Int. J. Control. Autom. Syst.* **2021**, *19*, 953–968. [CrossRef]
39. Majumdar, A.; Ahmadi, A.A.; Tedrake, R. Control design along trajectories with sums of squares programming. In Proceedings of the 2013 IEEE International Conference on Robotics and Automation, Karlsruhe, Germany, 6–10 May 2013; IEEE: Piscataway, NJ, USA, 2013; pp. 4054–4061.
40. Meyer-Delius, D.; Beinhofer, M.; Burgard, W. Occupancy grid models for robot mapping in changing environments. In Proceedings of the 26th AAAI Conference on Artificial Intelligence, Toronto, ON, Canada, 22–26 July 2012.
41. Dhulkefl, E.; Durdu, A.; Terzioğlu, H. Dijkstra algorithm using uav path planning. *Konya Mühendislik Bilimleri Dergisi* **2020**, *8*, 92–105.
42. Zhou, F.; Nie, H. Quick path planning based on shortest path algorithm for multi-uav system in windy condition. *Control. Syst. Eng.* **2021**, preprint.
43. Hong, Z.; Sun, P.; Tong, X.; Pan, H.; Zhou, R.; Zhang, Y.; Han, Y.; Wang, J.; Yang, S.; Xu, L. Improved a-star algorithm for long-distance off-road path planning using terrain data map. *ISPRS Int. J. Geo-Inf.* **2021**, *10*, 785. [CrossRef]
44. Belanová, D.; Mach, M.; Sinčák, P.; Yoshida, K. Path planning on robot based on D* lite algorithm. In *Procedings of the 2018 World Symposium on Digital Intelligence for Systems and Machines (DISA), Košice, Slovakia, 23–25 August 2018*; IEEE: Piscataway, NJ, USA, 2018; pp. 125–130.
45. Chen, X.; Zhang, J. The three-dimension path planning of UAV based on improved artificial potential field in dynamic environment. In Proceedings of the 2013 5th International Conference on Intelligent Human-Machine Systems and Cybernetics, Hangzhou, China, 26–27 August 2013; IEEE: Piscataway, NJ, USA, 2013; pp. 144–147.
46. Li, F.; Zlatanova, S.; Koopman, M.; Bai, X.; Diakité, A. Universal path planning for an indoor drone. *Autom. Constr.* **2018**, *95*, 275–283. [CrossRef]
47. González de Santos, L.M.; Frías Nores, E.; Martínez Sánchez, J.; González Jorge, H. Indoor path-planning algorithm for uav-based contact inspection. *Sensors* **2021**, *21*, 642. [CrossRef]
48. Xu, Z.; Zhang, L.; Ma, X.; Liu, Y.; Yang, L.; Yang, F. An anti-disturbance resilience enhanced algorithm for UAV 3D route planning. *Sensors* **2022**, *22*, 2151. [CrossRef]
49. Hao, K.; Zhao, J.; Yu, K.; Li, C.; Wang, C. Path planning of mobile robots based on a multi-population migration genetic algorithm. *Sensors* **2020**, *20*, 5873. [CrossRef]
50. Ajeil, F.H.; Ibraheem, I.K.; Azar, A.T.; Humaidi, A.J. Grid-based mobile robot path planning using aging-based ant colony optimization algorithm in static and dynamic environments. *Sensors* **2020**, *20*, 1880. [CrossRef] [PubMed]
51. Li, J.; Sun, T.; Huang, X.; Ma, L.; Lin, Q.; Chen, J.; Leung, V.C. A memetic path planning algorithm for unmanned air/ground vehicle cooperative detection systems. *IEEE Trans. Autom. Sci. Eng.* **2021**, *19*, 2724–2737. [CrossRef]
52. Liu, G.; Shu, C.; Liang, Z.; Peng, B.; Cheng, L. A modified sparrow search algorithm with application in 3d route planning for UAV. *Sensors* **2021**, *21*, 1224. [CrossRef]
53. Wang, X.; Pan, J.S.; Yang, Q.; Kong, L.; Snášel, V.; Chu, S.C. Modified mayfly algorithm for UAV path planning. *Drones* **2022**, *6*, 134. [CrossRef]
54. Li, S.; Zhang, H.; Li, Z.; Liu, H. An air route network planning model of logistics UAV terminal distribution in urban low altitude airspace. *Sustainability* **2021**, *13*, 13079. [CrossRef]
55. Yin, G.; Zhou, S.; Wu, Q. An improved RRT algorithm for UAV path planning. *Acta Electonica Sin.* **2017**, *45*, 1764
56. Pettersson, P.O.; Doherty, P. Probabilistic roadmap based path planning for an autonomous unmanned aerial vehicle. In Proceedings of the ICAPS-04 Workshop on Connecting Planning Theory with Practice, Whistler, Canada, 3–7 June 2004.
57. Gang, L.; Wang, J. PRM path planning optimization algorithm research. *Wseas Trans. Syst. Control.* **2016**, *11*, 81–86.

58. Chen, J.; Zhou, Y.; Gong, J.; Deng, Y. An improved probabilistic roadmap algorithm with potential field function for path planning of quadrotor. In Proceedings of the 2019 Chinese Control Conference (CCC), Guangzhou, China, 27–30 July 2019; IEEE: Piscataway, NJ, USA, 2019; pp. 3248–3253.
59. Mohanta, J.C.; Keshari, A. A knowledge based fuzzy-probabilistic roadmap method for mobile robot navigation. *Appl. Soft Comput.* **2019**, *79*, 391–409. [CrossRef]
60. Raheem, F.A.; Abdulakareem, M.I. Development of a* algorithm for robot path planning based on modified probabilistic roadmap and artificial potential field. *J. Eng. Sci. Technol* **2020**, *15*, 3034–3054.
61. Abdulakareem, M.I.; Raheem, F.A. Development of path planning algorithm using probabilistic roadmap based on ant colony optimization. *Eng. Technol. J.* **2020**, *38*, 343–351. [CrossRef]
62. Che, E.; Jung, J.; Olsen, M.J. Object recognition, segmentation, and classification of mobile laser scanning point clouds: A state of the art review. *Sensors* **2019**, *19*, 810. [CrossRef] [PubMed]
63. Rahul Kala's Source Codes. Available online: http://rkala.in/codes.php (accessed on 5 May 2022).
64. Point Cloud Library. Available online: https://pointclouds.org/ (accessed on 20 April 2022).

drones

MDPI

Article

Path Planning of Autonomous Mobile Robots Based on an Improved Slime Mould Algorithm

Ling Zheng [1,2,*], Yan Tian [1,2], Hu Wang [1], Chengzhi Hong [3] and Bijun Li [3]

1 School of Electronic Information and Communications, Huazhong University of Science and Technology, Wuhan 430074, China

2 Shenzhen Research Institute of Central China Normal University, Shenzhen 518052, China

3 State Key Laboratory of Information Engineering in Surveying, Mapping, and Remote Sensing, Wuhan University, Wuhan 430079, China

* Correspondence: lingzheng@whu.edu.cn

Abstract: Path planning is a crucial component of autonomous mobile robot (AMR) systems. The slime mould algorithm (SMA), as one of the most popular path-planning approaches, shows excellent performance in the AMR field. Despite its advantages, there is still room for SMA to improve due to the lack of a mechanism for jumping out of local optimization. This means that there is still room for improvement in the path planning of ARM based on this method. To find shorter and more stable paths, an improved SMA, called the Lévy flight-rotation SMA (LRSMA), is proposed. LRSMA utilizes variable neighborhood Lévy flight and an individual rotation perturbation and variation mechanism to enhance the local optimization ability and prevent falling into local optimization. Experiments in varying environments demonstrate that the proposed algorithm can generate the ideal collision-free path with the shortest length, higher accuracy, and robust stability.

Keywords: autonomous mobile robots (AMRs); path planning; slime mould algorithm; rotation transformation

1. Introduction

With the rapid development of artificial intelligence technology, the intelligence and automation of autonomous mobile robots (AMRs) have advanced significantly. AMRs are widely used in many high-tech aspects, such as smart homes, intelligent logistics, and autonomous vehicles. Path planning, which is a crucial component of the AMR system, aims to generate a feasible, safe, and smooth path from the starting point to the target point in an unknown or known environment [1]. However, with the increasing types and scales of AMR applications, existing path-planning approaches cannot fulfill all demands due to constraints such as complexity and nonlinearity [2]. As a challenging research hotspot, it has attracted several studies on performance improvement in terms of effectiveness and efficiency [3].

Many traditional path-planning methods do not consider complexity and randomness, resulting in locally optimal solutions [4]. Meta-heuristic algorithms, on the other hand, utilize global search to find the best approximate solution by exploring and exploiting the overall search space [5]. Meta-heuristic algorithms can be categorized into evolutionary algorithms, such as simulated annealing, and swarm intelligence algorithms (SIAs), which simulate the behavior of biological groups in nature and exchange information and experience to solve global optimization problems [6].

In recent years, researchers have used many algorithms based on swarm intelligence to solve the path-planning problem of robots. Commonly used meta-heuristic algorithms include the genetic algorithm (GA) [7], particle swarm optimization (PSO) [8], ant colony optimization (ACO) [9], monarch butterfly optimization (MBO) [10], whale optimization algorithm (WOA) [11], grey wolf optimization (GWO) [12], and slime mould algorithm

(SMA) [13]. SIAs are a combination of a stochastic algorithm and local search and perform well in solving highly nonlinear and multi-modal optimization problems [14]. However, SIAs still face some challenges, such as local optimization and slow convergence. GA simulates the evolution laws of the biological world. Although GA has great global optimization capability, it requires a large population and broad search space, which can result in local optimization and slow convergence in the search process [7]. In PSO, the flying process of particles is the individual's search process. The flying speed of particles can be dynamically adjusted based on the individual's historical best position and the population's historical best position. PSO is prone to early convergence when solving complex optimization problems [13]. ACO divides the search space into grids in the field of path planning and utilizes the state transition probability and pheromone updating method to solve the algorithm [15]. However, the convergence speed of ACO can be slow, resulting in an extensive calculation [16]. MBO, WOA, and SMA are also known to experience slow local convergence, while GWO exhibits poor population diversity and slow convergence speed in later stages. In the robot path-planning field, PSO, WOA, GWO, and SMA are generally used algorithms for comparison. This paper will also compare these algorithms, which are common in mainstream practice.

The SMA is a meta-heuristic algorithm that was first proposed by Nagataki in 2000 [17]. It has been successfully applied to various areas, such as the selection of traffic network nodes [18], economic emission dispatch [19], robot path planning [20], medical image classification [21], and feature selection [22]. In 2020, Li et al. [23] proposed an SMA optimization for studying the activity and dynamics of slime moulds. Due to the excessive randomness in the meta-heuristic optimization process, it is essential to strike a proper balance between exploration and exploitation to optimize the algorithm effectively. In various studies, it has been found that SMA achieves excellent exploitation, while maintaining a significant balance with exploration. In multiple benchmark functions, SMA has shown outstanding performance compared to other algorithms, such as WOA, GWO, and PSO [14]. However, there is still limited research on using the SMA algorithm for AMR navigation. Additionally, in some cases or functions, SMA may present defects, such as failure to converge to the global optimal or slow convergence, and it may lack an effective escape mechanism.

This paper builds upon the work of Agarwal and Bharti [20] and proposes a path-planning algorithm called Lévy flight-rotation SMA (LRSMA) that updates the variable neighborhood Lévy mechanism to improve the algorithm's convergence accuracy. It also introduces a real-time convergence stagnation monitoring strategy based on tolerance, which guides the population out of local optima using individual rotation perturbation and mutation mechanisms. The mobile robot considered in this paper is autonomous, a type of dynamic remotely operated navigation equipment (DRONE). Besides, it is easy to apply to other types of drones, such as aerial robots. The main contributions of this study are summarized as follows:

(a) A path-planning solution is proposed to generate the shortest no-collision curve path with robustness for AMRs.

(b) A perturbation mutation method based on rotation transformation is applied to increase the probability of finding the best solution near the suboptimal solution.

This paper is structured as follows. The first chapter introduces the research background, as mentioned above. The second chapter discusses related work. The third chapter introduces the SMA. The fourth chapter presents path planning based on the proposed LRSMA. The fifth chapter compares the LRSMA with other algorithms to verify its effectiveness in path planning. The last chapter provides the conclusion.

2. Related Work

Traditional path-planning algorithms, such as the A star algorithm [24], the artificial potential field method [25], the rapidly exploring random tree (RRT) algorithm [26], fuzzy logic [27], and neural networks [28], are widely used in robot path planning. However, these algorithms face challenges in complex environments due to poor adaptability and slow convergence. The A star algorithm requires prior information about the environment which can optimize the path but is inefficient under large-scale complex conditions [24]. The artificial potential field method is often used in real-time dynamic environments but can easily fall into planning stagnation and local optimization in complex environments with obstacles [25]. The RRT can quickly search an unknown space and solve complex constraints with few parameters through a simple structure, but there are still issues in improving search efficiency and obtaining the optimal path [26]. The fuzzy logic algorithm is applied to handle scenarios with uncertain environment information, but the rule database needs to be updated frequently during path planning, and it has a high computational cost under complex conditions [27]. The neural network method does not require prior information about the environment and has broad applicability to real-time dynamic scenarios, but in complex environments with obstacles, the processing of massive networks results in a significant computational burden [28]. Although traditional path-planning algorithms have shown excellent performance in solving path planning under specific conditions [29], they cannot adequately satisfy the requirements for path planning in complex environments due to inadequate adaption, slow convergence, and other factors [30].

The path-planning problem can be transformed into the problem of finding the optimal path with the minimum cost function value [31]. Thanks to the optimization advantages in solving discontinuous, non-smooth, and discrete variable problems, SIAs have been commonly applied to path planning. Researchers are continuously improving these algorithms and exploring ones with higher planning efficiency, better optimization ability, and greater robustness [32].

For example, Wang et al. [33] proposed an ant colony path-planning algorithm in a 3D environment, which improves the global pheromone and designs a heuristic function with a safe value. This solves the problems of easily falling into local optimization and having a long search time in 3D path planning. Teng et al. [34] proposed a GWO based on PSO and used nonlinear control parameters to balance the local and global search. They also introduced PSO to update the positions of grey wolves, preventing the algorithm from falling into local optimization. Fernandes et al. [35] proposed a Quantum-behaved Particle Swarm Optimization (QPSO) algorithm that can produce a diverse population of peak values, effectively avoiding stagnation in the optimization process and solving path planning in both static and dynamic scenarios. Dai et al. [11] proposed an optimized WOA, adopted adaptive technology to improve the convergence speed, and set virtual obstacles and improved potential field factors to enhance the dynamic obstacle avoidance ability of AMRs. The WOA has a strong search ability despite its slow computation.

Slime molds belong to the Polycephalum family. The Physarum polycephalum algorithm requires numerous iterations and initial parameters, making it redundant for complex problems. The SMA proposed by Li et al. [23] simulates the behavior and morphological changes of Physarum polycephalum myxomycetes during foraging. The front end of slime molds is fan-shaped, and the back is an interconnected venous network. When the venous network approaches food, the biological oscillator of the slime mold produces a diffusion wave that changes the flow of cytoplasm in the vein, and the slime mold moves to a higher-quality food. The SMA adjusts search modes according to the fitness value. Additionally, slime molds will shrink to the optimal position and separate into multiple individuals to explore other spaces.

The SMA has strong global optimization due to a multiple exploration mechanism. It has diverse applications, such as the selection of traffic network nodes [18], robot path planning [20], medical image classification [21], and feature selection [22]. However, the SMA has unstable optimization and can easily fall into local optimization due to its simple optimization mechanism. Achievable improvements are applied to solve this problem, which can divide into three categories. The first is optimizing the population position update mechanism. Yu et al. [36] applied the quantum rotation gate and water cycle strategies, keeping the population balanced between exploration and exploitation, thus solving local optimization and slow convergence problems in later iterations. Nguyen et al. [37] proposed an improved SMA that adjusted the weight coefficient and introduced a reverse learning strategy to enhance optimization performance when updating the positions of slime molds. The second category is expanding the search space of the population. Rizk-Allah et al. [38] adopted the chaos-opposition-enhanced strategy to expand the search space and avoid premature algorithm convergence. Houssein et al. [39] integrated improved opposition-based learning and orthogonal learning mechanisms to prevent the algorithm from falling into local optimization due to rapid population assimilation. The third approach is introducing mutation operators. Houssein et al. [40] employed an adaptive guided differential evolution algorithm to improve the population's local search ability, increase the population's diversity, and avoid premature convergence. Liu and Liu [41] used quasi-reverse and quasi-reflective learning to expand the search space and introduce an unscented transformation sigma point to guide the search of the SMA, thereby solving the problems of search stagnation and poor stability.

While previous research has improved the performance of the SMA, optimizing the multi-objective function in path planning remains unsolved. Additionally, few studies have reported on robot path planning with obstacle avoidance in various environments. In other words, it is necessary to develop practical applications and further improvements in path planning. Hence, this paper proposes the LRSMA to achieve the shortest and collision-free path with robustness and efficiency for AMRs. The effectiveness and efficiency of the LRSMA are compared with those of the GWO, WOA, PSO, and SMA to validate its performance.

3. Overview of the SMA

The SMA uses a mathematical model to simulate the slime molds' foraging behavior and morphological changes when approaching and surrounding food [23]. If slime molds find food while foraging, they generate interconnected vein networks of varying thicknesses between multiple food sources based on the quality and density of the food source. Slime molds spread into food, creating positive and negative feedback through diffusion waves, enabling them to search for higher-quality food. The whole process contains three stages.

- Approach food stage

Slime mold individuals approach food based on odor. Suppose the population of slime molds in a D-dimensional searching space is N. The updated position of a slime mold $X(t+1)$ at the tth iteration when approaching food can be expressed as follows:

$$X(t+1) = \begin{cases} X\_best(t) + vb \times (W \times X_{r1}(t) - X_{r2}(t)), & rand_1 < p, \\ vc \times X(t), & rand_1 \geq p, \end{cases} \tag{1}$$

where t is the current iteration index, $X(t)$ is the position of a slime mold, and $X(t+1)$ is the new updated position of the slime mold. $X\_best(t)$ is the position of the best solution at the tth iteration. $X_{r1}(t)$ and $X_{r2}(t)$ are the positions of two slime molds randomly selected from the population. vc is a control parameter that evaluates the changes in the use of historical data by slime molds, whose value decreases linearly from 1 to 0. $rand_1$ is a random number in the range [0, 1].

p is a control parameter that determines the position update mode of slime molds and can be obtained as follows:

$$p = tanh|S(i) - DF|, i \in \{1, 2, \ldots, N\}, \tag{2}$$

where i is the slime mold index, $S(i)$ is the fitness of the ith slime mold, and DF is the fitness of the best solution.

vb is a random number in the range $[-a, a]$ regarded as a control parameter; its value range decreases with the decrease in a. Parameter a, used to simulate the gradual and dynamic contraction of veins when slime molds approach food, is expressed as follows:

$$a = arctanh\left(-\frac{t}{IT_{max}} + 1\right), \tag{3}$$

where IT_{max} is the maximum number of iterations.

The weight coefficient W represents the oscillation frequency of the biological oscillator, which changes with food concentration when the slime molds approach food. If the food where the slime molds are located has a high concentration, there is positive feedback to the slime molds, and vice versa. W is defined as

$$\begin{cases} W(SIndex(i)) = \begin{cases} 1 + r_2 \times lg\left(\frac{bF-S(i)}{bF-wF} + 1\right), & condition, \\ 1 - r_2 \times lg\left(\frac{bF-S(i)}{bF-wF} + 1\right), & others, \end{cases} \\ SIndex(i) = sort(N), \end{cases} \tag{4}$$

where r_2 is a random number in the range [0, 1]; $SIndex(i)$ is the sort index of the slime mold, and bF and wF indicate the best fitness and the worst fitness, respectively. lg is used to reduce the change rate of the fitness, *condition* indicates slime molds with the fitness ranking of the first half of the population, and *others* denotes the remaining slime molds.

- Wrap food stage

When slime molds wrap around food, if the concentration of the searched food is greater than that of the current food, the oscillation wave of the biological oscillator becomes stronger, and the flow speed of the cytoplasm increases. Although slime molds may find the current best food sources, it is necessary to adjust search strategies and randomly assign a portion of the slime molds to explore other search spaces to locate better food sources. The random mechanism helps maintain the diversity of the slime mold population. The positions of slime molds are updated as follows:

$$X(t+1) = \begin{cases} rand \times (UB - LB) + LB, & rand < z, \\ X\_best(t) + vb \times (W \times X_{r1}(t) - X_{r2}(t)), & rand \geq z \text{ and } rand_1 < p, \\ vc \times X(t), & rand \geq z \text{ and } rand_1 \geq p, \end{cases} \tag{5}$$

where UB and LB indicate the upper and low bound constraints of the search space, respectively; $rand$ and $rand_1$ are two random numbers in the interval [0, 1]; and z represents the proportional parameters of randomly distributed slime mold individuals in the population.

- Oscillation stage

In the oscillation process, slime molds dynamically adjust the width of a vein based on the quality and density of the food. They then use the oscillation of the biological oscillator to adjust the flow speed of cytoplasm in the vein and adjust the oscillation mechanism through parameters W, vb, and vc. This completes the search for optimal food in the search space.

4. Path Planning Based on the LRSMA

4.1. Elite Learning Strategy Based on Variable Neighborhood Lévy Flight

Lévy flight can describe the foraging path of many creatures in nature [42]. The random walk mode of Lévy flight is special, mixing short and long distances. The random distance follows the Lévy distribution, and the Lévy distribution follows the power function distribution, which is given by

$$\begin{cases} Levy(\beta) = \frac{\mu}{|\nu|^{1/\beta}}, \\ \mu \sim N\left(0, \sigma_\mu^2\right), \\ \nu \sim N\left(0, \sigma_\nu^2\right), \\ \sigma_\mu = \left\{\frac{\Gamma(1+\beta) \times sin(\pi\beta/2)}{\Gamma[(1+\beta)/2] \times \beta \times 2^{(\beta-1)/2}}\right\}^{1/\beta}, \end{cases} \quad (6)$$

where $Levy(\beta)$ is the random distance processed by the Mantegna algorithm [43], $\Gamma(*)$ is a gamma function, and β is a constant. μ and ν are two normal stochastic variables with standard deviations σ_μ and σ_ν, $\sigma_\nu = 1$, respectively.

To ensure effective algorithm convergence without compromising optimization precision, the elite learning strategy based on variable neighborhood Lévy flight is applied. The elite slime mold is the global best solution, and variable neighborhood Lévy flight is adopted to generate the elite mutation. Selecting the positions of the elite before and after the Lévy flight provides a better position with more suitable fitness.

Variable neighborhood Lévy flight is given by

$$\begin{cases} X\_best(t)_l = \alpha \times \left(X_{best(t)} + stepsize \times Levy(\beta)\right), \\ \alpha = 1 + index \times r_l, \end{cases} \quad (7)$$

where t is the index of the iteration, α is the variable neighborhood coefficient, $X\_best(t)_l$ is the new position of the best solution after Lévy flight, and $X\_best(t)_i$ is the position of the best solution. $stepsize$ is a step factor whose goal is simply to adjust the random search range. $Levy(\beta)$ is the random Lévy flight distance. $index$=$\{0,1,2\}$. If the fitness of $X\_best(t)_l$ is less than the fitness of $X\_best(t)_i$, or if $index$ is equal to 2, the search of $index$ is abandoned, and Lévy flight is stopped. r_l is a random number in the range $[0, 0.5]$.

4.2. Tolerance-Based Rotation Perturbation Mutation Mechanism of Slime Mold Individuals

The SMA can obtain the optimization solution, but local exploration can be further improved. In the later stages of the iteration, the SMA converges slowly and may even stagnate. To address this issue, this paper proposes a rotation perturbation mutation mechanism based on the tolerance of slime mod individuals. First, a convergence stagnation monitoring strategy based on tolerance is proposed. Then, a perturbation mutation method based on rotation transformation is introduced to aid in the search for the best solution in the hypersphere zone for slime molds with higher fitness ranks. The probability of finding the best solution near the suboptimal solution is increased to enhance the algorithm's local search capabilities during the local search.

Tolerance parameter τ denotes the count variable for convergence state monitoring. After completing one iteration, the tolerance parameter τ can be updated according to the following formula:

$$\tau(t+1) = \begin{cases} \tau(t) + 1, if\ T(t) < Tmax\ and\ |\triangle F(t)| < Fmin, \\ 0, \quad if\ (T(t) == Tmax\ and\ |\triangle F(t)| < Fmin)\ or\ |\triangle F(t)| > Fmin, \end{cases} \quad (8)$$

where t is current iteration index and $\tau(t+1)$ is the newly updated tolerance parameter. $\tau(t)$ is the tolerance parameter at the tth iteration. Here, $t \in \{0, \ldots, Tmax\}$, and $Tmax = 2$. $|\triangle F(t)|$ is the absolute value of the best solution's fitness difference at the tth iteration and $(t-1)th$ iteration and is calculated by $|\triangle F(t)| = |F(t) - F(t-1)|$. The threshold value of fitness difference $Fmin$ is a fixed value.

When $\tau(t) \geq 0$, the algorithm may or may not be in the stagnation state. When $T(t) == Tmax$and$|\triangle F(t)| < Fmin$, it can be considered that the algorithm falls into local optimization. Furthermore, the slime mold needs to be redirected to improve the diversity of the distribution. Rotation transformation is used to conduct stochastic perturbation of slime mold individuals within the ω radius of the self-centered hypersphere, which is expressed as follows:

$$X(t)_r = X(t) + \omega \times \frac{1}{\|X(t)\|_2} \times (R \cdot X(t)), \tag{9}$$

where $X(t)$ is the position of the slime mold individual at the t-th iteration. $X(t)_r$ represents its new rotated position. The rotation transformation factor ω is in the range of $[0.1, 1]$. Let D be the dimension of the search space and R be a uniform random $1 \times D$ vector in $[-1, 1]$. $R \cdot X(t)$ is the dot product of the R vector and the position $X(t)$, while $\|X(t)\|_2$ represents the 2-norms (Euclidean norms).

Since there is no guarantee that the fitness values of slime molds after perturbation are more optimal, the greedy principle is adopted to select the positions of slime molds. When comparing the fitness values of slime molds before and after perturbation, if the fitness value after the perturbation is better, the position of the slime mold will be replaced with the new position after the perturbation. Otherwise, the position will not be replaced.

4.3. Elite Simulated Annealing Strategy

To enhance the population's diversity and fully utilize the evolutionary potential of the current best solution, the simulated annealing algorithm introduces the Metropolis criterion [44] to generate suboptimal solutions with probabilities. The elite slime mold currently possesses the current best solution. A new best solution is obtained through a perturbation mutation method based on rotation transformation. Let DF_r represent the new fitness of the best solution after rotation transformation, DF represent the fitness of the global best solution, and $rand_S$ be a random number within the range of [0, 1]. If DF_r is less than DF, the position of the global best solution is replaced with the new position. However, if DF_r is greater than or equal to DF, a random number $rand_S$ is generated. If $exp^{-(DF_r - DF)/T} < rand_S$, then DF_r is accepted, and the position of the global best solution is replaced with the new position. The default value for T is set to 1000.

4.4. Fitness Function Construction

Cubic spline interpolation is a classic method of piecewise interpolation that generates a smooth curve based on several interpolation point intervals defined by a cubic polynomial [45]. The resulting curve is smooth, which is ideal for the dynamic characteristics of AMRs. Therefore, this paper combines the LRSMA with the cubic spline interpolation method to solve optimal path-planning problems for AMRs. The turning point of each interpolation segment is the path node, and each slime mold individual represents all path nodes on one path. Assume that the coordinates of m path nodes are $(x_{n\_1}, y_{n\_1}), (x_{n\_2}, y_{n\_2}), \ldots, (x_{n\_m}, y_{n\_m})$. There is a start point (x_s, y_s) and an end point (x_e, y_e). By applying cubic spline interpolation in the x and y directions, the coordinates of n interpolation points are produced, i.e., $(x_1, y_1), (x_2, y_2), \ldots, (x_n, y_n)$. These start points, n interpolation points, and end points are then connected to form a smooth, continuous path.

The fitness values are used to evaluate paths and are determined by the fitness function. To search for the shortest collision-free path, the fitness function is defined as follows:

$$S = \eta_1 \times f_1 + \eta_2 \times f_2 \tag{10}$$

where S is the fitness function designed to find the minimum value in this work. f_1 is the path length, which represents the length of the curve connecting the start point and the end point by each interpolation point in sequence. η_1 is the path length penalty coefficient and is set as $1.f_2$ is the average distance between all interpolation points and all obstacles. If there is a collision in the path, $f_2 > 0$. If the path does not pass through the coverage area of obstacles, $f_2 = 0$. η_2 is the collision penalty coefficient and is set as 1000. Note that the higher the value of η_2, the lower the collision of the final path.

4.5. Path-Planning Process Based on the LRSMA

The flow diagram of LRSMA is shown in Figure 1. The detailed steps of the LRSMA are shown as follows.

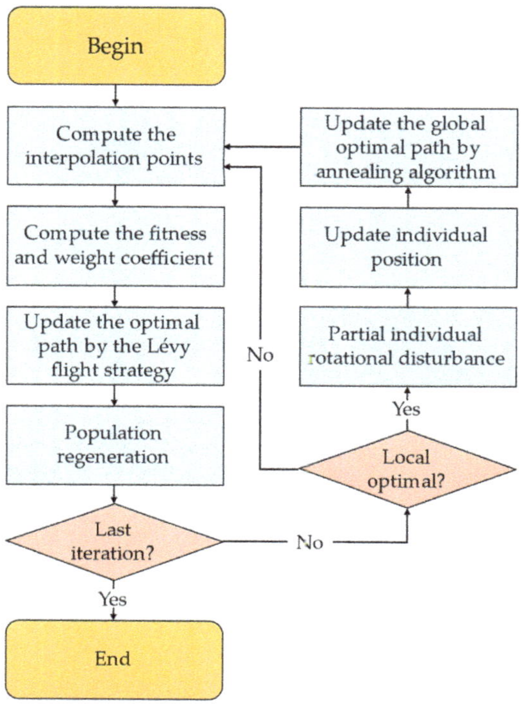

Figure 1. The diagram of the Lévy flight-rotation SMA (LRSMA).

(1) Initialize the algorithm parameters and population, including the population size N; the maximum number of iterations T_{max}; the upper and lower bound constraints of the search space UB and LB, respectively; the start point (x_s, y_s); the end point (x_e, y_e); the proportion of slime molds perturbed by the perturbation mutation method based on rotation transformation p_r; and the path nodes (the position of a slime mold) $(x_{n\_1}, y_{n\_1}), (x_{n\_2}, y_{n\_2}), \ldots, (x_{n\_m}, y_{n\_m})$.

(2) Obtain the coordinates of interpolation points $(x_1, y_1), (x_2, y_2), \ldots, (x_n, y_n)$ between the start point, path nodes, and end point using cubic spline interpolation.

(3) Calculate and rank the fitness of each slime mold. Determine the best fitness bF and the worst fitness wF.

(4) Update the weight coefficients of slime molds $W(SIndex)$ using Equation (4).

(5) Update the positions of all slime molds according to Equations (1)–(5).

(6) Apply the variable neighborhood Lévy flight strategy using Equations (6) and (7), and select and save the best solution.

(7) Monitor whether the population is in a state of convergence stagnation. If the population is stagnating, rotate the positions of the slime molds with higher fitness ranks using Equation (9) based on rotation transformation. If the new fitness of a slime mold after perturbation is less than its past fitness, update the position of the slime mold.

(8) Compare the new fitness of the best solution after perturbation with the past fitness, update the global best solution using the simulated annealing mechanism, and save the fitness and positions of the global best solution.

(9) Repeat Steps (2)–(8) until the maximum number of iterations is reached. Once reached, terminate the iteration, obtain the fitness and positions of the best solution, and output the optimal path.

The pseudocode for the LRSMA is shown in Algorithm 1.

Algorithm 1 LRSMA pseudocode

```
Initialize the population
while it < no. of iterations do
    for each slime mold individual do
        Calculate the fitness of each slime mold individual
        Calculate the local best by Equation (4)
    end for
    Update the local and global best by Equation (7)
    If T(t) == Tmax and |△F(t)| < Fmin then
        Update a portion of the population individuals by Equation (4)
        Update the global by annealing algorithm
    end if
    Update the population
end while
```

5. Simulation Results and Analysis

The goal of the LRSMA is to plan the shortest path without collision. To verify its performance, the path length and planning time were taken as the evaluation indicators, and the minimum, mean, and standard deviation of these indicators were compared.

5.1. Experimental Environment

To simplify the calculation, circles with different radii were used as obstacles, while the center points of the robot were used to represent the AMR. The size of the mobile was ignored during path planning. All simulations were carried out in Visual Studio Code1.68.1 on a laptop (Windows 10 64-bit, Intel Xeon E3-1231 v3 CPU, 16GB RAM). To verify the robustness of the algorithm, three scenarios with different numbers and distributions of obstacles were designed. Moreover, a light blue circle denotes obstacles in Figures 2–4. There were 12 obstacles in the first and second scenarios, and 24 in the third scenario. Detailed parameters are shown in Table 1.

Table 1. Type of scenario.

	Start Point	End Point	Obstacle Number	Obstacle Coordinate and Size (x, y, r)
Scenario 1	[−10, −10]	[10, 10]	12	Coordinate x: [−5.3, 7.6, 5.0, 3.5, −4.4, 0.0, −2.9, 0.0, 7.3, −3.6, 0.4, 2.4] Coordinate y: [−6.3, −7.0, 6.1, −1.4, −1.9, 2.6, −5.2, 6, 4.4, 2.3, −3.1, 1.4] Size r: [0.5, 1.0, 1.0, 0.5, 1.0, 1.0, 0.5, 0.5, 0.5, 1.0, 1.0, 1.0]
Scenario 2	[−10, −10]	[10, 10]	12	Coordinate x [−2.5, −7.6, 6.7, 6.6, −6.5, 0.0, −2.0, 0.0, 3.5, −3.6, 2.6, 2.4] Coordinate x [−6.4, −7.0, 3.5, −1.1, −1.1, 2.6, −2.2, 6, 7, 2.3, −3.8, 1.4] Size r: [1.0, 1.0, 2.0, 2.0, 1.0, 1.0, 2.0, 0.5, 2.0, 2.0, 1.0, 1.0]
Scenario 3	[−10, −10]	[10, 10]	24	Coordinate x: [−7.0, −6.7, −2.5, −2.5, 2.0, 1.6, 6.8, 7.1, −6.2, −5.9, 1.0, 0.8, −4.0, 4.2, 4.1, 5.1, 5.6, −3.7, −8.6, −8.1, −8.7, −4.5, −3.7, 6.0] Coordinate y: [−7.0, −6.7, −2.5, −2.5, 2.0, 1.6, 6.8, 7.1, −6.2, −5.9, 1.0, 0.8, −4.0, 4.2, 4.1, 5.1, 5.6, −3.7, −8.6, −8.1, −8.7, −4.5, −3.7, 6.0] Size r: [0.5, 0.5, 0.5, 1.0 0.5, 0.5, 0.5, 0.5, 0.5, 0.5, 0.5, 0.5, 1.0, 1.0, 1.0, 1.0, 1.0, 1.0, 1.0, 1.0, 1, 1.0, 1.0, 1.0]

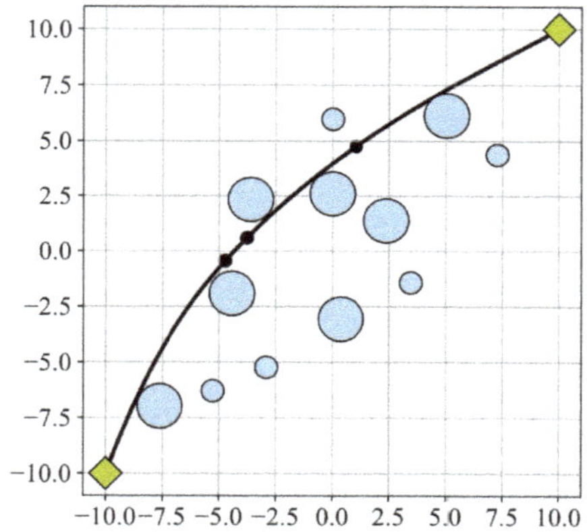

Cost = 29.15
(**a**) GWO.

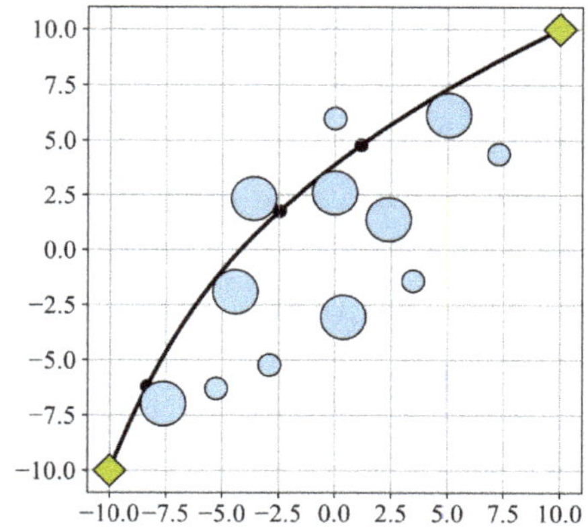

Cost = 29.10
(**b**) WOA.

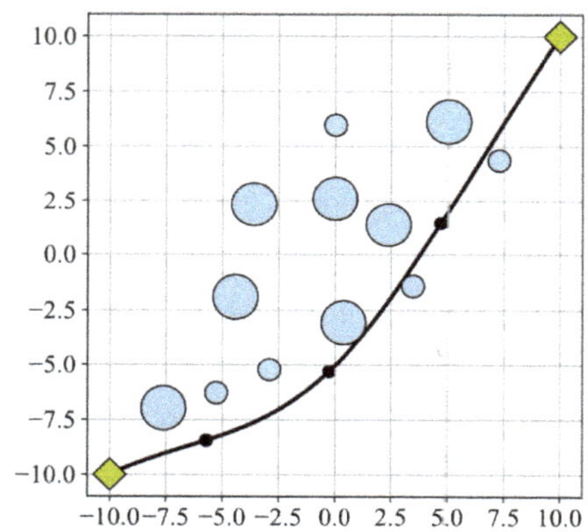

Cost = 29.37
(**c**) PSO.

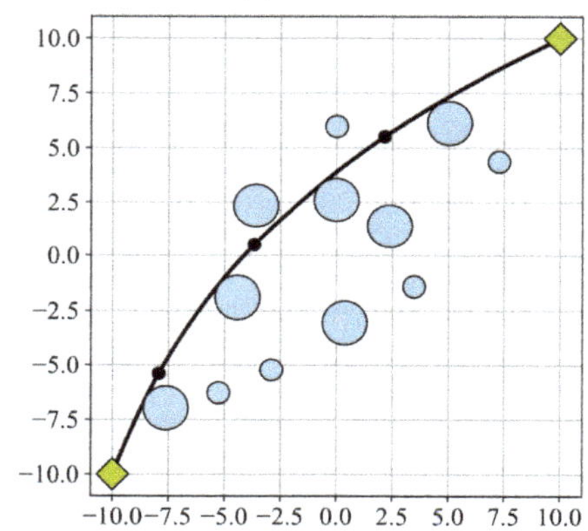

Cost = 29.10
(**d**) SMA.

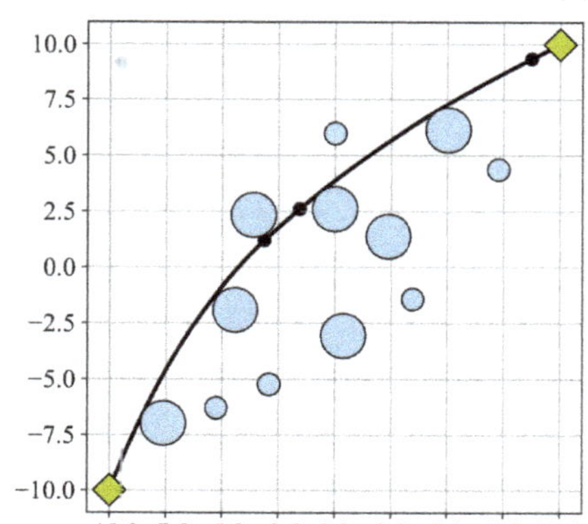

Cost = 29.11
(**e**) LRSMA.

Figure 2. Optimal paths for the first scenario: (**a**) GWO, (**b**) WOA, (**c**) PSO, (**d**) SMA, and (**e**) LRSMA.

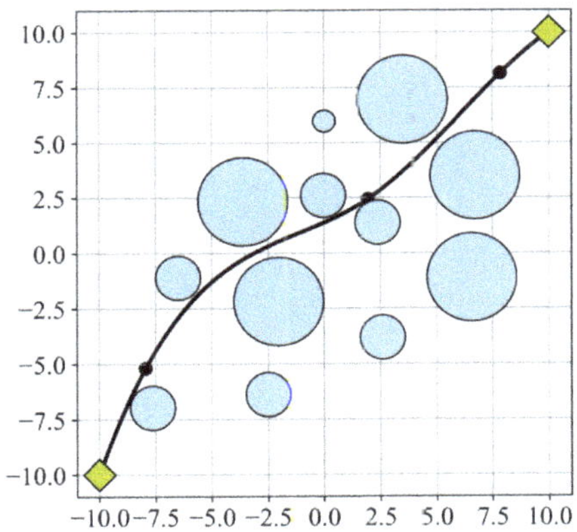

Cost = 30.03
(**a**) GWO.

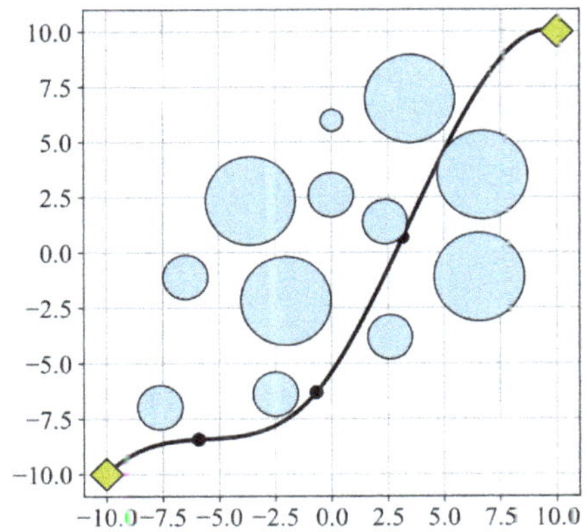

Cost = 30.51
(**b**) WOA.

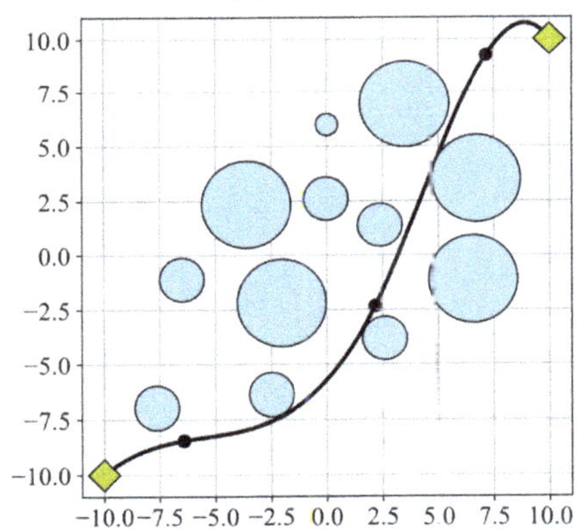

Cost = 31.52
(**c**) PSO.

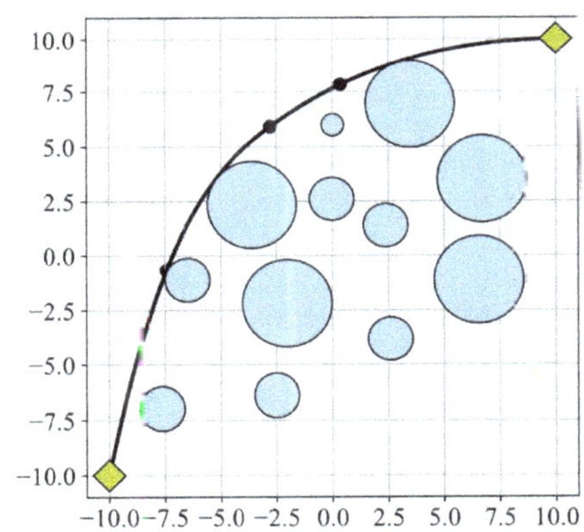

Cost = 31.52
(**d**) SMA.

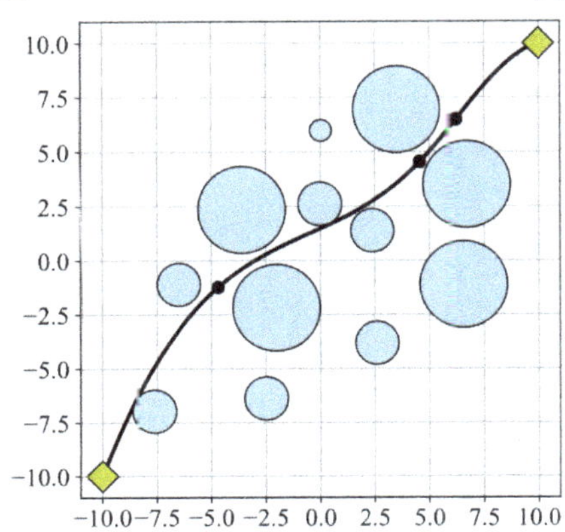

Cost = 29.11
(**e**) LRSMA.

Figure 3. Optimal paths for the second scenario: (**a**) GWO, (**b**) WOA, (**c**) PSO, (**d**) SMA, and (**e**) LRSMA.

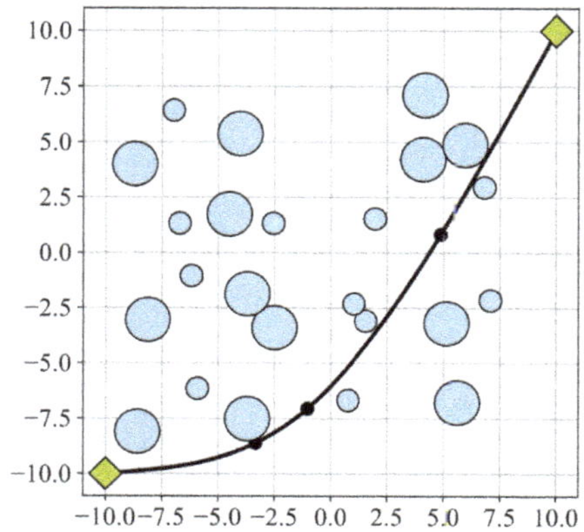

Cost = 29.99
(**a**) GWO.

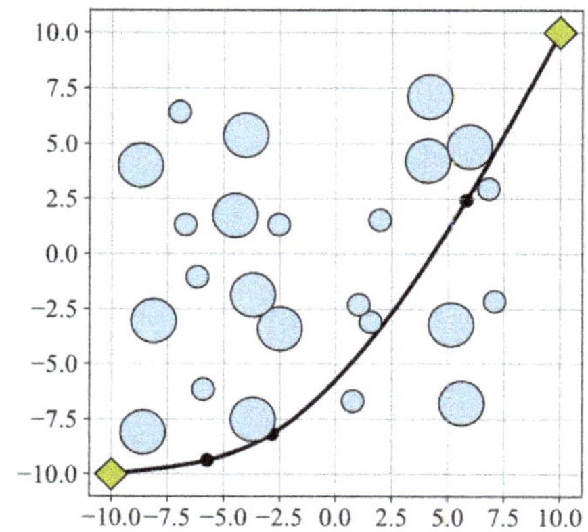

Cost = 29.90
(**b**) WOA.

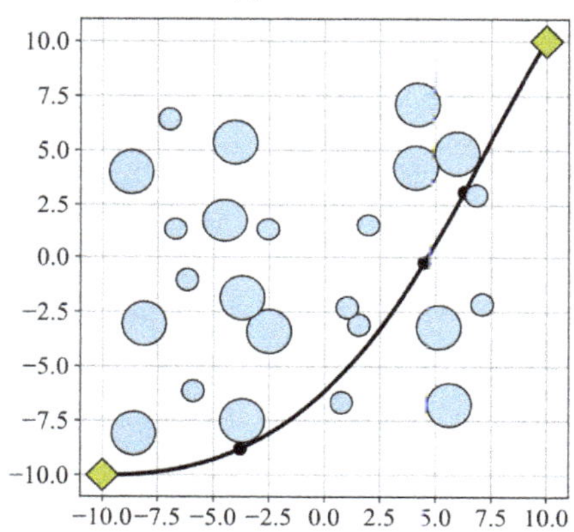

Cost = 30.10
(**c**) PSO.

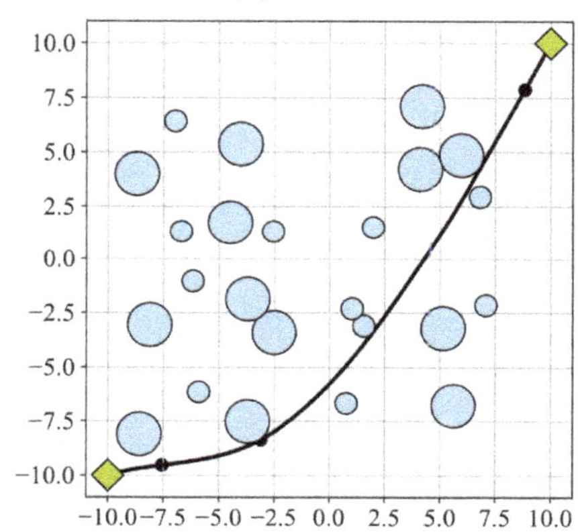

Cost = 29.90
(**d**) SMA.

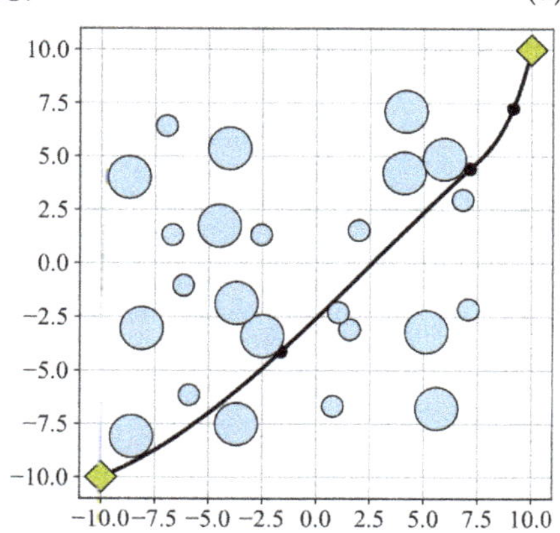

Cost = 28.91
(**e**) LRSMA.

Figure 4. Optimal paths for the third scenario: (**a**) GWO, (**b**) WOA, (**c**) PSO, (**d**) SMA, and (**e**) LRSMA.

5.2. Parameter Setting

The performance of the LRSMA was compared with those of the GWO [46], WOA [47], PSO [35], and SMA [20]. The parameter settings of the comparison algorithms were consistent with those in the reference papers. In GWO, a is a linear convergence factor. In WOA, a is a linearly decreasing convergence factor. For the PSO, the individual experience learning factor $c1$ and the group experience learning factor $c2$ determine the individual and group experience effect on particle trajectories, respectively. The number of path nodes is $m = 3$ for all of the algorithms All of the algorithms were run 100 times in each scenario and took the maximum number of iterations as the termination condition.

The detailed parameter settings are shown in Table 2.

Table 2. Detailed parameters of all algorithms.

	Population Size	Iteration Number	Parameter Value
GWO	30	30	$a_{max} = 2, a_{min} = 0$
WOA	30	30	$a_{max} = 2, a_{min} = 0$
PSO	30	30	$c1 = 0.4, c2 = 0.4, \omega = 1, m = 3$
SMA	30	30	$z = 0.3, m = 3$
LRSMA	30	30	$z = 0.3, Fmin = 1, \mathrm{m} = 3, p_r = 0.5$

5.3. Simulation Results and Discussion

Table 3 shows the overall performance of the four algorithms in three scenarios, including the minimum, mean, and standard deviation of the path length for the best solution, the planning time required to obtain the best optimal path, and the mean planning time. Each scenario and algorithm were tested 100 times.

Table 3. Performance of the four algorithms.

		Path Length			Planning Time	
		Minimum	Mean	Standard Deviation	Processing the Best Optimal Path	Mean
Scenario 1	GWO	29.15	30.14	0.71	2.00	2.00
	WOA	29.10	29.92	0.38	2.08	2.07
	PSO	29.37	30.25	0.37	1.73	1.76
	SMA	29.10	29.88	0.62	2.02	2.01
	LRSMA	29.11	29.81	0.55	2.16	2.18
Scenario 2	GWO	29.13	37.23	12.75	1.95	1.97
	WOA	30.51	33.43	15.70	2.08	2.09
	PSO	31.52	34.31	4.50	1.91	1.85
	SMA	31.52	32.50	1.04	1.97	1.97
	LRSMA	29.11	32.32	0.91	2.06	2.08
Scenario 3	GWO	30.03	42.52	22.78	2.80	2.81
	WOA	29.90	37.38	93.64	3.11	3.04
	PSO	30.10	37.61	9.46	2.81	2.68
	SMA	29.90	35.82	9.04	2.97	2.94
	LRSMA	28.91	34.58	4.81	2.97	2.98

Figures 2–4 show the collision-free shortest path results generated by each algorithm, indicating that all algorithms can generate collision-free paths. In these figures, blue circles represent obstacles, with larger radii indicating larger obstacles. Black dots represent the coordinates of path nodes, while yellow diamonds represent the starting and ending points. A comparison of path lengths between the LRSMA and other algorithms is shown in Tables 4 and 5.

Table 4. Comparison of the percentage change in the minimum path length for the best solution among the LRSMA and three other algorithms.

	GWO-LRSMA % Change in Minimum	WOA-LRSMA % Change in Minimum	PSO-LRSMA % Change in Minimum	SMA-LRSMA % Change in Minimum
Scenario 1	0.14	−0.03	0.82	−0.03
Scenario 2	0.07	4.59	7.65	7.65
Scenario 3	3.73	3.31	3.95	3.31

Table 5. Comparison of the percentage change in the mean path length for the best solution among the LRSMA and three other algorithms.

	GWO-LRSMA % Change in Mean	WOA-LRSMA % Change in Mean	PSO-LRSMA % Change in Mean	SMA-LRSMA % Change in Mean
Scenario 1	1.09	0.37	1.42	0.23
Scenario 2	13.19	3.32	5.80	0.55
Scenario 3	18.67	7.49	8.06	3.46

Figure 2a–d shows the best solutions generated by all the algorithms in the first scenario. In the first scenario, the LRSMA had a slight difference in the minimum and mean path length compared to the other algorithms. The PSO has the shortest solving time, followed by the SMA, WOA, GWO, and LRSMA. This is because the LRSMA's variable neighborhood flight consumes time. In simple scenarios, the LRSMA has little advantage over the other three algorithms. The SMA is less likely to fall into local optimization in simple scenarios, so the LRSMA's rotation perturbation mechanism has difficulty functioning effectively. The LRSMA's planning time is greater than that of the SMA due to its need to monitor the population state and perturb it if it is stagnating.

Figure 3a–d shows the best solution generated by all the algorithms in the second scenario. In the second scenario, the LRSMA reduces the minimum of the path length by 0.07% compared with the GWO, 4.59% compared with the WOA, 7.65% compared with the PSO, and 7.65% compared with the SMA. Meanwhile, the LRSMA reduces the mean of the path length by 13.19% compared with the GWO, 3.32% compared with the WOA, 5.80% compared with the PSO, and 0.55% compared with the SMA. Moreover, the LRSMA is better than the GWO, WOA, PSO, and SMA in terms of the standard deviation of the path length. In addition, the planning time of the GWO, SMA, PSO, and LRSMA differs by 0.1 and is less than that of the WOA. There is a difference of 0.2 between the planning time of the best optimal path and the mean planning time for all algorithms. In complex scenarios, the GWO, WOA, PSO, and SMA take longer as the scenario complexity increases. The strategy proposed by LRSMA accelerates the algorithm's convergence, making the planning times similar across all algorithms. Furthermore, the time consumed by the LRSMA is relatively stable in all scenarios. This indicates that the proposed rotation perturbation mechanism in this paper can improve the algorithm's global search and local exploitation capabilities, therefore improving the convergence efficiency of the algorithm and finding shorter paths.

Figure 4a–d shows the best solution generated by all the algorithms in the third scenario. In the third scenario, the LRSMA reduces the minimum path length by 3.37% compared with the GWO, 3.31% compared with the WOA, 3.95% compared with the PSO, and 3.31% compared with the SMA. Meanwhile, the LRSMA reduces the mean of the path length by 18.67% compared with the GWO, 7.49% compared with the WOA, 8.06% compared with the PSO, and 3.46% compared with the SMA. Moreover, the standard deviation of path length for the best solution obtained by the LRSMA is 4.81, which is less than those of the GWO, WOA, SMA and PSO. Moreover, the LRSMA and SMA have longer planning times than the GWO and PSO and shorter planning times than WOA. In addition, there is a difference of 0.1 between the planning time of the best optimal path and the mean planning time for all algorithms. In complex scenarios, the GWO, WOA, PSO, and SMA take longer time as the scenario complexity increases. The strategy proposed by LRSMA accelerates the algorithm's convergence, making the planning times similar across all algorithms. Furthermore, the time consumed by the LRSMA is relatively stable in all scenarios. This is because the Lévy algorithm enhances the global search capability of the algorithm, while the rotation perturbation increases the local development capability of the algorithm. This indicates that the path modification of LRSMA is the most stable, and the rotation perturbation mechanism proposed in this paper can monitor the algorithm state and jump out of local optima, finding a shorter path.

Overall, the LRSMA produces more stable, collision-free, and shorter paths in both simple and complex scenarios compared with the other algorithms, especially in complex scenarios with obstacles on the path from the starting point to the destination.

Figure 5 shows the iteration curves of the four algorithms in the third scenario, revealing that LRSMA convergence requires more iterations than the SMA. GWO stopped at the 26th iteration. The WOA suddenly finds the most optimal solution at 25 iterations. The PSO has difficulty improving its results after 17 iterations, while SMA has difficulty improving after 19 iterations. By contrast, the LRSMA slowly converges from the beginning to the later stages of iteration, indicating that the strategy proposed by the LRSMA can detect local optimal traps and achieve slow convergence.

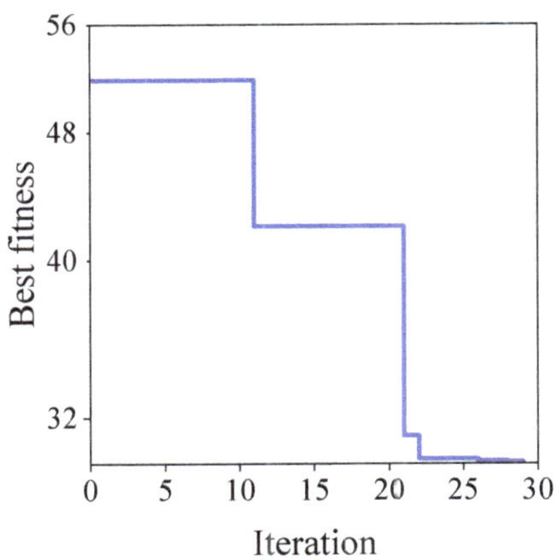

(**a**) GWO.

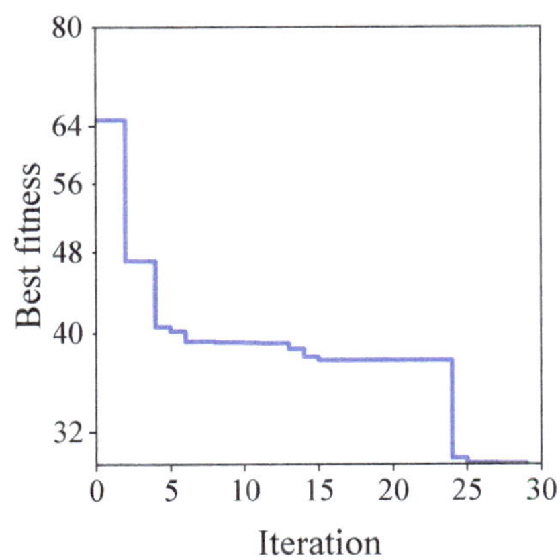

(**b**) WOA.

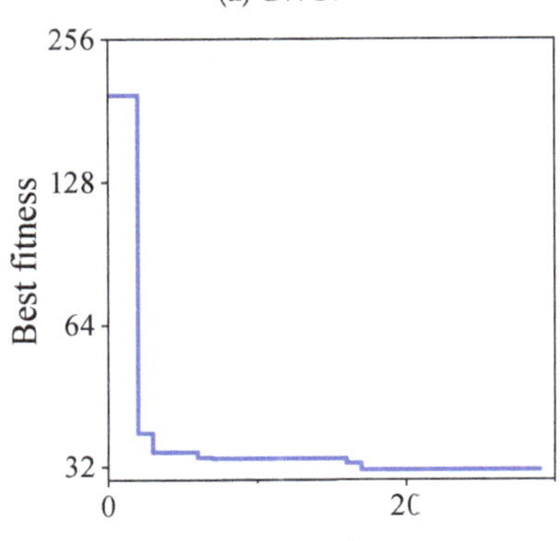

(**c**) PSO.

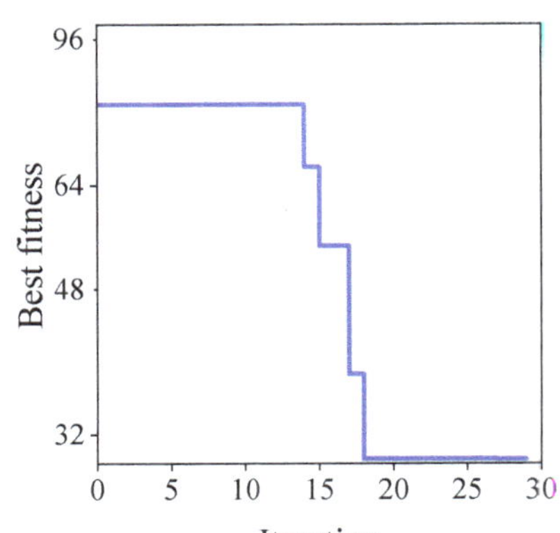

(**d**) SMA.

Figure 5. *Cont.*

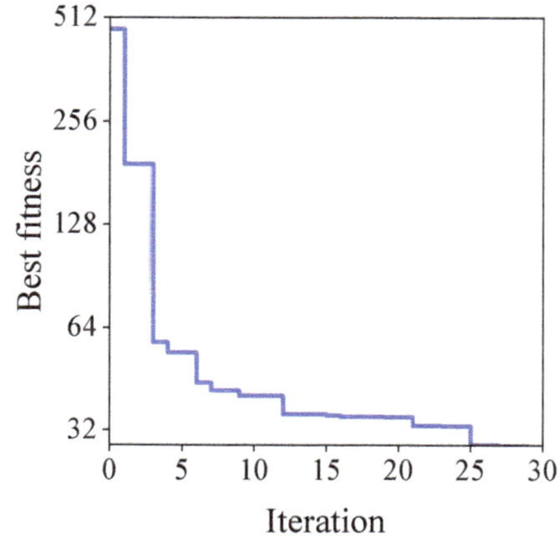

(**e**) LRSMA.

Figure 5. Iteration curves of (**a**) GWO, (**b**) WOA, (**c**) PSO, (**d**) SMA, and (**e**) LRSMA in the third scenario.

Figure 6 illustrates the iterative path planning curve obtained using the LRSMA in both the first and second scenario. The LRSMA demonstrates a gradual evolution throughout. Moreover, the convergence times in the first scenario are shorter than those in the second scenario. As per the LRSMA's definition, it requires time to converge after detecting that the population has entered the local optimum. The convergence then takes place after a sufficient number of iterations. Figures 5 and 6 suggest that 30 iterations can achieve satisfactory performance across all scenarios.

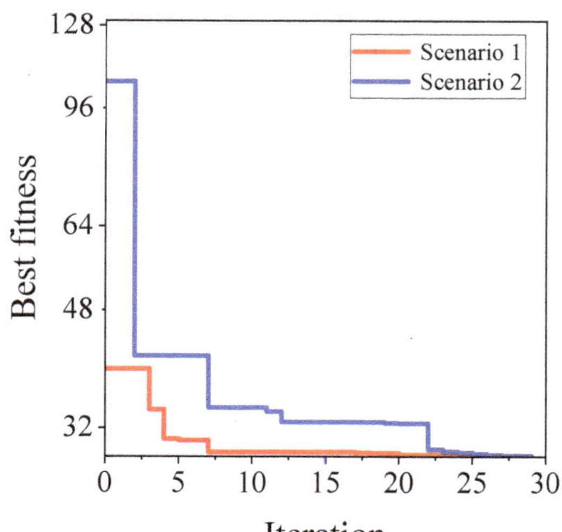

Figure 6. Iterative curves of the LRSMA in the first scenario and the second scenario.

6. Conclusions

This paper proposes a path-planning algorithm using the improved SMA for AMRs, achieving shorter and more stable path solutions. The proposed algorithm addresses the issues of local optima trapping and the lack of an effective escape mechanism in SMA path planning. By introducing an elite learning strategy utilizing variable neighborhood Lévy flight and an individual rotational perturbation and variation mechanism based on tolerance, the proposed algorithm enhances the SMA's global search and local exploitation capabilities and planning capabilities. The experiments conducted on three different obstacle distributions demonstrate that the collision-free path generated by our algorithm has the shortest length, highest accuracy, and greatest stability. However, the LRSMA does not outperform the WOA, PSO, and SMA in terms of time consumption, especially in optimizing the best position variation. The computational cost must be further improved. Additionally, path planning based on SMA in dynamic environments is also an important research direction. Moreover, the algorithm needs to be tested in multi-robot scenarios.

In terms of real-world applications, this algorithm can be applied to autonomous vehicles or slow-moving robots in situations without no road constraints. For autonomous vehicles, it can be used for mapping specific locations, monitoring, or filming scenes. For robots, it can be used in logistics for sorting delivery routes, guiding underground parking in transportation, and other applications. Since this algorithm requires convergence time, it is suitable for situations that do not require urgent and quick responses. For instance, in rescue missions, faster algorithms are needed.

Author Contributions: L.Z. and Y.T. conceived the framework of the paper. L.Z. designed the algorithm and wrote the paper. L.Z. and H.W. designed the experimental scene and performed the experiment. Y.T. and B.L. were responsible for the paper writing guidance and reviewed the paper. C.H. was responsible to the writing refinement. All authors have read and agreed to the published version of the manuscript.

Funding: This work was supported by China Postdoctoral Science Foundation(2021M691114), the Fundamental Research Funds for the Central Universities (2042022kf0049) and the Central Funds Guiding the Local Science and Technology Development (2021Szvup045).

Data Availability Statement: The data that support the findings of this study are available from the corresponding author upon reasonable request.

Conflicts of Interest: The authors declare no conflict of interest.

Abbreviations

Acronyms	Definitions
3D	3-Dimension
AMR	Autonomous Mobile Robot
ACO	Ant Colony Optimization
DRONE	Dynamic Remotely Operated Navigation Equipment
GA	Genetic Algorithm
GWO	Gray Wolf Optimization
LRSMA	Lévy Flight-Rotation Slime Mould Algorithm
MBO	Monarch Butterfly Optimization
NISI	Naturally Inspired Swarm Intelligence
PSO	Particle Swarm Optimization
QPSO	Quantum-behaved Particle Swarm Optimization
RRT	Rapidly-Exploring Random Tree
SIA	Swarm Intelligence Algorithm
SMA	Slime Mould Algorithm
WOA	Whale Optimization Algorithm

References

1. Cui, P.; Yan, W.; Cui, R.; Yu, J.; Ju, Z. Smooth path planning for robot docking in unknown environment with obstacles. *Complex* **2018**, *2018*, 4359036. [CrossRef]
2. Zhao, M.; Lu, H.; Yang, S.; Guo, Y.; Guo, F. A fast robot path planning algorithm based on bidirectional associative learning. *Comput. Ind. Eng.* **2021**, *155*, 107173. [CrossRef]
3. Patle, B.K.; Babu L, G.; Pandey, A.; Parhi, D.R.K.; Jagadeesh, A. A review: On path planning strategies for navigation of mobile robot. *Def. Technol.* **2019**, *15*, 582–606. [CrossRef]
4. Barraquand, J.; Langlois, B.; Latombe, J.-C. Numerical potential field techniques for robot path planning. In Proceedings of the Fifth International Conference on Advanced Robotics 'Robots in Unstructured Environments, Pisa, Italy, 19–22 June 1991; Volume 1012, pp. 1012–1017.

5. Hao, K.; Zhao, J.; Yu, K.; Li, C.; Wang, C. Path planning of mobile robots based on a multi-population migration genetic algorithm. *Sensors* **2020**, *20*, 5873. [CrossRef] [PubMed]
6. Mu, Y.; Li, B.; An, D.; Wei, Y. Three-dimensional route planning based on the beetle swarm optimization algorithm. *IEEE Access.* **2019**, *7*, 117804–117813. [CrossRef]
7. Wang, H.J.; Fu, Z.J.; Zhou, J.J.; Fu, M.Y.; Ruan. L. Cooperative collision avoidance for unmanned surface vehicles based on improved genetic algorithm. *Ocean Eng.* **2021**, *222*, 25. [CrossRef]
8. Ajeil, F.H.; Ibraheem, I.K.; Sahib, M.A.; Humaidi. A.J. Multi-objective path planning of an autonomous mobile robot using hybrid PSO-MFB optimization algorithm. *Appl. Soft Comput.* **2020**, *89*, 106076. [CrossRef]
9. Sun, Y.; Wang, H.P. A novel A* method fusing bio-inspired algorithm for mobile robot path planning. *Eai Endorsed Trans. S* **2022**, *9*, 12. [CrossRef]
10. Feng, Y.H.; Deb, S.; Wang, G.G.; Alavi, A.H. Monarch butterfly optimization: A comprehensive review. *Expert Syst. Appl.* **2021**, *168*, 11. [CrossRef]
11. Dai, Y.; Yu, J.; Zhang, C.; Zhan, B.; Zheng, X. A novel whale optimization algorithm of path planning strategy for mobile robots. *Appl. Intell.* **2022**. [CrossRef]
12. Luo, J.; Liu, Z.W. Novel grey wolf optimization based on modified differential evolution for numerical function optimization. *Appl. Intell.* **2020**, *50*, 468–486. [CrossRef]
13. Zhang, L.; Zhang, Y.J.; Li, Y.F. Mobile robot path planning based on improved localized particle swarm optimization. *IEEE Sens. J.* **2021**, *21*, 6962–6972. [CrossRef]
14. Zhang, Y.; Gong, D.-W.; Zhang, J.-H. Robot path planning in uncertain environment using multi-objective particle swarm optimization. *Neurocomputing* **2013**, *103*, 172–185. [CrossRef]
15. Wu, L.; Huang, X.D.; Cui, J.G.; Liu, C.; Xiao, W.S. Modified adaptive ant colony optimization algorithm and its application for solving path planning of mobile robot. *Expert Syst. Appl.* **2023**, *215*, 22. [CrossRef]
16. Liu, C.; Wu, L.; Huang, X.D.; Xiao, W.S. Improved dynamic adaptive ant colony optimization algorithm to solve pipe routing design. *Knowledge-Based Syst.* **2022**, *237*, 13. [CrossRef]
17. Nakagaki, T.; Yamada, H.; Ueda, T. Interaction between cell shape and contraction pattern in the physarum plasmodium. *Biophys. Chem.* **2000**, *84*, 195–204. [CrossRef]
18. Cai, Z.; Xiong, Z.; Wan, K.; Xu, Y.; Xu, F. A node selecting approach for traffic network based on artificial slime mold. *IEEE Access.* **2020**, *8*, 8436–8448. [CrossRef]
19. Hassan, M.H.; Kamel, S.; Abualigah, L.; Eid, A. Development and application of slime mould algorithm for optimal economic emission dispatch. *Expert Syst. Appl.* **2021**, *182*, 28. [CrossRef]
20. Agarwal, D.; Bharti, P.S. Implementing modified swarm intelligence algorithm based on slime moulds for path planning and obstacle avoidance problem in mobile robots. *Appl. Soft Comput.* **2021**, *107*, 107372. [CrossRef]
21. Naik, M.K.; Panda, R.; Abraham, A. An entropy minimization based multilevel colour thresholding technique for analysis of breast thermograms using equilibrium slime mould algorithm. *Appl. Soft Comput.* **2021**, *113*, 107955. [CrossRef]
22. Abdel-Basset, M.; Mohamed, R.; Chakrabortty, R.K.; Ryan, M.J.; Mirjalili, S. An efficient binary slime mould algorithm integrated with a novel attacking-feeding strategy for feature selection. *Comput. Ind. Eng.* **2021**, *153*, 107078. [CrossRef]
23. Li, S.; Chen, H.; Wang, M.; Heidari, A.A.; Mirjalili, S. Slime mould algorithm: A new method for stochastic optimization. *Future Gener. Comput. Syst.* **2020**, *111*, 300–323. [CrossRef]
24. Dai, X.; Long, S.; Zhang, Z.; Gong, D. Mobile robot path planning based on ant colony algorithm with A* heuristic method. *Front. Neurorob.* **2019**, *13*. [CrossRef] [PubMed]
25. Szczepanski, R.; Bereit, A.; Tarczewski, T. Efficient local path planning algorithm using artificial potential field supported by augmented reality. *Energies* **2021**, *14*, 6642. [CrossRef]
26. Qi, J.; Yang, H.; Sun, H. Mod-RRT*: A sampling-based algorithm for robot path planning in dynamic environment. *IEEE Trans. Ind. Electron.* **2021**, *68*, 7244–7251. [CrossRef]
27. Chang, Y.-C.; Shi, Y.; Dostovalova, A.; Cao, Z.; Kim, J.; Gibbons, D.; Lin, C.-T. Interpretable fuzzy logic control for multirobot coordination in a cluttered environment. *IEEE Trans. Fuzzy Syst.* **2021**, *29*, 3676–3685. [CrossRef]
28. Qu, H.; Yang, S.X.; Willms, A.R.; Yi, Z. Real-time robot path planning based on a modified pulse-coupled neural network model. *IEEE Trans. Neural Netw.* **2009**, *20*, 1724–1739. [CrossRef]
29. Nazarahari, M.; Khanmirza, E.; Doostie, S. Multi-objective multi-robot path planning in continuous environment using an enhanced genetic algorithm. *Expert Syst. Appl.* **2019**, *115*, 106–120. [CrossRef]
30. Zhao, Z.; Jin, M.; Lu, E.; Yang, S.X. Path planning of arbitrary shaped mobile robots with safety consideration. *IEEE Trans. Intell. Transp. Syst.* **2022**, *23*, 16474–16483. [CrossRef]
31. Cui, Y.; Hu, W.; Rahmani, A. A reinforcement learning based artificial bee colony algorithm with application in robot path planning. *Expert Syst. Appl.* **2022**, *203*, 117389. [CrossRef]
32. Das, P.K.; Jena, P.K. Multi-robot path planning using improved particle swarm optimization algorithm through novel evolutionary operators. *Appl. Soft Comput.* **2020**, *92*, 106312. [CrossRef]
33. Wang, L.; Kan, J.; Guo, J.; Wang, C. 3D path planning for the ground robot with improved ant colony optimization. *Sensors* **2019**, *19*, 815. [CrossRef] [PubMed]
34. Teng, Z.-j.; Lv, J.-l.; Guo, L.-w. An improved hybrid grey wolf optimization algorithm. *Soft Comput.* **2019**, *23*, 6617–6631. [CrossRef]

35. Fernandes, P.B.; Oliveira, R.C.L.; Fonseca Neto, J.V. Trajectory planning of autonomous mobile robots applying a particle swarm optimization algorithm with peaks of diversity. *Appl. Soft Comput.* **2022**, *116*, 108108. [CrossRef]
36. Yu, C.; Heidari, A.A.; Xue, X.; Zhang, L.; Chen, H.; Chen, W. Boosting quantum rotation gate embedded slime mould algorithm. *Expert Syst. Appl.* **2021**, *181*, 115082. [CrossRef]
37. Nguyen, T.-T.; Wang, H.-J.; Dao, T.-K.; Pan, J.-S.; Liu, J.-H.; Weng, S. An improved slime mold algorithm and its application for optimal operation of cascade hydropower stations. *IEEE Access.* **2020**, *8*, 226754–226772. [CrossRef]
38. Rizk-Allah, R.M.; Hassanien, A.E.; Song, D. Chaos-opposition-enhanced slime mould algorithm for minimizing the cost of energy for the wind turbines on high-altitude sites. *ISA Trans.* **2022**, *121*, 191–205. [CrossRef]
39. Houssein, E.H.; Helmy, B.E.-d.; Rezk, H.; Nassef, A.M. An efficient orthogonal opposition-based learning slime mould algorithm for maximum power point tracking. *Neural Comput. Appl.* **2022**, *34*, 3671–3695. [CrossRef]
40. Houssein, E.H.; Mahdy, M.A.; Blondin, M.J.; Shebl, D.; Mohamed, W.M. Hybrid slime mould algorithm with adaptive guided differential evolution algorithm for combinatorial and global optimization problems. *Expert Syst. Appl.* **2021**, *174*, 114689. [CrossRef]
41. Liu, Y.; Liu, S. Unscented sigma point guided quasi-opposite slime mould algorithm and its application in engineering problem. *Appl. Res. Comput.* **2022**, *39*, 2709–2716. [CrossRef]
42. Wang, J.; Elia, N. Distributed averaging under constraints on information exchange: Emergence of lévy flights. *IEEE Trans. Automat. Contr.* **2012**, *57*, 2435–2449. [CrossRef]
43. Mantegna, R.N. Fast, accurate algorithm for numerical simulation of lévy stable stochastic processes. *Phys. Rev. E* **1994**, *49*, 4677–4683. [CrossRef] [PubMed]
44. Hwang, C.-R. Simulated annealing: Theory and applications. *Acta Appl. Math.* **1988**, *12*, 108–111. [CrossRef]
45. Lian, J.; Yu, W.; Xiao, K.; Liu, W. Cubic spline interpolation-based robot path planning using a chaotic adaptive particle swarm optimization algorithm. *Math. Probl. Eng.* **2020**, *2020*, 1–20. [CrossRef]
46. Sang-To, T.; Le-Minh, H.; Mirjalili, S.; Wahab, M.A.; Cuong-Le, T. A new movement strategy of grey wolf optimizer for optimization problems and structural damage identification. *Adv. Eng. Softw.* **2022**, *173*, 31. [CrossRef]
47. Mirjalili, S.; Lewis, A. The whale optimization algorithm. *Adv. Eng. Softw.* **2016**, *95*, 51–67. [CrossRef]

drones

MDPI

Article

An Efficient Framework for Autonomous UAV Missions in Partially-Unknown GNSS-Denied Environments

Michael Mugnai *,†, Massimo Teppati Losé †, Edwin Paúl Herrera-Alarcón, Gabriele Baris, Massimo Satler and Carlo Alberto Avizzano

Institute of Mechanical Intelligence, Sant'Anna School of Advanced Studies, 56127 Pisa, Italy; massimo.teppatilose@santannapisa.it (M.T.L.); edwinpaul.herreraalarcon@santannapisa.it (E.P.H.-A.); gabriele.baris@santannapisa.it (G.B.); massimo.satler@santannapisa.it (M.S.); carloalberto.avizzano@santannapisa.it (C.A.A.)
* Correspondence: michael.mugnai@santannapisa.it
† These authors contributed equally to this work.

Abstract: Nowadays, multirotors are versatile systems that can be employed in several scenarios, where their increasing autonomy allows them to achieve complex missions without human intervention. This paper presents a framework for autonomous missions with low-cost Unmanned Aerial Vehicles (UAVs) in Global Navigation Satellite System-denied (GNSS-denied) environments. This paper presents hardware choices and software modules for localization, perception, global planning, local re-planning for obstacle avoidance, and a state machine to dictate the overall mission sequence. The entire software stack has been designed exploiting the Robot Operating System (ROS) middleware and has been extensively validated in both simulation and real environment tests. The proposed solution can run both in simulation and in real-world scenarios without modification thanks to a small sim-to-real gap with PX4 software-in-the-loop functionality. The overall system has competed successfully in the Leonardo Drone Contest, an annual competition between Italian Universities with a focus on low-level, resilient, and fully autonomous tasks for vision-based UAVs, proving the robustness of the entire system design.

Keywords: UAV; MAV; mission planning; collision avoidance; navigation in partially-known environments; visual-based navigation; GNSS-denied

1. Introduction

Recent technological advances have raised interest in autonomous Micro-Aerial Vehicles (MAVs) for different applications from agriculture to industry. MAVs assist human operators in repetitive or dangerous tasks, such as powerline maintenance [1] and Search and Rescue (SAR) activities. The fields where autonomous drones are reducing the complexity and risk of different human chores are increasing daily. Current challenges in MAV research involve using only local sensing and pushing the efficiency of the algorithms further in order to improve the agent's self-sufficiency. Anyway, the design of a fully autonomous framework for a small platform with limited payload and sensing capabilities poses significant challenges. The design of such systems is a process that requires the continuous improvement of both the hardware and the software components to find the optimal trade-off between performance and flight duration.

For example, from 2018 to 2021 international agencies such as DARPA funded the Subterranean Challenge (SubT) Search and Rescue competition. The challenge aimed to boost the research and the development of technologies that can support operations in complex and diverse underground settings which pose significant risks and challenges for military and civilian first responders. In this challenge, the teams had to address perception, networking, and navigation problems that arise when using autonomous robots in unpredictable subsurface environment conditions in order to obtain a map of the scenario. The teams that participated in the challenge were Team CoStar [2], Team CSIRO Data61 [3], Team CTU-CRAS-NORLAB [4], Team Cerberus [5], Team Explorer (www.cmu.edu/news/stories/archives/2021/september/darpa-subt-finals.html (accessed on 17 July 2023)), Team Marble (www.colorado.edu/today/2021/09/24/engineers-take-home-500000-international-underground-robotics-competition (accessed on 17 July 2023)), Team Coordinated Robotics, and Team Robotika. These teams implemented different heterogeneous multi-robot systems for the SAR exploration task that can be split into two categories. On one side, there was a significant variety of ground robots, including legged robots, crawling hexapods, and both wheeled and tracked rovers. On the other side, the flying robots consisted of custom-built multirotors, which are optimal for autonomous operations relying only on onboard sensing due to their maneuverability, vertical take-off and landing capability, small factor size, and low cost.

The flying platforms of the DARPA Subterranean Challenge present different frameworks towards full aerial autonomy, for which environment perception sensors strongly rely on 3D LiDAR technology due to the lack of good visibility in certain areas. Nevertheless, most MAVs used cameras, as the subterranean exploration mission was oriented towards identifying and localizing specific objects within the environment, which was incrementally mapped as the robots navigate in it.

Depending on the scenario, lighting conditions, and tasks to be accomplished, autonomous platforms could generally operate based on camera systems only regardless of whether they used a 3D LiDAR. LiDAR technology is fast and accurate, and the amount of information obtained at each sample time is huge. Therefore, its usage in an autonomous system requires an onboard computer with the computational capacity to handle a large amount of information in real-time. In addition, its price with respect to other onboard sensors is significantly higher. Vision-based navigation relying only on onboard sensors such as cameras and inertial measurement units (IMU) is a plausible option. Cameras are promising sensors for small drones because they can gather rich information and span wide fields of view for comparable mass. An example of a vision-based MAV autonomous stack is Ingenuity (https://mars.nasa.gov/technology/helicopter/ (accessed on 17 July 2023)) [6], a 1.8 kg multirotor designed to autonomously fly for up to 90 s in the atmosphere of Mars. Ingenuity is the first aircraft to accomplish autonomous exploration on a planet relying entirely on onboard sensors. Its sensors include inertial measurement units for accelerations and angular rates, a laser rangefinder to measure its distance to the ground, and a navigation camera to take pictures.

New challenges related to vision-based tasks in autonomous drones have appeared in recent years. For instance, the MBZIRC 2020 competition's first challenge was intended for an autonomous UAV to localize, track, and neutralize an intruder UAV in an outdoor open space. The International Conference of Unmanned Aircraft Systems (ICUAS) 2022 proposed a UAV competition in which the drone had to quickly explore an environment, detect a static target, and precisely deliver an object to the target.

1.1. Competition Details and Proposed System Overview

Leonardo S.p.A., an Italian aerospace and defense company, created the Drone Contest challenge (https://www.leonardo.com/it/innovation-technology/open-innovation/drone-contest (accessed on 17 July 2023)) in 2019, a yearly event among Italian universities to solve vision-based tasks in an indoor GNSS-denied environment using a UAV. In the 2022 competition, the environment was partially unknown: teams knew the position of many obstacles, while others were randomly placed before each round. Each round consisted of two phases: first, the drone had to explore the environment looking for a moving target, then it had to be tracked for at least 10 s. In the second phase, a sequence of actions was provided to the team, with each action having a score. The teams decided on the execution order of the actions. Actions could be of two types: precision landing, or taking a picture of a marker. The winner was the team with the highest score over the three rounds of the competition. In detail, the autonomous platform had to navigate autonomously in an environment of dimensions 20 m × 10 m with a maximum height of 3 m while relying only on vision-based sensors. Within the environment, the drone had to search for a mobile ground robot and track it for a minimum amount of time. Then, a sequence of actions was provided to the UAV. These actions were divided between precision landing and taking pictures of fiducial markers positioned on the vertical faces of the obstacles.

This paper introduces a ROS framework capable of performing autonomous missions in a partially unknown GNSS-denied environment and presents a solution for the Drone Contest challenge with a vision-based autonomous MAV based on the open-source PX4 firmware. The proposed solution was tested in a real environment during competition. (a video of the competition can be found at https://www.youtube.com/watch?v=HiNrlqCKWlQ&t=3133s (accessed on 17 July 2023)). While within the Drone Contest challenge a partial map of the environment is available, this work considers the possibility of variations within the environment such as the presence of new unknown obstacles, extending the global planning strategy to work alongside a local re-planning module. In previous works, we proposed an object-oriented exploration and volumetric mapping of an unknown environment [7] and a 3D low-dimensional topological graph generated on previously explored environments, which is useful for global path planning [8]. In this work, the navigation graph is extended as the entry point of a trajectory planner able to guide UAVs in partially known environments through an obstacle avoidance module that overrides desired UAV velocities in order to evade unmapped obstacles. The overall system is handled by a mission planner that drives the UAV during the entire challenge using a finite-state machine.

1.2. Related Works

This section reviews works presented in the literature regarding fully autonomous frameworks, then works concerning their evolution towards vision-based platforms, and finally provides a comparison between the presented hardware and software modules with other state-of-the-art systems with similar characteristics.

Over the last decade, the research topic of aerial navigation has driven towards the challenging problem of employing solely onboard sensors. The earliest overall frameworks, such as Shen et al. [9], proposed a platform based on an IMU, monocular camera, and 3D LiDAR as unique sensors for the autonomous UAV. Tomic et al. [10] introduced a quadcopter with navigation based on a combination of Visual Odometry from a stereo camera and 3D LiDAR scanning. These works focused on providing greater robustness for both indoor and outdoor SAR scenarios. Both frameworks were validated through experiments that highlighted navigation performance through evaluations of the pose estimation and loop closure in different indoor and outdoor scenarios. On the other side, Fraundorfer et al. [11] presented a quadrotor with navigation that depended solely on visual information, using a front-facing stereo camera and a down-facing optical flow camera. Their work emphasized accurate perception of the environment by using two different sensors to analyze different planes. The navigation was validated by testing it with an onboard planning and mapping algorithms, while the loop closure was done offboard.

Nowadays, the most common architecture for autonomous custom-built drones uses two boards: a companion computer for high-level modules and a low-level attitude controller board. For the latter, Pixhawk (https://pixhawk.org (accessed on 17 July 2023)) hardware has become an open-source standard. Up to its original work on autonomous vision-based MAVs, Pixhawk [12] has become a flexible research module to standardize the low-level controller and the state estimation. An important advantage of Pixhawk boards is that they are supported by the most popular autopilots for autonomous vehicles: PX4 [13] and ArduPilot (https://ardupilot.org/ (accessed on 17 July 2023)).

Another option for low-level control and estimation is the board used by Loianno et al. [14] with the Qualcomm® Snapdragon™ Flight™, which aims to achieve lightweight visual–inertial navigation. This board features a downward-facing VGA camera with 160° field of view, a VGA stereo camera, and a 4 K camera, all packed into a board of dimensions 58 mm × 40 mm. Ge et al. [15] used this setup in a MAV swarm system for decentralized vision-based detection and tracking strategies. On the same topic, Thakur et al. [16] instead used the successor of the previously mentioned board. These works depict examples of lightweight vision-based platforms using smartphone processors with weights under 450 g. These platforms are lightweight enough to be both agile and resistant in case of collisions, and as such are a good prototype for testing new algorithms. However, they are currently not designed to fulfill long high-level missions due to their lack of high computational power and loe level of flight autonomy.

The Size, weight, and power (SWaP) constraints perfectly describe the trade-offs for designing MAV platforms. For this reason, autonomous flying system developers often have to accurately choose the number of sensors to use, then estimate the computational cost needed for processing their information and the minimum accuracy they want to reach in terms of both the navigation error and the trajectory tracking performance. Liu et al. [17] highlighted the evolution of technologies for autonomous multirotors as well as the need for bigger sensors, more computational power, and better batteries for GNSS-denied environments. On one side, platforms under 450 g (wheelbase under 0.35 m from tip to tip with 3-inch propellers) continue to see reduced form factor and battery size while increasing their flight time and performance. On the other side, the evolution of larger prototypes (wheelbase over 0.75 m) has significantly improved in terms of endurance, sensor carrying technology, and computational power. However, their weight remains over 3.5 kg.

Moon et al. [18] described the challenges involved in Autonomous Drone Racing (ADR) and analyzed the technologies and results in the IROS competitions in 2016 and 2017. ADR is a vision-based competition in which the drone has to navigate in a cluttered indoor environment while detecting and crossing through gates and relying only on onboard resources. Several of the platforms used in this competition are based on commercial off-the-shelf drones [19,20] modified for the competition rules, while other teams build their own custom drones [21,22]. Other published platforms relate to specific competitions or programs, such as the drone from the AlphaPilot [23] ADR Competition (www.lockheedmartin.com/en-us/news/events/ai-innovation-challenge.html (accessed on 14 March 2023)), the team from Politecnico di Milano [24] in the Leonardo Drone Contest, and Mohta et al. [25,26] from the DARPA Fast Lightweight Autonomy program.

Several projects have been published with the aim of reproducibility and to encourage open-sourcing. For example, the Open Vision Computer (https://osrf.github.io/ovc/ (accessed on 17 July 2023)) project (OVC) was tested on a drone [27]. Oleynikova et al. [28] and Kompis et al. [29] have published both the hardware and software setup of their fully autonomous projects. Finally, other projects have published their integrated frameworks to display their performance in accomplishing specific problems, such as SAR tasks [30], sensor placement [31], and using computational-constrained platforms [32].

Table 1 summarizes the overall specifications and hardware information of the previously mentioned frameworks. All of the mentioned platforms navigate autonomously while relying only on vision sensors. These platforms were presented to solve different problems, such as racing, high-speed navigation, exploration in unknown cluttered environments, dense mapping and reconstruction of real-world scenarios, and sensor allocation. Compared to the other solutions, our proposed platform shows a fair trade-off between the available computational power of the companion computer and its small size, making it particularly suitable for performing complex navigation and exploration tasks in cluttered environments while relying primarily on vision. The main disadvantages are the limited flight time and the limited angular range to detect obstacles due to the lack of LiDAR sensors.

For autonomous navigation and obstacle avoidance in GNSS-denied environments, the platform must continuously preserve information about its pose with different localization strategies. Different works in the literature have focused on the development of several individual modules in simulation [33,34] without developing a comprehensive solution in real-world experiments. In this work, we present a complete framework that enables a drone to autonomously navigate in partially known environments even in the absence of a reliable position reference such as a GNSS.

Table 1. Summarized characteristics of autonomous vision-based middle-size platforms described in the literature.

Reference	Wheelbase/ Weight	Flight Time	Companion Computer	FCU/ Autopilot	Sensors
Rojas-Perez et al. [19]	47 cm ≈560 g	<20 min	Odroid XU4	Parrot Board Bebop Autopilot	Sensors from Parrot Bebop 2
Kaufmann et al. [21]	49 cm 0.950 g	N.S.	Aaeon UP Board	Qualcomm Snapdragon Flight N.S.	Intel RS R200 or Front-facing Camera
Li et al. [20]	51 cm ≈400 g	≈12 min	Parrot P7 dual-core CPU Cortex A9	Parrot Board Paparazzi	Sensors from Parrot Bebop 1
Jung et al. [22]	73 cm 2.26 kg	12 min	NVIDIA Jetson TK1	In-house	ZED Stereo Camera LiDAR Lite V3 PX4FLow
Mohta et al. [25,26]	75 cm ≈2.8 kg	≈5 min	Intel NUC i7	Pixhawk 4 PX4	2 × FLIR Chamaleon VN-100 IMU LiDAR Lite V3 Hokuyo UTM-30LX *
Quigley et al. [27]	40 cm ≈1.3 kg	15 min	NVIDIA Jetson TX2	Pixhawk N.S.	FPGA with: 2 × 1300 Python Cameras 4 × Fish-eye Cameras
AlphaPilot [23]	70 cm 3.4 kg	N.S.	NVIDIA Jetson Xavier AGX	N.S. N.S.	2 × Leopard IMX 264 IMU Bosch BMI088 LiDAR Lite V3
Sandino et al. [30]	73 cm Under 2 kg	8 min	Aaeon UP Squared Board	Pixhawk 4 PX4	Intel RS D435 Intel RS T265 HBV-1615 RGB camera TFMini Plus
Oleynikova et al. [28]	70 cm ≈2.5 kg	15 min	Intel NUC i7	mRo Pixhawk PX4	VI Sensor ([35]) Intel RS D435
Kompis et al. [29]	75 cm ≈1.6 kg **	15 min	Intel NUC i5/i7 or NVIDIA Jetson TX2	Pixhawk 4 PX4	Stereo VI Units ***: Intel RS D455 Alphasense Core Skybotix VI-Sensor ZED 2 Structure Core Venus RGB BlueFox
Campos-Macias et al. [32]	59 cm 1.3 kg	<20 min	Intel Atom x7-Z8750	STM32F427V N.S.	Intel RS D435 Intel RS T265
Roggi et al. [24]	53 cm 2.85 kg	≈20 min	NVIDIA Jetson Xavier NX	Pixhawk 4 PX4	ZED Stereo Camera 2 × OpenMV H7 Plus LiDAR Lite V3
Stephens et al. [31]	60 cm 1.9 kg	≈20 min	Intel NUC i7	Pixhawk Pixracer PX4	Intel RS D435i Intel RS T265
our quadcopter	54 cm 1.5 kg	≈12 min	Lattepanda Alpha 864s	Pixhawk Durandal PX4	Intel RS D435i Intel RS T265 TFMini-S

N.S.: Not Specified. * The LiDAR sensor was used only for dense mapping. ** Weight without the VI camera and its respective supports. *** The platform was tested with different VI cameras, not simultaneously.

The rest of this paper is organized as follows. Section 1.2 reviews fully autonomous UAV frameworks employed in applications similar to the proposed one. Section 2 presents the proposed custom UAV's architecture, including both the hardware and software modules. In Section 3, the navigation stack is exposed in depth. Simulations and real experiments are described in Section 4, and the conclusions are presented in Section 5.

2. System Overview

This section describes the UAV architecture, consisting of hardware components and software modules, that guarantees autonomous flight while relying only on onboard vision-based sensing. The competition rules forbid the use of LiDARs and GNSS-based positioning systems.

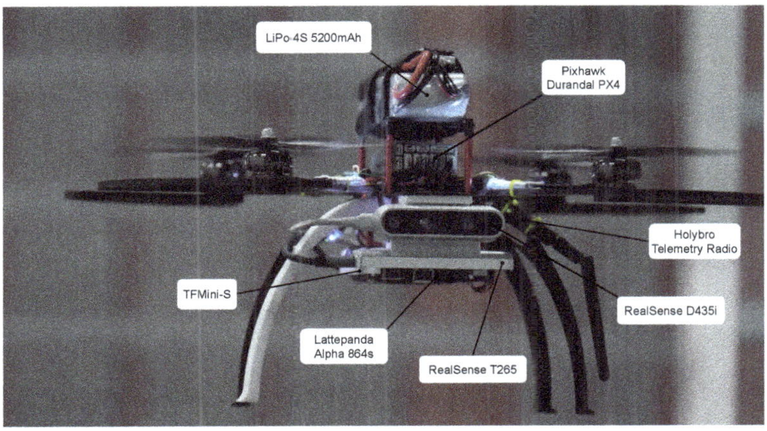

Figure 1. One of the prototypes used in the Leonardo Drone Contest 2022 competition, with 7-inch propellers (wheelbase 51.5 cm from tip to tip) and a LattePanda Alpha 864s for onboard processing. The low-level control board is a Pixhawk Durandal. The drone is equipped with a front-facing camera for stereo depth and an RGB channel and a down-facing fisheye camera for VIO and object tracking.

2.1. Hardware Overview

The MAV employed in this work in Figure 1 is an evolution of the experimental platform presented in our previous work [7], a custom quadrotor built with commercially available off-the-shelf products. The chosen frame is a carbon fiber quadcopter with a wheelbase without propellers of 36 cm, intended for propellers up to 8 inches (maximum wheelbase of up to 54 cm from tip to tip).

The perception task is accomplished by two stereo-cameras, one front-facing for obstacle avoidance and one down-facing for Visual–Inertial Odometry (VIO). In particular, the former is an Intel Realsense D435i camera able to provide depth-maps through stereo-vision information, while the latter is an Intel Realsense T265 camera able to provide MAV's pose in a local reference frame. Both the depth and VIO are computed onboard these devices through proprietary firmware.

Because the environment is only partially known and front pictures must be taken as a task during the mission, it is necessary to employ a front-facing camera for the local planner module. A down-facing camera can work instead as an optical flow module; thanks to its wide fisheye's Field Of View (FOV), the T265 can observe the UAV's surroundings ($\approx$5 m wide at an altitude of 1.5 m) both for visual navigation and mission tasks such as ground patrolling. The chosen altimeter is a Benewake TFmini-S that allows distances up to 12 m to be measured in a small form-factor. The flight controller is a Pixhawk Durandal board with PX4 Autopilot firmware v1.12.3.

2.2. Software Overview

The UAV estimates its 3D pose and velocity relative to a local starting position through Visual Inertial Odometry (VIO). The resulting pose estimation is combined with the Pixhawk's IMUs and altimeter through an Extended Kalman Filter (EKF) implemented in PX4 and computed inside the flight controller. In particular, the EKF fuses the on-board IMU data (two accelerometers, two gyros, and one magnetometer) with external pose estimates, which in our case are computed by the VIO and the drone altitude sensor. The flight controller board runs multiple instances of the Extended Kalman Filter. By comparing estimate consistencies of the EKF instances, an EKF selector is able to determine which filter output is the more confident, thereby monitoring sensor faults. This enables detection and isolation of faults such as sudden changes in IMU bias, saturation, and stuck data.

As shown in Figure 2, the estimated pose is provided to both the mission planner and global planner. The mission planner, described in depth in Section 3.6, dictates the overall behavior of the UAV during the mission through a Finite State Machine that spans from the initial takeoff to each mission task needed to fulfill the competition or challenge. It requests goal poses in a map frame for the global planner, which is divided into two components. The global planner first performs a fast exploration of a precomputed navigation graph, generated as specified in Section 3.1. With the obtained feasible sequence of intermediate waypoints, the current pose is connected to the goal pose through rectilinear segments, which are proven to lie in the obstacle-free region of the known map. In Section 3.2, the trajectory reference is generated from the interpolation of the provided waypoints while obeying dynamic constraints. Collisions with unknown obstacles or due to trajectory tracking errors are handled by the local planner, which steers the MAV by altering the trajectory reference velocity vector, as explained in Section 3.3. The output trajectory is tracked with velocity commands, generated as in Section 3.4; these are forwarded to the Pixhawk's micro-controller, which converts the center-of-mass desired velocities as commands for the actuator drivers.

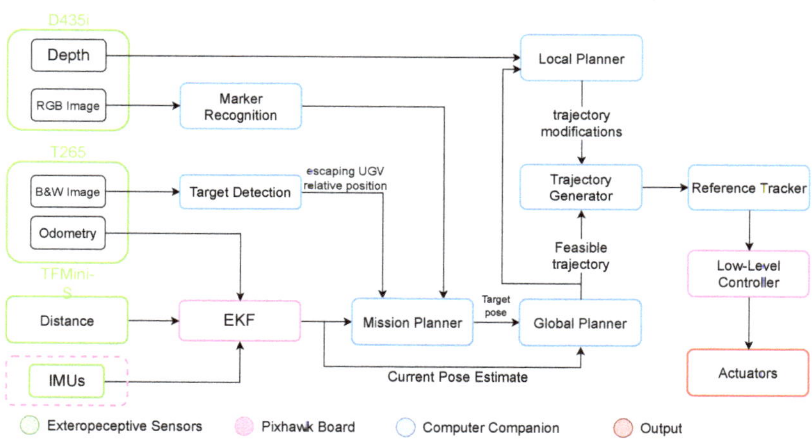

Figure 2. Overall architecture of the system. Data flows from the sensor's inputs (green boxes) to the main modules of the system in their respective boards (pink boxes for the MCU, blue boxes for the companion computer) up to the controlled output (red box).

3. Core Modules of the Navigation Stack

In this section, a detailed overview of the navigation stack is presented. Figure 3 depicts the information flow that composes the overall structure of the navigation stack. Each time a goal end pose is supplied, a pathfinder queries an offline-computed topological graph to find intermediate waypoints that connect the current UAV pose to the desired goal pose. In this passage, the selected crossing points as well as the path that connects them lie in the obstacle-free region of the pre-acquired map. This path is used as a nominal reference for the UAV trajectory, while unpredicted obstacles are handled online by the influence of the obstacle avoidance module on the actual trajectory to be tracked. The waypoints selected from the topological graph are interpolated by a dynamically feasible trajectory that drives the UAV in the obstacle-free region of the previously known map. Unpredicted obstacles are handled by the local planner, which alters the nominal trajectory by steering the UAV along free regions.

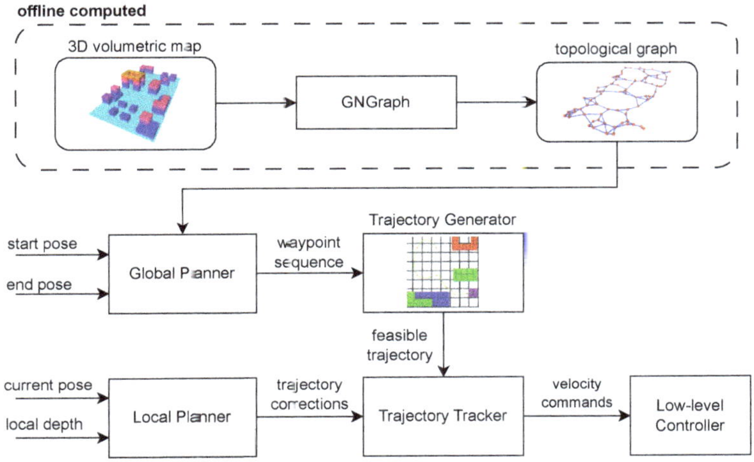

Figure 3. Core elements of the navigation stack. A navigation graph, computed offline over a given 3D map, is queried to produce a feasible trajectory that can be modified by the local planner for avoidance of unknown obstacles.

3.1. *Offline Path Planning*

The path planning is composed of two successive parts: an offline precomputing of the topological graph and an online trajectory generation and tracking. The former is approached using a graph-based solution, which can model the essential notions of navigational knowledge using route graphs, as presented by Werner et al. [36].

Our previous work, GNGraph [8], was a map-based general method that computed a low-dimensional topological graph of the environment for path planning. This solution exploits an unsupervised Hebbian learning algorithm, and is devised for a fast run-time global planning query on low-computing power embedded systems. The resulting graph is extracted from the environment's Euclidean Signed Distance Field (ESDF) map, and represents a clearance roadmap of the collision-free surroundings close to its 3D Generalized Voronoi Diagram (GVD). A clearance roadmap is a suitable solution for the challenge in question: considering the partially-known environment, the GVD maintains the safest distance from the known obstacles; a collision-avoidance module can then operate in the presence of unknown obstacles.

For this work, the shortest path search algorithm was changed from A* to Dijkstra's algorithm. This decision was taken because A* searches for the optimal path through a heuristic function, while in the case of the contest environment, considering its content size and topology, it is more feasible to run Dijkstra, which guarantees optimality without the need to define a heuristic.

The waypoint sequence between two extreme points, that is, the starting position and goal position, is obtained by first finding the closest vertices of the graph to the starting and goal positions in the environment considering the Euclidean distance between each of them. For this, it is assumed that the segment that connects starting and goal positions to their nearest graph vertices relies on the obstacle-free space. Then, the Dijkstra algorithm finds the shortest path that connects both nodes. It is guaranteed that at least one path is found, as GNGraph ensures the graph's connectivity using the spectral analysis included in its stopping learning criteria.

3.2. Online Trajectory Planning

The second step of the path planning is achieved online, working with the information computed by the method in Section 3.1. While navigating in the environment, the pathfinding algorithm is applied in the topological graph in order to obtain a feasible sequence of waypoints that allows the UAV to move from the current (starting) point to any desired final position. Note that the topological graph dictates only desired 3D positions of the UAV, while the orientation is left free and can be programmed in order to fulfill a secondary task, such as pointing the front-facing camera in any desired direction.

From the obtained sequence of positions, a feasible trajectory that obeys UAV dynamic constraints such as maximum velocities and accelerations must be computed. Time-Optimal Path Parametrization (TOPP) is a class of methods that has the aim of generating a feasible trajectory in the configuration space that interpolate an ordered sequence of desired configurations. The planned trajectory lies in the flat space that is the outcome of the application of the Differential Flatness theory [37] to the quadrotor dynamics. As commonly used in the literature, the chosen flat states in this work are the position of the UAV's center of mass and heading angle, both of which are taken with respect to a fixed frame. Indeed, the 3D position and orientation cannot be controlled independently; the UAV's roll and pitch angles used to obtain the desired position and yaw are constrained by its dynamics. These two elements generate a four-dimensional space that composes the hyperplane in which the trajectory is planned.

The feasible first- and second-order pose derivatives constrain the allowed set, while the ordered sequence of waypoints obtained by the exploration of the GNGraph imposes intermediate cross-points as position-only constraints. The current UAV position $\boldsymbol{p}_{ref}(0)$, together with the M desired waypoints $\boldsymbol{w}_i$ and the final position $\boldsymbol{p}_{ref}(t_f)$ are interpolated with a cubic spline $\mathcal{S}$ that defines the UAV trajectory as $\boldsymbol{p}_{ref}(t) = \mathcal{S}(t)$ in $t \in [0, t_f]$.

The heading angle can be imposed as a function of the current mission; during the exploration phase, i.e., while the UAV is navigating in previously unknown environments, the heading can be fixed towards the direction of motion by imposing the desired yaw angle ψ as

$$\psi_{ref}(t) = \arctan \frac{v_y(t)}{v_x(t)}, \tag{1}$$

where (v_x, v_y) are the horizontal components of the reference UAV velocity with respect to the map reference frame. This allows the front camera to detect unattended obstacles. During scanning phases, i.e., while the quadrotor is seeking some element of interest in a known direction, the desired heading can instead be directly imposed for each waypoint $\boldsymbol{w}_i$.

Among TOPP algorithms, TOPP-RA [38] exploits Reachability Analysis, a linear control systems theory that allows for the identification of feasible states starting from constrained inputs. In the studied case, feasible trajectories for the UAV position and heading angle are generated as time-dependent parametric curves using this method, where velocity and acceleration constraints are enforced over a sampling grid. Together with their time derivatives, the computed trajectories are sampled to generate the time-varying setpoint employed as a reference for the low-level controller, which is discussed in Section 3.4.

3.3. Local Planning for Obstacle Avoidance

The developed local planner module allows the robot to navigate in environments where the map is partially known a priori. The unknown obstacles that compose the environment are represented by columns; these are placed at random points in the environment in a scattered manner such that their composition does not generate a concave configuration. The module interacts with the global planner and sends addictive speed references to change the planned trajectory in real-time based on local sensing sensors. The perception of obstacles is carried out by the obstacle detector system (https://github.com/tysik/obstacle_detector (accessed on 17 July 2023)), which reads the depth information from the RGB-D camera and performs a clusterization to generate point cloud groups that identify the obstacles. Because the size range of the obstacles is known a priori, the clusterization algorithm fits the point cloud data with the expected dimension to detect obstacles.

After the obstacles have been detected within the camera's field of view, the respective centroids are used as reference points for the obstacle avoidance algorithm.

Artificial Potential Field (APF) [39] represents a simple approach for implementing an obstacle avoidance module able to interface easily with a generic global planner. The method is a control strategy for autonomous systems based on the artificial potential function that guides the motion of the system. The APF has been shown to be effective in a wide range of applications, including robotic navigation, formation control, and multi-agent coordination. Our implemented local planner is based on this method and allows the autonomous platform to achieve real-time performance with unpredicted obstacles, showing robustness in the presence of location uncertainty while exploiting solely local sensing information. In addition, the implemented approach only requires a few parameters to be tuned.

The planar position $\boldsymbol{r}_i = [x_i,\ y_i]^T$ of each i-th detected obstacle is expressed with respect to the local reference system attached to the quadrotor. For each i-th obstacle, the corresponding planar velocity

$$\boldsymbol{v}_i = \begin{cases} -\left[\frac{k_r}{x_i^2},\ \frac{k_r}{y_i^2}\right]^T & \text{if } \|\boldsymbol{r}_i\| \leq R \\ [0,\ 0]^T & \text{otherwise} \end{cases} \tag{2}$$

is computed only if the obstacle is detected within a certain threshold of the distance R from the geometric center of the drone. The gain k_r is the tuning parameter that modulates the motion aggressiveness in avoiding obstacles. The vector modulus of each velocity $\boldsymbol{v}_i$ is saturated if a certain threshold v_{max} in magnitude is exceeded; in this way, the contribution of the repulsion velocity is limited and the safety of autonomous navigation can be ensured by tuned limits. Then, the average speed $\bar{\boldsymbol{v}}$ is computed on the actual repulsive contributions to calculate the overall repulsive velocity for N obstacles

$$\bar{\boldsymbol{v}} = \frac{1}{N}\sum_{i=1}^{N} \boldsymbol{v}_i. \tag{3}$$

The average operator ensures that the generated repulsive velocity does not exceed the established threshold v_{max} for any distribution of detected obstacles. The computed overall repulsive velocity is sent to the global planner's reference tracking control in order to obtain feasible trajectories that evade unknown obstacles in the environment, as described in the next Subsection.

3.4. Trajectory Reference Tracking and Low-Level Control

The desired trajectory that the UAV has to track, as computed in Section 3.2, can be expressed with the desired position $\boldsymbol{p}_{\text{ref}}(t)$ and heading angle $\psi_{\text{ref}}(t)$, the function of the time $t \in [t_0, t_f]$, and their time derivatives $v_{\text{ref}}(t)$ and $\dot{\psi}_{\text{ref}}(t)$.

The lower level control of the UAV is handled by the PX4 autopilot, which communicates desired propeller velocities to the ESCs with several nested control loops. The chosen entry point for the position and velocity references designated by the high-level control is the UAV's center of mass velocity. Here, a PID controller implemented in the PX4 is able to map the desired UAV velocities to actuation reference signals.

Therefore, it is sufficient to produce a desired UAV center-of-mass velocity $\boldsymbol{v}_d$ together with a desired heading rate $\dot{\psi}_d$ to feed to the control loops of the PX4. The position and velocity references produced by the trajectory planning in any time instant t are converted to the desired center of mass velocities as

$$\boldsymbol{v}_d = \dot{\boldsymbol{p}}_{\text{ref}}(t) + k_P \boldsymbol{p}_{\text{ref}}(t)\,, \tag{4a}$$

$$\dot{\psi}_d = \dot{\psi}_{\text{ref}}(t) + k_\psi \psi_{\text{ref}}(t). \tag{4b}$$

As computed in Section 3.3, the additional contribution of the local planner can be added to Equation (4a) after each position and velocity vector is written in the reference system fixed to the UAV frame. Indeed, velocity variations due to the avoidance of obstacles are identified via local sensing, as they are already expressed in the UAV frame and are not impacted by navigation errors.

3.5. UGV Tracking

The tracking problem is stated as an unknown target moving freely on the ground constrained inside the partially known environment. The target's state and kinematics are entirely unknown; therefore, its dynamics is assumed to be a double integrator in discrete time with a sampling period Δ_t.

The target UGV position is measured using the intrinsic camera matrix to project the position from the pixel 2D image plane to the 3D camera plane and then estimated with a Kalman filter. The state vector is composed of positions and velocities in the horizontal plane, such as $\boldsymbol{x} = [p_x \quad p_y \quad \dot{p}_x \quad \dot{p}_y]^T$, while the state transition matrix $\boldsymbol{A}$ and the observation matrix $\boldsymbol{H}$ are

$$\boldsymbol{A} = \begin{bmatrix} 1 & 0 & \Delta_t & 0 \\ 0 & 1 & 0 & \Delta_t \\ 0 & 0 & 1 & 0 \\ 0 & 0 & 0 & 1 \end{bmatrix}, \quad \boldsymbol{H} = \begin{bmatrix} 1 & 0 & 0 & 0 \\ 0 & 1 & 0 & 0 \end{bmatrix}. \tag{5}$$

In the prediction step

$$\begin{aligned} \widehat{\boldsymbol{x}}_{k|k-1} &= \boldsymbol{A}\widehat{\boldsymbol{x}}_{k-1}, \\ \boldsymbol{P}_{k|k-1} &= \boldsymbol{A}\boldsymbol{P}_{k-1}\boldsymbol{A}^T + \boldsymbol{Q} \end{aligned} \tag{6}$$

the Kalman filter produces the a priori prediction state $\widehat{\boldsymbol{x}}_{k|k-1}$ and state estimation error covariance $\boldsymbol{P}_{k|k-1}$ considering the covariance $\boldsymbol{Q}$ of the process noise.

In the update step

$$\begin{aligned} \boldsymbol{K}_k &= \boldsymbol{P}_{k|k-1}\boldsymbol{H}^T\left(\boldsymbol{H}\boldsymbol{P}_{k|k-1}\boldsymbol{H}^T + \boldsymbol{R}\right)^{-1}, \\ \boldsymbol{P}_k &= (\boldsymbol{I}_{4\times 4} - \boldsymbol{K}_k\boldsymbol{H})\boldsymbol{P}_{k|k-1}, \\ \widehat{\boldsymbol{x}}_k &= \widehat{\boldsymbol{x}}_{k|k-1} + \boldsymbol{K}_k\left(\boldsymbol{y}_k - \boldsymbol{H}\widehat{\boldsymbol{x}}_{k|k-1}\right) \end{aligned} \tag{7}$$

the current a priori prediction is combined with the current observation $\boldsymbol{y}_k$ to obtain the a posteriori state estimate $\widehat{\boldsymbol{x}}_k$ and the estimate covariance matrix $\boldsymbol{P}_k$ considering the covariance of the observation noise $\boldsymbol{R}$.

3.6. Mission Planner

The high-level mission logic is managed by means of a Finite State Machine (FSM), the representation of which is shown in Figure 4. The FSM dictates the overall UAV behavior while taking on all the challenge tasks exposed in Section 1.1.

The mission starts in the `Idle` state. When the UAV is set to the PX4 offboard mode and receives the takeoff command, the system moves to the `Hover` state until the `Start Search` command is received, at which point the system moves into the `Search` state. In the latter state, the drone starts exploring the environment and looking for the moving target in the down-facing camera stream. When the target is found, the system moves into the `Tracking` state. The tracking of the target must last for 10 s; if the target is lost, the drone starts exploring again (i.e., the system moves again into the `Search` state). If the tracking is performed continuously for 10 s, the system moves into the `Wait for Mission` state and the drone stays in hover.

At this point, the Ground Control Station (GCS) can send a mission to the drone. A mission consists of a sequence of actions: either land, or take a photo. In both cases, a target coordinate is specified as a cell of the competition field. The system automatically picks up the first action in the list and moves to the desired location. If the action is a land request, then the drone lands, waits 10 s, and takes off again; if the action is a photo request, the drone starts an exploration phase looking for the target that must be shot. During the challenge, the environment is partially known in terms of obstacles; however, we only have an approximate location of the target to be shot. For this reason, a series of waypoints is used as a reference for the exploration phase. When the target is found, the drone takes a picture and the operator at the GCS accepts the picture or sends the retry command.

When an action is completed, the system again enters the `Select Action` state and the loop repeats until all actions have been executed. Finally, the drone moves to a safe location (away from obstacles) and performs a final landing before returning to the `Idle` state.

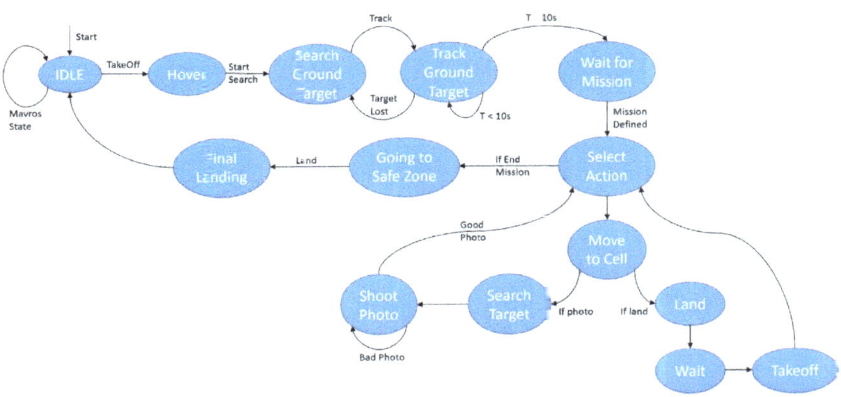

Figure 4. Schematic representation of the FSM managing the whole mission logic. The blue circles represent states, while the arrows represent transitions.

4. Evaluation and Experiments

The effectiveness of the different modules described in this work was first tested in simulated environments, then validated in real-world scenarios. The PX4 Software In The Loop (PX4-SITL) allows testing and comparison of the proposed algorithm in simulation and in the real world without any modification. PX4-SITL is based on the RotorS simulator [40], a Gazebo-based simulation environment that provides reliable multirotor models. The simulated quadrotor, mimicking the real one, is equipped with a face-forward stereo camera that provides real-time color images and depths of the simulated environment as well as with a downfacing camera.

The partially known 3D map of the Drone Contest is reproduced in Blender and imported in Gazebo with a dynamical spawning of the additional unknown obstacles in random positions; these are regulated as stated in Section 1.1. The escaping UGV is implemented in simulation with a Turtlebot 3 (https://www.turtlebot.com/turtlebot3 (accessed on 17 July 2023)), whereas during the competition the real robot is a Roomba (https://www.irobot.it/roomba (accessed on 17 July 2023)) with original firmware. Both are constrained in rectangular subsections of the map's ground.

4.1. Global Trajectory Planner

Example trajectories are shown in Figure 5, where a top view of the competition field is represented. In Figure 5a, the computed trajectory moves the UAV from the starting cell `I12` to the left side of the map in order to look for the escaping UGV. The trajectory is generated through the method described in Section 3.2 by interpolation of the red waypoints. The ordered list is obtained prepending the user-defined sequence `G8-G6-D6-D8`, with the waypoint sequence selected from the GNGraph obtained as in Section 3.1, which connects the starting point (I12 to the first user-defined search position (G8).

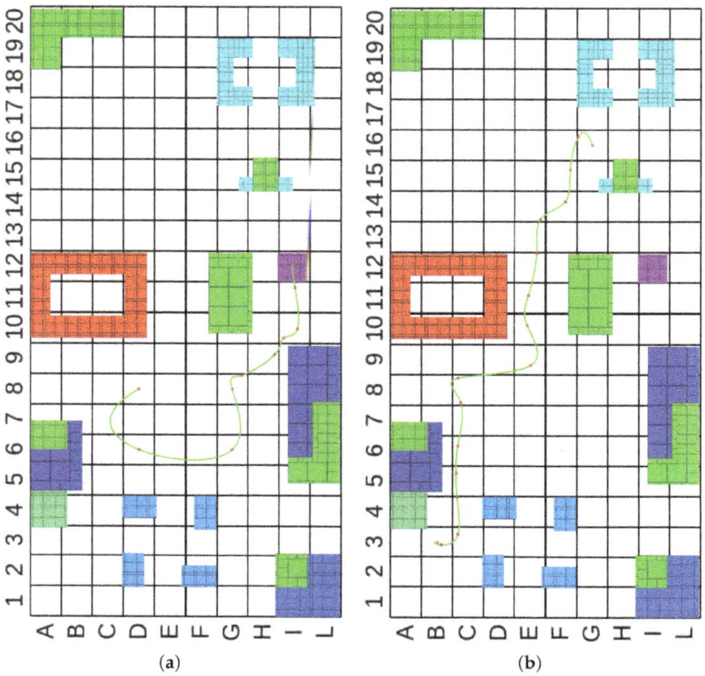

Figure 5. UAV planned trajectories, in red, for (**a**) beginning UGV search performed when taking off from the starting square I12 and (**b**) long-distance navigation from B3 to G16. The red dots are the intermediate waypoints selected from the GNGraph, that the trajectory has to interpolate.

Figure 5b shows the planned trajectory from the start position B3 to the target G16. The entire list of intermediate waypoints is inferred from the GNGraph algorithm. At any time instant, the heading angle is set tangent to the green curve in order to always monitor the direction of motion for unknown obstacles with the forward-facing camera.

These examples show the versatility of the algorithm with a different number of waypoints that can be explicitly requested, inferred from the topological graph, or both. Short (<5 m) and long (up to 5 m) trajectories are planned on the onboard companion computer in less than 700 ms.

4.2. Local Planner

The corrections of a simulated obstacle avoidance experiment are shown in Figures 6. The trajectory replanning is due to the presence of two unknown cylindrical obstacles, represented in red in the top-down view and in white in the gazebo environment. Obstacles are placed in the simulation such that the desired trajectory (in green) intersects them. When obstacles are detected, the local planner module makes the drone deflect the reference trajectory, sending additive speed commands to the global planner. The parameters of the local planning module are tuned in simulation to guarantee UAV readiness in avoiding the detected obstacles and at the same time to ensure compliance with the desired kinematic constraints in terms of the maximum speed limits of the vehicle. From the plots in Figure 7, it is apparent that the vehicle does not violate the kinematic constraints of the global planner (± 0.5 m/s) when no obstacles are detected; in the presence of obstacles, the vehicle must not exceed the sum of the limits of the global and local planner (± 1.1 m/s).

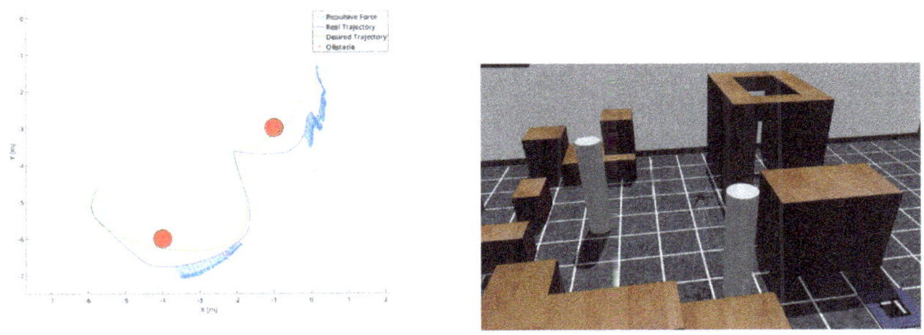

(**a**) (**b**)

Figure 6. Local Planner module performing obstacle avoidance starting from the local position [0, −1] in simulation. The desired trajectory in green is the reference provided to the quadrotor, while the real (corrected by obstacle avoidance) trajectory is in blue. (**a**) Trajectory due to obstacle avoidance, top view and (**b**) scenario viewed in Gazebo Simulator.

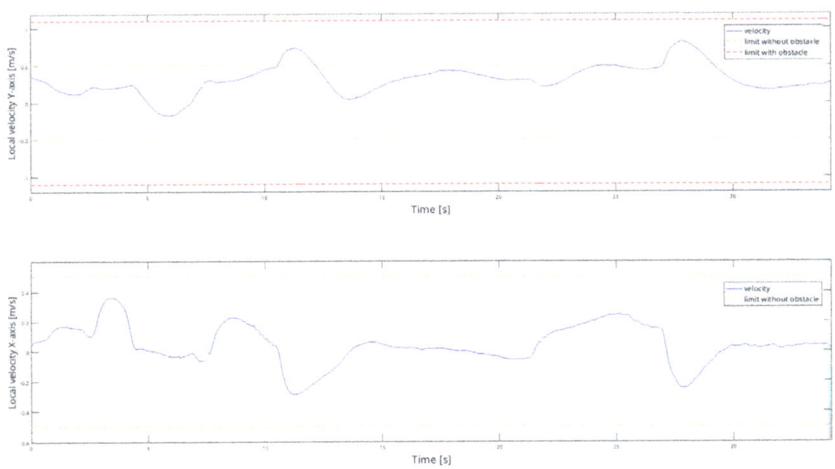

Figure 7. Vehicle speed expressed in the local coordinate frame during the obstacle avoidance experiment in Gazebo Simulator. The vehicle respects the speed thresholds set both in the absence and in the presence of obstacles.

4.3. Overall System in Real Environment

The proposed framework was used during the competition in the indoor environment of the 2022 Leonardo Drone Contest. With the help of the Rviz tool, we developed a graphical user interface (Figure 3) to monitor the progress of the mission and the status of the drone in real time. On the right of the GUI, the operator can see the streams from the forward-facing camera and the undistorted down-facing camera, which are enabled for searching markers. In the center it can be seen where the drone is located with respect to the map, which is known a priori. For each planning, the selected nodes of the roadmap are colored in red while the computed trajectory of the global planner is shown in green. The cyan sphere indicates the current reference position for the low-level tracking control. The unknown obstacle mapped by the detector module is represented by concentric red and green cylinders. On the left panel, the operator has the possibility of deciding which box the drone has to land in by typing the coordinates on the interface.

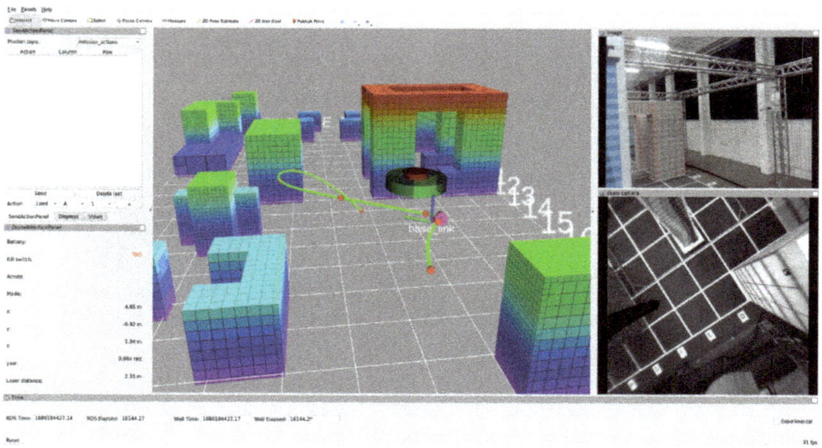

Figure 8. RVIZ GUI used to monitor the drone's mission in real-life experiments. In the right column, both camera FOVs are shown. In the central window, the environment's known obstacles are represented with voxels and an unknown obstacle is depicted with a dark red and green cylinder. The drone's trajectory and reference points are shown in light green and red, respectively.

The expected performance was verified in the real scenarios in terms of both readiness to avoid unknown obstacles and ability to return to the desired trajectory when no more obstacles were detected. Figure 9 show a sequence of manoeuvres along the planned trajectory in the presence of an unknown obstacle.

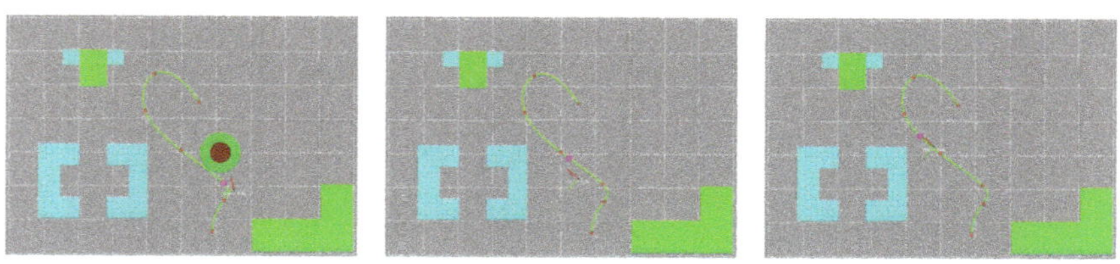

(**a**) (**b**) (**c**)

Figure 9. A sequence of images during the tracking of a reference trajectory in the presence of an unknown obstacle. The UAV detects the obstacle (**a**), sends speed commands to the global planner to change the planned trajectory (**b**), and returns to the previous trajectory after avoiding the obstacle (**c**).

5. Conclusions

This paper has presented a versatile solution for addressing complex autonomous tasks, specifically focusing on the case of the 2022 Leonardo Drone Contest. The proposed vision-based framework demonstrates the ability to navigate efficiently in partially unknown GNSS-denied real-world environments. By leveraging an offline-computed topological graph and online obstacle avoidance, the system effectively handles both planned and unpredicted obstacles.

The complexity of the mission tasks was tackled by using a Finite State Machine that effectively activates the needed modules of the stack. The entire software stack was designed, developed, and tested entirely in the ROS framework, with the middleware exploited on the drone companion computer to run and arrange the communication among all the software modules.

Localization and navigation capabilities were initially tested in simulation and their robustness was subsequently validated in real-world scenarios, specifically, the Leonardo Drone Contest challenge, further confirming the effectiveness of the proposed system. The proposed framework and the choices related to the hardware and the sensors represent a widely tested innovative solution that ensures autonomous flight in real-world GNSS-denied environments even in partially unknown scenarios.

In summary, this work provides a comprehensive and adaptable solution for autonomous tasks that has demonstrated its efficacy in challenging environments, thereby setting a foundation for further advances in the field of UAV autonomy.

Author Contributions: Conceptualization, M.M., M.T.L., E.P.H.-A., G.B. and M.S.; methodology, M.M., M.T.L., E.P.H.-A. and G.B.; software, M.M., M.T.L., E.P.H.-A. and G.B.; validation, M.M., M.T.L., E.P.H.-A. and G.B.; resources, E.P.H.-A., M.S. and C.A.A.; writing—original draft preparation, M.M., M.T.L., E.P.H.-A., G.B. and M.S.; writing—review and editing, M.M., M.T.L. and M.S.; supervision, M.S.; project administration, M.S.; funding acquisition, M.S. and C.A.A. All authors have read and agreed to the published version of the manuscript.

Funding: This research received no external funding.

Data Availability Statement: The data presented in this study are available on request from the corresponding author. The data are not yet publicly available at the time of writing but they will

Conflicts of Interest: The authors declare no conflict of interest

Abbreviations

The following abbreviations are used in this manuscript:

UAV	Unmanned Aerial Vehicle
MAV	Micro-Aerial Vehicle
UGV	Unmanned Ground Vehicle
GNSS	Global Navigation Satellite System
ROS	Robot Operating System
RViz	ROS visualization tool
SAR	Search And Rescue
FSM	Finite State Machine
FCU	Flight Control Unit
FPGA	Field-Programmable Gate Array
RS	RealSense
LiDAR	Light Detection and Ranging
IMU	Inertial Measurement Unit
VIO	Visual Inertial Odometry
FOV	Field Of View
EKF	Extended Kalman Filter
TOPP	Time-Optimal Path Parametrization
APF	Artificial Potential Field
ESC	Electronic Speed Controller
GCS	Ground Control Station
GUI	Graphical User Interface

References

1. Silano, G.; Baca, T.; Penicka, R.; Liuzza, D.; Saska, M. Power Line Inspection Tasks with Multi-Aerial Robot Systems via Signal Temporal Logic Specifications. *IEEE Robot. Autom. Lett.* **2021**, *6*, 4169–4176. [CrossRef]
2. Agha, A.; Otsu, K.; Morrell, B.; Fan, D.D.; Thakker, R.; Santamaria-Navarro, A.; Kim, S.K.; Bouman, A.; Lei, X.; Edlund, J.; et al. NeBula: Quest for Robotic Autonomy in Challenging Environments; TEAM CoSTAR at the DARPA Subterranean Challenge. *arXiv* **2021**. [CrossRef]
3. Hudson, N.; Talbot, F.; Cox, M.; Williams, J.; Hines, T.; Pitt, A.; Wood, B.; Frousheger, D.; Lo Surdo, K.; Molnar, T.; et al. Heterogeneous Ground and Air Platforms, Homogeneous Sensing: Team CSIRO Data61's Approach to the DARPA Subterranean Challenge. *Field Robot.* **2022**, *2*, 595–636. [CrossRef]
4. Rouček, T.; Pecka, M.; Čížek, P.; Petříček, T.; Bayer, J.; Šalanský, V.; Azayev, T.; Heřt, D.; Petrlík, M.; Báča, T.; et al. System for multi-robotic exploration of underground environments CTU-CRAS-NORLAB in the DARPA Subterranean Challenge. *arXiv* **2021**. [CrossRef]

5. Tranzatto, M.; Dharmadhikari, M.; Bernreiter, L.; Camurri, M.; Khattak, S.; Mascarich, F.; Pfreundschuh, P.; Wisth, D.; Zimmermann, S.; Kulkarni, M.; et al. Team CERBERUS Wins the DARPA Subterranean Challenge: Technical Overview and Lessons Learned. *arXiv* **2022**. [CrossRef]
6. Balaram, B.; Canham, T.; Duncan, C.; Grip, H.F.; Johnson, W.; Maki, J.; Quon, A.; Stern, R.; Zhu, D. Mars helicopter technology demonstrator. In Proceedings of the 2018 AIAA Atmospheric Flight Mechanics Conference, Kissimmee, FL, USA, 8–12 January 2018; p. 0023.
7. Alarcón, E.P.H.; Ghavifekr, D.B.; Baris, G.; Mugnai, M.; Satler, M.; Avizzano, C.A. An Efficient Object-Oriented Exploration Algorithm for Unmanned Aerial Vehicles. In Proceedings of the 2021 International Conference on Unmanned Aircraft Systems (ICUAS), Athens, Greece, 15–18 June 2021; pp. 330–337. [CrossRef]
8. Herrera-Alarcón, E.; Satler, M.; Vannucci, M.; Avizzano, C. GNGraph: Self-Organizing Maps for Autonomous Aerial Vehicle Planning. *IEEE Robot. Autom. Lett.* **2022**, *7*, 10721–10728. [CrossRef]
9. Shen, S.; Michael, N.; Kumar, V. Autonomous multi-floor indoor navigation with a computationally constrained MAV. In Proceedings of the 2011 IEEE International Conference on Robotics and Automation, Shanghai, China, 9–13 May 2011; pp. 20–25. [CrossRef]
10. Tomic, T.; Schmid, K.; Lutz, P.; Domel, A.; Kassecker, M.; Mair, E.; Grixa, I.; Ruess, F.; Suppa, M.; Burschka, D. Toward a fully autonomous UAV: Research platform for indoor and outdoor urban search and rescue. *IEEE Robot. Autom. Mag.* **2012**, *19*, 46–56. [CrossRef]
11. Fraundorfer, F.; Heng, L.; Honegger, D.; Lee, G.H.; Meier, L.; Tanskanen, P.; Pollefeys, M. Vision-based autonomous mapping and exploration using a quadrotor MAV. In Proceedings of the IEEE International Conference on Intelligent Robots and Systems, Vilamoura-Algarve, Portugal, 7–12 October 2012; pp. 4557–4564. [CrossRef]
12. Meier, L.; Tanskanen, P.; Heng, L.; Lee, G.H.; Fraundorfer, F.; Pollefeys, M. PIXHAWK: A micro aerial vehicle design for autonomous flight using onboard computer vision. *Auton. Robot.* **2012**, *33*, 21–39. [CrossRef]
13. Meier, L.; Honegger, D.; Pollefeys, M. PX4: A node-based multithreaded open source robotics framework for deeply embedded platforms. In Proceedings of the IEEE International Conference on Robotics and Automation, Seattle, WA, USA, 26–30 May 2015; pp. 6235–6240. [CrossRef]
14. Loianno, G.; Brunner, C.; McGrath, G.; Kumar, V. Estimation, Control, and Planning for Aggressive Flight With a Small Quadrotor With a Single Camera and IMU. *IEEE Robot. Autom. Lett.* **2017**, *2*, 404–411. [CrossRef]
15. Ge, R.; Lee, M.; Radhakrishnan, V.; Zhou, Y.; Li, G.; Loianno, G. Vision-based Relative Detection and Tracking for Teams of Micro Aerial Vehicles. In Proceedings of the 2022 IEEE/RSJ International Conference on Intelligent Robots and Systems (IROS), Kyoto, Japan, 23–27 October 2022; pp. 380–387. [CrossRef]
16. Thakur, D.; Tao, Y.; Li, R.; Zhou, A.; Kushleyev, A.; Kumar, V. Swarm of Inexpensive Heterogeneous Micro Aerial Vehicles. In *Proceedings of the Experimental Robotics*; Siciliano, B., Laschi, C., Khatib, O., Eds.; Springer International Publishing: Cham, Switzerland, 2021; pp. 413–423.
17. Liu, X.; Chen, S.W.; Nardari, G.V.; Qu, C.; Ojeda, F.C.; Taylor, C.J.; Kumar, V. Challenges and Opportunities for Autonomous Micro-UAVs in Precision Agriculture. *IEEE Micro* **2022**, *42*, 61–68. [CrossRef]
18. Moon, H.; Martinez-Carranza, J.; Cieslewski, T.; Faessler, M.; Falanga, D.; Simovic, A.; Scaramuzza, D.; Li, S.; Ozo, M.; De Wagter, C.; et al. Challenges and implemented technologies used in autonomous drone racing. *Intell. Serv. Robot.* **2019**, *12*, 137–148. [CrossRef]
19. Rojas-Perez, L.O.; Martinez-Carranza, J. Metric monocular SLAM and colour segmentation for multiple obstacle avoidance in autonomous flight. In Proceedings of the 2017 Workshop on Research, Education and Development of Unmanned Aerial Systems (RED-UAS), Linköping, Sweden, 3–5 October 2017; pp. 234–239. [CrossRef]
20. Li, S.; Ozo, M.M.; De Wagter, C.; de Croon, G.C. Autonomous drone race: A computationally efficient vision-based navigation and control strategy. *Robot. Auton. Syst.* **2020**, *133*, 103621. [CrossRef]
21. Kaufmann, E.; Gehrig, M.; Foehn, P.; Ranftl, R.; Dosovitskiy, A.; Koltun, V.; Scaramuzza, D. Beauty and the Beast: Optimal Methods Meet Learning for Drone Racing. In Proceedings of the 2019 International Conference on Robotics and Automation (ICRA), Montreal, QC, Canada, 20–24 May 2019; pp. 690–696. [CrossRef]
22. Jung, S.; Cho, S.; Lee, D.; Lee, H.; Shim, D.H. A direct visual servoing-based framework for the 2016 IROS Autonomous Drone Racing Challenge. *J. Field Robot.* **2018**, *35*, 146–166. [CrossRef]
23. Foehn, P.; Brescianini, D.; Kaufmann, E.; Cieslewski, T.; Gehrig, M.; Muglikar, M.; Scaramuzza, D. AlphaPilot: Autonomous drone racing. *Auton. Robot.* **2022**, *46*, 307–320. [CrossRef]
24. Roggi, G.; Meraglia, S.; Lovera, M. Leonardo Drone Contest 2021: Politecnico di Milano team architecture. In Proceedings of the 2022 International Conference on Unmanned Aircraft Systems, ICUAS 2022, Dubrovnik, Croatia, 21–24 June 2022; pp. 191–196. [CrossRef]
25. Mohta, K.; Watterson, M.; Mulgaonkar, Y.; Liu, S.; Qu, C.; Makineni, A.; Saulnier, K.; Sun, K.; Zhu, A.; Delmerico, J.; et al. Fast, autonomous flight in GPS-denied and cluttered environments. *J. Field Robot.* **2018**, *35*, 101–120. [CrossRef]
26. Mohta, K.; Sun, K.; Liu, S.; Watterson, M.; Pfrommer, B.; Svacha, J.; Mulgaonkar, Y.; Taylor, C.J.; Kumar, V. Experiments in Fast, Autonomous, GPS-Denied Quadrotor Flight. In Proceedings of the 2018 IEEE International Conference on Robotics and Automation (ICRA), Brisbane, QLD, Australia, 21–25 May 2018; pp. 7832–7839. [CrossRef]

27. Quigley, M.; Mohta, K.; Shivakumar, S.S.; Watterson, M.; Mulgaonkar, Y.; Arguedas, M.; Sun, K.; Liu, S.; Pfrommer, B.; Kumar, V.; et al. The open vision computer: An integrated sensing and compute system for mobile robots. In Proceedings of the IEEE International Conference on Robotics and Automation, Montreal, QC, Canada, 20–24 May 2019; pp. 1834–1840. [CrossRef]
28. Oleynikova, H.; Lanegger, C.; Taylor, Z.; Pantic, M.; Millane, A.; Siegwart, R.; Nieto, J. An open-source system for vision-based micro-aerial vehicle mapping, planning, and flight in cluttered environments. *J. Field Robot.* **2020**, *37*, 642–666. [CrossRef]
29. Kompis, Y.; Bartolomei, L.; Chli, M. Fully Autonomous Live 3D Reconstruction with an MAV: Hardware- and Software-Setup, 2021-12-02. In Proceedings of the 9th International Conference on 3D Vision (3DV 2021), Online, 1–3 December 2021. [CrossRef]
30. Sandino, J.; Vanegas, F.; Maire, F.; Caccetta, P.; Sanderson, C.; Gonzalez, F. UAV framework for autonomous onboard navigation and people/object detection in cluttered indoor environments. *Remote Sens.* **2020**, *12*, 3386. [CrossRef]
31. Stephens, B.; Nguyen, H.N.; Hamaza, S.; Kovac, M. An Integrated Framework for Autonomous Sensor Placement With Aerial Robots. *IEEE/ASME Trans. Mechatron.* **2022**, *28*, 38-49. [CrossRef]
32. Campos-Macías, L.; Aldana-López, R.; de la Guardia, R.; Parra-Vilchis, J.I.; Gómez-Gutiérrez, D. Autonomous navigation of MAVs in unknown cluttered environments. *J. Field Robot.* **2021**, *38*, 307–326.
33. Ko, C.; Han, S.; Choi, M.; Kim, K.S. Integrated path planning and tracking control of autonomous vehicle for collision avoidance based on model predictive control and potential field. In Proceedings of the 2020 20th International Conference on Control, Automation and Systems (ICCAS), Busan, Republic of Korea, 13–16 October 2020; pp. 956–961.
34. Nieuwenhuisen, M.; Droeschel, D.; Beul, M.; Behnke, S. Autonomous navigation for micro aerial vehicles in complex GNSS-denied environments. *J. Intell. Robot. Syst.* **2016**, *84*, 199–216. [CrossRef]
35. Nikolic, J.; Rehder, J.; Burri, M.; Gohl, P.; Leutenegger, S.; Furgale, P.T.; Siegwart, R. A synchronized visual-inertial sensor system with FPGA pre-processing for accurate real-time SLAM. In Proceedings of the 2014 IEEE International Conference on Robotics and Automation (ICRA), Hong Kong, China, 31 May–7 June 2014; pp. 431–437. [CrossRef]
36. Werner, S.; Krieg-Brückner, B.; Herrmann, T. Modelling navigational knowledge by route graphs. In *Spatial Cognition II*; Springer: Berlin/Heidelberg, Germany, 2000; pp. 295–316.
37. Mellinger, D.; Kumar, V. Minimum snap trajectory generation and control for quadrotors. In Proceedings of the IEEE International Conference on Robotics and Automation, Shanghai, China, 9–13 May 2011; pp. 2520–2525. [CrossRef]
38. Pham, H.; Pham, Q.C. A New Approach to Time-Optimal Path Parameterization Based on Reachability Analysis. *IEEE Trans. Robot.* **2018**, *34*, 645–659. [CrossRef]
39. Khatib, O. Real-time obstacle avoidance for manipulators and mobile robots. In Proceedings of the 1985 IEEE International Conference on Robotics and Automation, St. Louis, MO, USA, 25–28 March 1985; Volume 2, pp. 500–505.
40. Furrer, F.; Burri, M.; Achtelik, M.; Siegwart, R. Rotors—A modular gazebo mav simulator framework. *RObot Oper. Syst. Complet. Ref.* **2016**, *1* , 595–625.

drones

MDPI

Article

Integration of Deep Sequence Learning-Based Virtual GPS Model and EKF for AUV Navigation

Peng-Fei Lv [1], Jun-Yi Lv [2], Zhi-Chao Hong [1,3] and Li-Xin Xu [1,3,*]

1 Ocean College, Jiangsu University of Science and Technology, Zhenjiang 212003, China; lvpengfei@just.edu.cn (P.-F.L.); hongzhichao@just.edu.cn (Z.-C.H.)
2 Shenzhen Institute for Advanced Study, University of Electronic Science and Technology of China, Shenzhen 518110, China; 202222280319@std.uestc.edu.cn
3 Jiangsu Marine Technology Innovation Center, Nantong 226199, China
* Correspondence: lxu@just.edu.cn

Abstract: To address the issue of increasing navigation errors in low-cost autonomous underwater vehicles (AUVs) operating without assisted positioning underwater, this paper proposes a Virtual GPS Model (VGPSM) based on deep sequence learning. This model is integrated with an Extended Kalman Filter (EKF) to provide a high-precision navigation solution for AUVs. The VGPSM leverages the time-series characteristics of data from sensors such as the Attitude and Heading Reference System (AHRS) and the Doppler Velocity Log (DVL) while the AUV is on the surface. It learns the relationship between these sensor data and GPS data by utilizing a hybrid model of Long Short-Term Memory (LSTM) and Bidirectional Long Short-Term Memory (Bi-LSTM), which are well-suited for processing and predicting time-series data. This approach constructs a virtual GPS model that generates virtual GPS displacements updated at the same frequency as the real GPS data. When the AUV navigates underwater, the virtual GPS displacements generated using the VGPSM in real-time are used as measurements to assist the EKF in state estimation, thereby enhancing the accuracy and robustness of underwater navigation. The effectiveness of the proposed method is validated through a series of experiments under various conditions. The experimental results demonstrate that the proposed method significantly reduces cumulative errors, with navigation accuracy improvements ranging from 29.2% to 69.56% compared to the standard EKF, indicating strong adaptability and robustness.

Keywords: autonomous underwater vehicle; underwater navigation; virtual GPS model; deep sequence learning; extended Kalman filter

1. Introduction

With the rapid development of marine science and technology, the demand for autonomous underwater vehicles (AUVs) in civil, military, and scientific research fields is increasing [1]. As the core component of AUV autonomy and intelligence, the positioning accuracy of the navigation system directly affects the success and efficiency of AUV underwater missions [2,3]. However, the special characteristics of the underwater environment bring great challenges to the navigation and positioning of AUVs [4,5]. Therefore, it is of great research significance to improve the navigation accuracy of underwater vehicles.

Currently, AUV underwater navigation sensors mainly include the inertial navigation system (INS) [6], Doppler Velocimetry (DVL) [7], a Depth System (PS) [8], and a Global Positioning System (GPS) [9]. In general, INS can provide AUVs with acceleration, angular velocity, angle, velocity, and position information. However, since the calculation of velocity and position relies on the integral accumulation of acceleration and angular velocity, it requires the high accuracy of gyroscopes and accelerometers. Conventional AUVs generally use an INS that cannot provide velocity and position information, such as inertial measurement units (IMUs) and attitude and heading reference systems (AHRSs). DVL provides velocity relative to the bottom through acoustic Doppler frequency shift, which assists inertial navigation in generating position information. PS provides the AUV with depth information between the current position and the water surface. GPS can provide accurate position information on the water surface, but the signal cannot be transmitted in underwater environments [10]. An ultra-short baseline (USBL) [11] can provide underwater positioning; however, it is expensive and relies on external equipment. Therefore, the integrated navigation technology of INS/DVL/PS/GPS has become a research hotspot for low-cost AUV underwater navigation [2,3], which achieves a higher accuracy of navigation through fusing data from multiple sensors. However, it is still a challenge to optimize the sensor data fusion algorithm in the underwater complex environment.

In practical applications, commonly used navigation algorithms include dead reckoning (DR) [12] and the extended Kalman filter (EKF) [13]. DR estimates the position of the AUV by continuously calculating its motion state, but is easily affected by cumulative errors. EKF effectively reduces the cumulative error and provides more accurate navigation information through multiple filtering and the fusion of sensor data. However, the performance of EKF depends on accurate sensor models and environmental models, and the accuracy is still limited without the assistance of external position information [14]. In recent years, machine learning [15,16], deep learning [17,18], and other neural network-based methods in AUV navigation [19] have gained widespread attention.

More and more scholars are exploring the impact of machine learning and deep learning methods on improving the accuracy of AUV navigation systems [20–23]. He et al. summarized how deep neural networks (DNN) can suppress the drift error of navigation systems by calibrating inertial sensor noise and fusing navigation sensor information [24]. Xie et al. proposed a neural network-based approach to explore the time-varying relationship between the acceleration and attitude angle, using acoustic localization and a neural network to estimate the pitch angle and reduce the impact of waves on inertial sensors [25]. Huang et al. proposed a multi-error model-assisted extended Kalman filter method for attitude estimation, which improves the performance of MEMS sensors and enhances the accuracy of underwater glider navigation systems [26]. Li et al. constructed a nonlinear regressive neural network model for the supervised learning of the velocity increments of DVL and INS, assisting the navigation when the DVL fails [27]. Sabet et al. proposed a dynamic surge identification model based on low-cost MEMS to assist the dead reckoning, which improves the navigation accuracy by reducing the effects of external acceleration and electromagnetic interference on attitude [28]. Topini et al. used four deep learning models, including MLP [29], LSTM [30], CNN [31], and CNN-LSTM [32], to estimate DVL velocity and combined dead reckoning for navigation [33] The above methods are mainly based on supervised learning models outputting information such as attitude and velocity to assist in improving navigation and positioning accuracy.

Since GPS can provide the navigation system with position data without cumulative errors when the AUV is sailing on the surface, the divergence of the navigation system error is suppressed. However, the GPS information is not available due to the attenuation of radio signals underwater, and the USBL data update rate is low, which makes it impossible to control the error dispersion of the navigation system. Scholars have begun to study the direct relationship between the inertial navigation system, the Doppler velocity log, the depth sensor, and GPS positioning data. The most common method is to achieve end-to-end position output through a neural network [34]. Lv et al. proposed an adaptive GPS filtering method to address the GPS drift caused by the GPS antenna being easily covered when the AUV is sailing near the surface. By comparing the length of time without GPS data and the distance difference between GPS position and navigation position with the threshold value, the GPS drift data are filtered out and the confidence of GPS position is improved [35]. Then, an AUV navigation dataset with GPS correction is constructed, a hybrid recurrent neural network is selected to learn the AUV navigation data, and a position correction model is constructed for the position output of the navigation system [36]. Guo et al. estimated the AUV position based on an online process model by reconstructing state information between valid data from an ultrashort baseline positioning system [37]. Mu et al. trained GPS and other navigation data directly through LSTM and bidirectional long- and short-term memory, and obtained the AUV position data [38] through an end-to-end approach [39].

Although the above navigation technologies have suppressed the divergence rate of navigation system errors to a certain extent, they still have certain limitations [40]. For example, deep learning methods rely on a large amount of high-quality data and computing resources. To overcome the above problems, this paper proposes a navigation method that combines the traditional EKF and the virtual GPS model (VGPSM) based on deep sequence learning to provide high-precision and high-update rate navigation information for an autonomous underwater vehicle. The main contributions of this paper are summarized as follows:

1. According to the high-quality navigation sensor data, when GPS data are available on the surface, based on the time-series characteristics of the navigation system data, a deep sequence learning model integrating LSTM and Bi-LSTM is constructed to learn the relationship between the GPS data and the DVL and AHRS data in the surface environment, and forms a virtual GPS model, which solves the problem of no GPS signal underwater.
2. Observation vectors are constructed for two cases, depending on whether the AUVs have positional aids when performing the mission. The virtual GPS model is updated at the same frequency as the real GPS data. When there is an update in the surface GPS or the underwater VGPSM, the corresponding position is added to the observation vector. Otherwise, the observation vector does not contain position information, which effectively solves the limitations of traditional navigation methods when there is no external position assistance.
3. Through experiments on real navigation data, the performance and practical application effects of the proposed method in different situations are verified, which provides new ideas and references for the development of underwater vehicle navigation technology.

The structure of this paper is organized as follows: Section 2 introduces the basic knowledge of navigation system. Section 3 describes the specific implementation process of a virtual GPS model based on deep sequence learning and the proposed methodology. Section 4 conducts experimental design and results, followed by Section 5, which offers an in-depth analysis and discussion of the experimental findings. Finally, Section 6 summarizes the key conclusions and contributions of this study. Through these contents, this study hopes to provide new theoretical and technical support for the development of AUV navigation systems.

2. Navigation System

The AUV navigation system contains functions such as sensor data acquisition and navigation positioning. The data collected using each sensor will be sent to the data center (MOOSDB). The navigation algorithm module, after acquiring the sensor data at the current moment from the MOOSDB, calculates a stable and reliable position, velocity, attitude, and other information in real time, and then sends it to the MOOSDB for the other systems of the AUV to acquire in order to perform the task more accurately. This section focuses on the two coordinate systems and the main onboard sensors of the AUV navigation system in this paper.

2.1. Coordinate System for AUV Navigation

The navigation system in this paper uses both the navigation coordinate system and the body coordinate system, as shown in Figure 1. In general, the position, velocity, attitude, acceleration, and angular velocity of the AUV come from the navigation coordinate system by default, but the velocity and acceleration of the AUV during the control system's manipulation of the AUV's movement come from the AUV coordinate system.

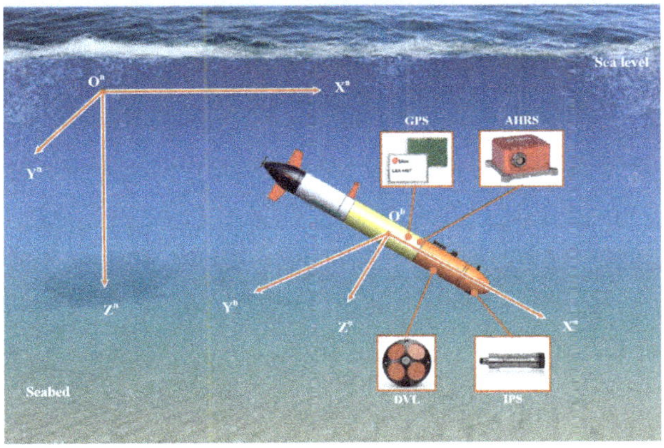

Figure 1. The coordinate system and onboard sensors for the AUV navigation system.

The letter n is generally used to denote the navigation coordinate system, whose origin O^n is a point fixed to the Earth, generally starting from a stable position obtained after the navigation system is switched on. The O^nX^n axis points geographical north along the tangent direction of the reference ellipsoid meridian, the O^nY^n axis points geographical east along the normal direction of the reference ellipsoid, and the O^nZ^n axis forms a right-handed coordinate system with the O^nX^n axis and the O^nY^n axis in the local horizontal plane, which is also known as the North-East-Down (NED) coordinate system due to its orientation.

The letter b is generally used to denote the carrier coordinate system, whose origin O^b is fixed at the center of mass and moves with the movement of the AUV. The O^bX^b axis is in the same direction as the roll axis of the AUV's angular motion and points to the front of the AUV. The O^bY^b axis is in the same direction as the pitch axis of the AUV's angular motion and points to the right of the AUV. The O^bZ^b axis is in the same direction as the heading axis of the AUV's angular motion, and forms a right-handed coordinate system with the O^bX^b axis and the O^bY^b axis, also known as the Forward-Right-Down (FRD) coordinate system.

2.2. Onboard Sensors for AUV Navigation

This section focuses on the onboard sensors of the navigation system as shown in Figure 1, as well as their performance parameters and output data types, which are shown in Table 1.

Table 1. Performance parameters and output data types of onboard sensors.

Sensors	Data Type	Parameter	Value
AHRS (Ellipse-A)	w_x^b, w_y^b, w_z^b, a_x^b, a_y^b, a_z^b, $\xi_x^b, \xi_y^b, \xi_z^b$	Heading Pitch Roll Update Rate	1° RMS 0.2° RMS 0.2° RMS 10 Hz (up to 200 Hz)
DVL (NavQuest 600)	v_x^b, v_y^b, v_z^b, a_z^b	Velocity Accuracy Altitude Range Maximum Velocity Update Rate	1 m/s 0.3–110 m 20 knot 1 Hz
GPS (LEA-M8T)	lat, lon	Position Accuracy Acquisition Update Rate Cold Starts	2.5 m CEP GPS & BeiDou 1 Hz 25 s
IPS (Valeport)	p_z^n	Accuracy Update Rate	0.01 % Range 1 Hz

The attitude and heading reference system (AHRS), model Ellipse-A AHRS, provides heading (ξ_z^b), pitch (ξ_y^b), roll (ξ_x^b), three-axis acceleration (a_x^b, a_y^b, a_z^b,), and three-axis angular velocity (w_x^b, w_y^b, w_z^b) for the navigation system. The data of AHRS are mainly obtained based on three-axis gyroscopes, three-axis accelerometers, and three-axis magnetometers. Due to the use of the magnetometer, the AHRS needs to be calibrated for the local magnetic field every time the AUV is tested in a new environment.

The Doppler velocity log (DVL) sensor, model NavQuest 600 micro DVL, is used to provide three-dimensional velocity (v_x^b, v_y^b, v_z^b) and bottom height (al_z^b) for the navigation system. The transducer array of DVL adopts a four-beam JANUS configuration, which is able to reduce the adverse effects of AUV motion on DVL measurement, and has a high accuracy in the velocity measurement. The bottom height can assist in determining the validity of DVL data.

The global positioning system (GPS), model LEA-M8T GPS, provides latitude and longitude with no accumulated errors for the navigation system. This sensor can provide high-precision data near the water surface where its antenna is not covered, but it is unable to provide a position underwater due to the rapid decay of radio signals, and this paper focuses on improving this.

The intelligent pressure system (IPS), model miniIPS, provides depth (p_z^n) for navigation system. The sensor obtains the current depth with high accuracy based on the effect of water pressure on the resistance value, and can provide it to the AUV continuously.

3. Proposed Method for AUV Navigation

As a commonly used method for AUV navigation, EKF can efficiently handle nonlinear systems with good real-time performance and robustness, but it still inevitably suffers from accumulated errors. Sensor noise and model uncertainty lead to gradual deviations in state estimation over time, resulting in cumulative errors. Therefore, these errors need to be corrected by periodic correction to ensure the long-term accuracy and reliability of the navigation system. The navigation system can use GPS position data for error correction while sailing on the surface, but it cannot be performed underwater. We propose a virtual GPS model based on deep sequence learning to address this issue, which outputs the virtual GPS position during AUV underwater navigation for error correction. This section will describe the proposed method in detail, focusing on the proposed virtual/real GPS-assisted EKF and the virtual GPS model.

3.1. Proposed Navigation Method Based on Virtual/Real GPS

The flowchart of the proposed navigation method based on virtual/real GPS is shown in Figure 2. First, a virtual GPS model is constructed and trained based on sensor data, which is consistent with the update frequency of the GPS data. Then, when the AUV navigates near the water surface, its position is assigned to the measurement vector $Z_2k \in \mathbb{R}^9$ if there is a GPS data update, corresponding to "Case2" in Figure 2; otherwise, the measurement vector $Z_1k \in \mathbb{R}^7$ has no two-dimensional (2D) position, corresponding to "Case1". When the AUV has no GPS data underwater, the VGPSM position is assigned to the measurement vector $Z_2k \in \mathbb{R}^9$ if there is an update of the virtual GPS model; otherwise, there is no 2D position in the measurement vector $Z_1k \in \mathbb{R}^7$. Finally, the AUV performs navigation positioning based on the prediction and update steps of the EKF.

The motion model and measurement model of AUV are shown in Equation (1) and Equation (2), respectively.

$$X_k = f(X_{k-1}, U_k, W_k) \tag{1}$$

$$Z_k = h(X_k, V_k) \tag{2}$$

$$W_k \sim N(0, Q_k) \tag{3}$$

$$V_k \sim N(0, R_k) \tag{4}$$

where $f(\cdot)$ and $h(\cdot)$ represent the motion function and measurement function, respectively. $X_k \in \mathbb{R}^9$ is the state vector at time k as shown in Equation (5), $W_k \in \mathbb{R}^9$ is Gaussian noise at time k with a mean zero and covariance matrix Q_k as shown in Equation (3), $U_k \in \mathbb{R}^6$ is the control vector at time k as shown in Equation (6), V_k is Gaussian noise at time k with a mean zero and covariance matrix R_k as shown in Equation (4), and Z_k is the measurement vector at time k. When there is no update from GPS or VGPSM, Z_k is expressed as Z_{1k} as shown in Equation (7); otherwise, Z_k is expressed as Z_{2k}, as shown in Equation (8).

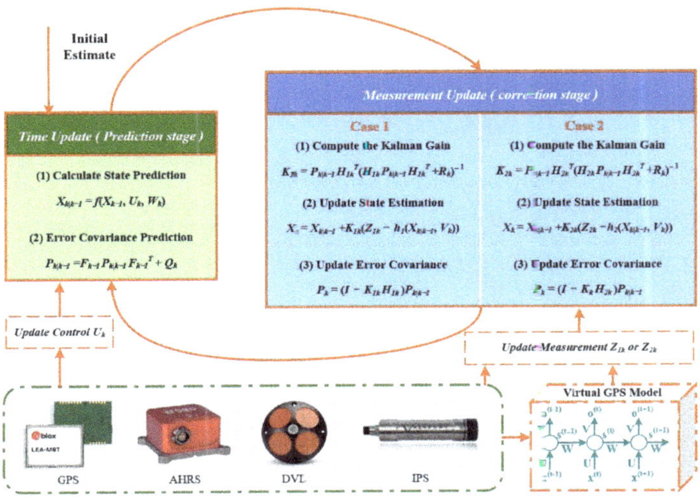

Figure 2. The framework of the proposed navigation method for AUVs.

$$X_k = \left[p^n_{x,k}, p^n_{y,k}, p^n_{z,k}, v^b_{x,k}, v^b_{y,k}, v^b_{z,k}, \xi_{x,k}, \xi_{y,k}, \xi_{z,k}\right]^T \quad (5)$$

$$U_k = \left[a^b_{x,k}, a^b_{y,k}, a^b_{z,k}, w_{x,k}, w_{y,k}, w_{z,k}\right]^T \quad (6)$$

$$Z_{1k} = \left[p^{IPS}_{z,k}, v^{DVL}_{x,k}, v^{DVL}_{y,k}, v^{DVL}_{z,k}, \xi^{AHRS}_{x,k}, \xi^{AHRS}_{y,k}, \xi^{AHRS}_{z,k}\right]^T \quad (7)$$

$$Z_{2k} = \left[p^{GPS/VGPSM}_{x,k}, p^{GPS/VGPSM}_{y,k}, p^{IPS}_{z,k}, v^{DVL}_{x,k}, v^{DVL}_{y,k}, v^{DVL}_{z,k}, \xi^{AHRS}_{x,k}, \xi^{AHRS}_{y,k}, \xi^{AHRS}_{z,k}\right]^T \quad (8)$$

where

$p^n_{x,k}, p^n_{y,k}, p^n_{z,k}$: the position of the x, y, and z axes in the navigation coordinate system.
$v^b_{x,k}, v^b_{y,k}, v^b_{z,k}$: the velocity of the x, y, and z axes in the body coordinate system.
$\xi_{x,k}, \xi_{y,k}, \xi_{z,k}$: the three-dimensional (3D) attitude of the AUV for roll, pitch, and heading.
$a^b_{x,k}, a^b_{y,k}, a^b_{z,k}$: the 3D acceleration of the AUV in the body coordinate system.
$w_{x,k}, w_{y,k}, w_{z,k}$: the 3D angular velocities of the AUV.
$p^{GPS/VGPSM}_{x,k}$, $p^{GPS/VGPSM}_{y,k}$: the position of the x and y axes in the navigation coordinate system from GPS or VGPSM.
$p^{IPS}_{z,k}$: the position of the z axes in the navigation coordinate system from IPS.
$v^{DVL}_{x,k}, v^{DVL}_{y,k}, v^{DVL}_{z,k}$: the 3D velocity in the body coordinate system from DVL.
$\xi^{AHRS}_{x,k}$, y, k^{AHRS}, $\xi^{AHRS}_{z,k}$: the 3D attitude of the AUV for roll, pitch, and heading from AHRS.

The pseudo-code of the proposed method is shown in Algorithm 1. When the navigation system is initialized, the state vector X_0 and covariance matrix P_0 are set up at the initial moment. At this stage, the current GPS position (lat, lon) is recorded and converted into Universal Transverse Mercator (UTM) coordinates $(p_{x,0}^{UTM}, p_{y,0}^{UTM})$. During the AUV navigation process, the system primarily relies on two stages: the time update (prediction) and the measurement update (correction) stages. The correction stage is divided into two scenarios: one where the measurement vector $Z_1k \in \mathbb{R}^7$ does not include the two-dimensional position, corresponding to Case 1, and another where the measurement vector $Z_2k \in \mathbb{R}^9$ includes the two-dimensional position, corresponding to Case 2. The detailed procedures for these two scenarios are described in Sections 3.1.1 and 3.1.2, respectively.

Algorithm 1: The pseudo-code of proposed method

1 Initialize state vector: X_0, covariance matrix:P_0, initial position: $(p_{x,0}^{UTM}, p_{y,0}^{UTM})$;
2 **while** $\exists$ *new data* **do**
3 *** **Time Update Step (Prediction)** *** ;
4 $X_{k|k-1} = f(X_{k-1}, U_k, 0)$;
5 $P_{k|k-1} = F_{k-1}P_{k-1}F_{k-1}^T + Q_k$;
6 *** **Measurement Update Step (Correction)** *** ;
7 **if** *AUV on the surface* **then**
8 **if** *the GPS data are updated (Case 2)* **then**
9 $p_{x,k}^{GPS} = p_{x,k}^{UTM} - p_{x,0}^{UTM}$;
10 $p_{y,k}^{GPS} = p_{y,k}^{UTM} - p_{y,0}^{UTM}$;
11 $Z_{2k} = \left[p_{x,k}^{GPS}, p_{y,k}^{GPS}, p_{z,k}^{IPS}, v_{x,k}^{DVL}, v_{y,k}^{DVL}, v_{z,k}^{DVL}, \zeta_{x,k}^{AHRS}, \zeta_{y,k}^{AHRS}, \zeta_{z,k}^{AHRS}\right]^T$;
12 $Z_{2k|k-1} = H_{2k}X_{k|k-1}$;
13 $H_{2k} = I_9 = diag(1, 1, \ldots, 1)$;
14 $K_k = P_{k|k-1}H_{2k}^T[H_{2k}P_{k|k-1}H_{2k}^T + R_k]^{-1}$;
15 $X_k = X_{k|k-1} + K_k[Z_{2k} - Z_{2k|k-1}]$;
16 $P_k = [I - K_kH_{2k}]P_{k|k-1}$;
17 **else**
18 $Z_{1k} = \left[p_{z,k}^{IPS}, v_{x,k}^{DVL}, v_{y,k}^{DVL}, v_{z,k}^{DVL}, \zeta_{x,k}^{AHRS}, \zeta_{y,k}^{AHRS}, \zeta_{z,k}^{AHRS}\right]^T$;
19 $Z_{1k|k-1} = H_{1k}X_{k|k-1}$;
20 $H_{1k} = [0_{7\times2}, I_7]$;
21 $K_k = P_{k|k-1}H_{1k}^T[H_{2k}P_{k|k-1}H_{1k}^T + R_k]^{-1}$;
22 $X_k = X_{k|k-1} + K_k[Z_{1k} - Z_{1k|k-1}]$;
23 $P_k = [I - K_kH_{1k}]P_{k|k-1}$;
24 **end**
25 **else**
26 **if** *the VGPSM is updated (Case 2)* **then**
27 $p_{x,k}^{VGPSM} = p_{x,k-m}^{n} + \delta p_{x,k}^{VGPSM}$;
28 $p_{y,k}^{VGPSM} = p_{y,k-m}^{n} + \delta p_{y,k}^{VGPSM}$;
29 $Z_{2k} = \left[p_{x,k}^{VGPSM}, p_{y,k}^{VGPSM}, p_{z,k}^{IPS}, v_{x,k}^{DVL}, v_{y,k}^{DVL}, v_{z,k}^{DVL}, \zeta_{x,k}^{AHRS}, \zeta_{y,k}^{AHRS}, \zeta_{z,k}^{AHRS}\right]^T$;
30 Repeat steps 12-16;
31 **else**
32 Repeat steps 18-23;
33 **end**
34 **end**
35 **end**
36 return

3.1.1. The Prediction Stage of Proposed Navigation Method

The prediction and update of the state vector $X_{k|k-1}$ and the covariance matrix $P_{k|k-1}$ are performed at each time step k according to the update frequency of the navigation system. The prediction steps are consistent with Figure 2, as shown in Equations (9) and (10).

$$X_{k|k-1} = f(X_{k-1}, U_k, 0) = \begin{bmatrix} p^n_{x,k-1} + C_{11}\delta p^b_{x,k} + C_{12}\delta p^b_{y,k} + C_{13}\delta p^b_{z,k} \\ p^n_{y,k-1} + C_{21}\delta p^b_{x,k} + C_{22}\delta p^b_{y,k} + C_{23}\delta p^b_{z,k} \\ p^n_{z,k-1} + C_{31}\delta p^b_{x,k} + C_{32}\delta p^b_{y,k} + C_{33}\delta p^b_{z,k} \\ v^b_{x,k-1} + a^b_{x,k}\delta t \\ v^b_{y,k-1} + a^b_{y,k}\delta t \\ v^b_{z,k-1} + a^b_{z,k}\delta t \\ \xi_{x,k-1} + w_{x,k}\delta t \\ \xi_{y,k-1} + w_{y,k}\delta t \\ \xi_{z,k-1} + w_{z,k}\delta t \end{bmatrix} \tag{9}$$

$$P_{k|k-1} = F_{k-1}P_{k-1}F^T_{k-1} + Q_k \tag{10}$$

where $\delta p^b_{x,k}, \delta p^b_{y,k}, \delta p^b_{z,k}$ denote the displacement of the x, y, and z axes in the body coordinate system between time steps $k-1$ and k as shown in Equation (11), δt denotes the time between time step $k-1$ and k, F_{k-1} is the Jacobi matrix of the motion function $f(\cdot)$ on the state vector X_{k-1} as shown in Equation (12), and $C_{ij}(1 \le i \le 3, 1 \le j \le 3)$ denotes the data in the i-th row and j-th column of the rotation matrix C^n_b that transform from the body coordinate system to the navigation coordinate system, as shown in Equation (13).

$$\Delta P^b_k = \begin{bmatrix} \delta p^b_{x,k} \\ \delta p^b_{y,k} \\ \delta p^b_{z,k} \end{bmatrix} = \begin{bmatrix} v^b_{x,k-1} + \frac{1}{2}a^b_{x,k}\delta t^2 \\ v^b_{y,k-1} + \frac{1}{2}a^b_{y,k}\delta t^2 \\ v^b_{z,k-1} + \frac{1}{2}a^b_{z,k}\delta t^2 \end{bmatrix} \tag{11}$$

$$F_{k-1} = \frac{\partial f(X_{k-1}, U_k, 0)}{\partial X_{k-1}} \tag{12}$$

$$C^n_b = \begin{bmatrix} \cos\xi_y \cos\xi_z & -\cos\xi_y \sin\xi_z & \sin\xi_y \\ \cos\xi_x \sin\xi_y \cos\xi_z + \cos\xi_x \sin\xi_z & -\sin\xi_x \sin\xi_y \sin\xi_z + \cos\xi_x \cos\xi_z & -\sin\xi_y \cos\xi_y \\ -\cos\xi_x \sin\xi_y \cos\xi_z + \sin\xi_x \sin\xi_z & \cos\xi_x \sin\xi_y \sin\xi_z + \sin\xi_x \cos\xi_z & \cos\xi_x \cos\xi_y \end{bmatrix} \tag{13}$$

3.1.2. The Correction Stage of Proposed Navigation Method

After the prediction stage, the measurement update stage follows. When the AUV is sailing on the surface, if the GPS data are updated at this time, the current GPS position $(lat^{GPS}_k, lon^{GPS}_k)$ is converted to the UTM coordinate position $(p^{UTM}_{x,k}, p^{UTM}_{y,k})$; then, the measured position $(p^{GPS}_{x,k}, p^{GPS}_{y,k})$, as shown in Equation (14), is obtained and assigned to Z_{2k}, and the measured values of IPS, DVL, and AHRS are also assigned to Z_{2k}, as shown in Equation (15).

$$\begin{bmatrix} p^{GPS}_{x,k} \\ p^{GPS}_{y,k} \end{bmatrix} = \begin{bmatrix} p^{UTM}_{x,k} - p^{UTM}_{x,0} \\ p^{UTM}_{y,k} - p^{UTM}_{y,0} \end{bmatrix} \tag{14}$$

$$Z_{2k} = \left[p^{GPS}_{x,k}, p^{GPS}_{y,k}, p^{IPS}_{z,k}, v^{DVL}_{x,k}, v^{DVL}_{y,k}, v^{DVL}_{z,k}, \xi^{AHRS}_{x,k}, \xi^{AHRS}_{y,k}, \xi^{AHRS}_{z,k}\right]^T \tag{15}$$

The measurement value $Z_{2k|k-1}$, as shown in Equation (16), is predicted based on the measurement function $h_2(\cdot)$, where H_{2k} is a nine dimensional identity matrix as shown in Equation (17).

$$Z_{2k|k-1} = h_2(X_{k|k-1}, 0) = H_{2k}X_{k|k-1} \tag{16}$$

$$H_{2k} = \frac{\partial h_2(X_{k|k-1}, 0)}{\partial X_{k|k-1}} \tag{17}$$

The Kalman gain K_k at time step k can be calculated using Equation (18). Then the state vector X_k and its covariance Matrix P_k are corrected as shown in Equations (19) and (20).

$$K_k = P_{k|k-1}H_{2k}^T[H_{2k}P_{k|k-1}H_{2k}^T + R_k]^{-1} \tag{18}$$

$$X_k = X_{k|k-1} + K_k[Z_{2k} - Z_{2k|k-1}] \tag{19}$$

$$P_k = [I - K_kH_{2k}]P_{k|k-1} \tag{20}$$

If the virtual GPS model is updated when the AUV is sailing underwater, the measurement position $(p_{x,k}^{VGPSM}, p_{y,k}^{VGPSM})$, as shown in Equation (21), is obtained based on the displacement output by VGPSM $(\delta p_{x,k}^{VGPSM}, \delta p_{y,k}^{VGPSM})$ and the AUV position $(p_{x,k-m}^n, p_{y,k-m}^n)$ at time step $k - m$, which is assigned to Z_{2k} together with the measurement data of IPS, DVL, and AHRS, as shown in Equation (22).

$$\begin{bmatrix} p_{x,k}^{VGPSM} \\ p_{y,k}^{VGPSM} \end{bmatrix} = \begin{bmatrix} p_{x,k-m}^n + \delta p_{x,k}^{VGPSM} \\ p_{y,k-m}^n + \delta p_{y,k}^{VGPSM} \end{bmatrix} \tag{21}$$

$$Z_{2k} = \left[p_{x,k}^{VGPSM}, p_{y,k}^{VGPSM}, p_{z,k}^{IPS}, v_{x,k}^{DVL}, v_{y,k}^{DVL}, v_{z,k}^{DVL}, \xi_{x,k}^{AHRS}, \xi_{y,k}^{AHRS}, \xi_{z,k}^{AHRS}\right]^T \tag{22}$$

where m is the number of time steps within one GPS update cycle. Then the updated state vector X_k and its covariance matrix P_k can be obtained based on Equations (18)–(20).

The measurement vector Z_{1k}, as shown in Equation (7), does not contain a 2D position if there is no GPS data on the surface and no VGPSM data underwater, which is consistent with the situation of "Case 1" in Figure 2. The measurement value $Z_{1k|k-1}$, as shown in Equation (23), is predicted based on the measurement function $h_1(\cdot)$, where the 7×9 matrix H_{1k}, as shown in Equation (24), is the Jacobian matrix of the measurement function $h_1(\cdot)$ with respect to the predicted state vector $X_{k|k-1}$.

$$Z_{1k|k-1} = h_1(X_{k|k-1}, 0) = H_{1k}X_{k|k-1} \tag{23}$$

$$H_{1k} = \frac{\partial h_1(X_{k|k-1}, 0)}{\partial X_{k|k-1}} \tag{24}$$

The Kalman gain K_k at time step k can be calculated using Equation (25). Then the state vector X_k and its covariance Matrix P_k are corrected, as shown in Equations (26) and (27).

$$K_k = P_{k|k-1}H_{1k}^T[H_{2k}P_{k|k-1}H_{1k}^T + R_k]^{-1} \tag{25}$$

$$X_k = X_{k|k-1} + K_k[Z_{1k} - Z_{1k|k-1}] \tag{26}$$

$$P_k = [I - K_kH_{1k}]P_{k|k-1} \tag{27}$$

3.2. Virtual GPS Model

Aiming at the problem of an increasing cumulative error in the navigation system without underwater positioning assistance, we propose to use a virtual GPS model, as shown in Figure 3, for the underwater position assistance of the AUV navigation system, in order to slow down the increase in cumulative error.

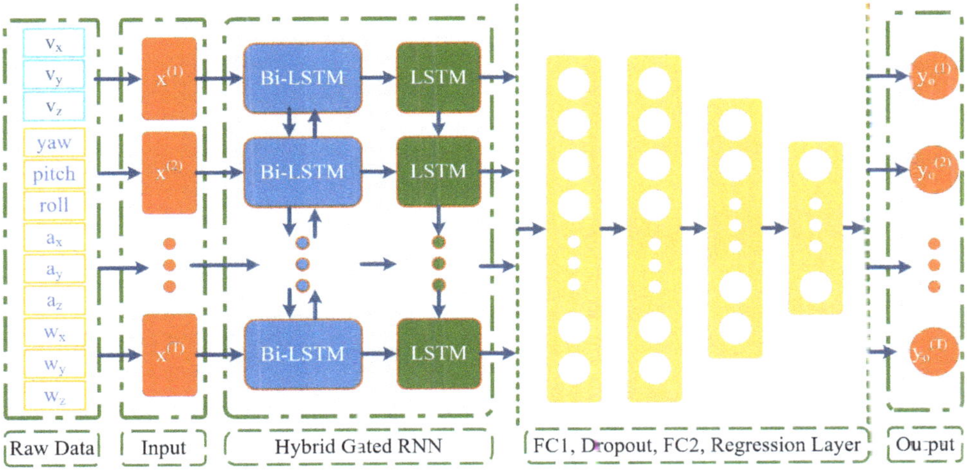

Figure 3. The flowchart of virtual GPS model based on deep sequence learning.

3.2.1. The Construction of Virtual GPS Model

VGPSM is trained and generated based on deep sequence learning, which comprehensively utilizes long short-term memory and bidirectional long short-term memory to effectively capture long-term dependencies in time-series navigation data, and, when combined with a two-layer neural network to learn the nonlinear relationship between navigation sensor data and GPS data, generates virtual GPS displacement.

The construction process of the virtual GPS model is as follows:

- First, the construction of high-quality datasets. Deep sequence learning models can efficiently handle the effects of nonlinearity and noise, but their training process relies on a large amount of high-quality data. Therefore, based on the data from AHRS, DVL, and GPS, we combine improved incremental smoothing and mapping (iSAM) and an adaptive GPS filter to generate position data that are consistent with the update frequency of the navigation system, and then construct them into a high-quality navigation dataset.
- Then, the structural design of a deep sequence model. Based on the time-series characteristics of navigation data, LSTM, which can capture long-term dependencies, and Bi-LSTM, which considers both forward and backward sequence data, are selected to process navigation sequence data. After that, data integration prediction is carried out by combining a 2-layer fully-connected layer, a Dropout layer, which prevents overfitting, and a Regression layer. Finally, the predicted virtual GPS position can be obtained using the output layer.
- Finally, the training and prediction process of the virtual GPS model. In total, 80% of the navigation dataset constructed above is used for training and 20% is used for testing. The input and output of the dataset are normalized before the training begins. The training process uses the mean square error (MSE) as the loss function, the optimizer is selected as Adam, and the initial learning rate is set to 0.001. After the DVL and AHRS data are processed with the same normalization parameters, they are input into the trained virtual GPSM model to generate the virtual GPS displacement.

Before constructing the navigation dataset, it is necessary to collect and save the data of the same AUV under various trajectories when there is an update of GPS on the surface, including the 3D attitude of DVL, the 2D position of GPS, the 3D acceleration, the 3D angular velocity, and the 3D attitude of AHRS. The update frequency of the navigation system is 10 Hz, and the update frequency of the GPS is 1 Hz. Since the GPS antenna may be covered by waves on the water surface causing drift errors, the above data are put through improved iSAM and the adaptive GPS filter for the navigation computation, and the navigation data with the update frequency of 10 Hz are generated uniformly.

Based on the described procedures, we can construct a dataset 'NavD' specifically for AUV navigation. Considering the navigation update frequency, the dataset is sampled at 0.1s intervals. Taking the j-th row as an example, the input vector I_j^n shown in Equation (28) consists of sensor data including DVL, AHRS, IPS, and GPS with a feature set of 15 in every time step j.

$$I_j^n = \left[v_{x,j}^{DVL}, v_{y,j}^{DVL}, v_{z,j}^{DVL}, \zeta_{x,j}^{AHRS}, \zeta_{y,j}^{AHRS}, \zeta_{z,j}^{AHRS}, a_{x,j}^{AHRS}, a_{y,j}^{AHRS}, a_{z,j}^{AHRS}, w_{x,j}^{AHRS}, w_{y,j}^{AHRS}, w_{z,j}^{AHRS}, p_{z,j}^{IPS}, lat_j^{GPS}, lon_j^{GPS}\right] \quad (28)$$

The output vector O_j^n of 'NavD' shown in Equation (29) includes the northward and eastward positions $(p_{x,j}^n, p_{y,j}^n)$, as well as the northward and eastward displacements $(p_{x,j}^n - p_{x,j-1}^n, p_{y,j}^n - p_{y,j-1}^n)$ at time step j.

$$O_j^n = \left[p_{x,j}^n, p_{y,j}^n, p_{x,j}^n - p_{x,j-1}^n, p_{y,j}^n - p_{y,j-1}^n\right] \quad (29)$$

Based on the existing dataset 'NavD', the relevant portions are selected to construct a specialized dataset "VGD" for the training and testing of the virtual GPS model. As shown in Figure 3, the input sequence $x^{(i)}$ at time step i is shown in Equation (30).

$$x^{(i)} = \left[I_{m\times(i-1)+1}^n(1:12), I_{m\times(i-1)+2}^n(1:12), I_{m\times(i-1)+3}^n(1:12), \ldots, I_{m\times i}^n(1:12)\right] \quad (30)$$

where $I_{i\times m}^n(1:12)$ is the DVL and AHRS data at time step $i \times m$, and m is the number of time steps within one GPS update cycle.

The output vector $y^{(i)}$ of 'VGD' shown in Equation (31) is the northward and eastward displacements of m time steps in the navigation coordinate system, and its output frequency is consistent with GPS.

$$y^{(i)} = \left[O_{m\times i}^n(1:2) - O_{m\times(i-1)}^n(1:2)\right] \quad (31)$$

The training of the virtual GPS model is based on the input $x^{(i)}$ and output $y^{(i)}$ from the dataset 'VGD'. The model starts with an input layer, where each sequence contains 10 time steps, with 12 features per time step. The hidden layer is a hybrid gated recurrent neural network (HGRNN) composed of an LSTM and a bidirectional LSTM (Bi-LSTM). The Bi-LSTM layer has 128 units, followed by a unidirectional LSTM layer with 64 units. The output is then passed through a fully connected layer (FC1) with 64 units, followed by a Dropout layer with a dropout rate of 0.5 to prevent overfitting. Afterward, the data go through another fully connected layer (FC2) with 32 units, and finally, the regression layer outputs two features. The model is trained with a batch size of 64, a learning rate of 0.001, and for 100 epochs. The loss function used is Mean Squared Error (MSE), optimized using the Adam optimizer.

The virtual GPS model is imported into the AUV navigation system for the virtual GPS displacement output $y_o^{(i)}$, as shown in Equation (32), during navigation.

$$y_o^{(i)} = \left[\delta p_{x,i}^{VGPSM}, \delta p_{y,i}^{VGPSM}\right] \quad (32)$$

where $\delta p_{x,i}^{VGPSM}$ and $\delta p_{y,i}^{VGPSM}$ denote the northward and eastward displacement over m time steps.

3.2.2. Bi-LSTM and LSTM

A hybrid recurrent gated neural network consisting of Bi-LSTM and LSTM is used in the virtual GPS model. The Bi-LSTM, as shown in Figure 4, consisting of forward and backward LSTM, can learn the bidirectional long-term dependencies of sequence navigation data. The structure of the LSTM cell can integrate long-term memory and short-term memory based on gating with different functions.

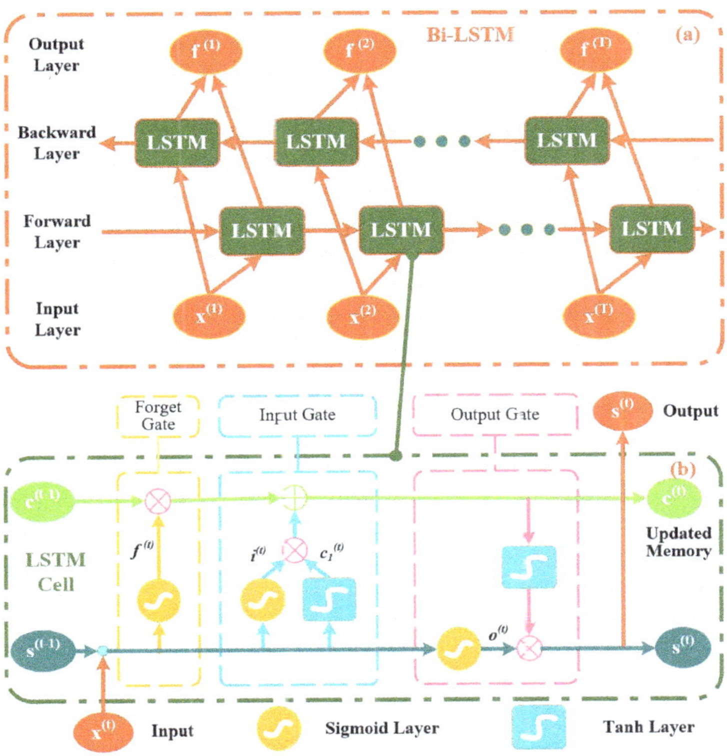

Figure 4. The structure of Bi-LSTM and LSTM. (**a**) Bi-LSTM. (**b**) LSTM.

LSTM is a recurrent neural network (RNN) that can effectively solve the gradient vanishing problem in traditional RNN, and is suitable for time-series-based data prediction. As shown in Figure 4, the LSTM cell consists of an input gate, a forget gate, and an output gate.

As described in the yellow dashed box in Figure 4, the forget gate processes the output $s^{(t-1)}$ of the previous moment and the input $x^{(t)}$ of the current moment through a sigmoid layer to obtain the activation vector $f^{(t)}$, shown in Equation (33).

$$f^{(t)} = \sigma(W_f s^{(t-1)} + U_f x^{(t)} + b_f) \tag{33}$$

As shown in the blue dashed box, the input gate combines $s^{(t-1)}$ and $x^{(t)}$ through the sigmoid layer to obtain the input gate's activation vector $i^{(t)}$, as shown in Equation (34), and through the tanh layer to obtain the cell input activation vector i_t, as shown in Equation (35).

$$i^{(t)} = \sigma(W_i s^{(t-1)} + U_i x^{(t)} + b_i) \tag{34}$$

$$c_1^{(t)} = tanh(W_c s^{(t-1)} + U_c x^{(t)} + b_c) \tag{35}$$

The light green line represents long-term memory. The forget gate determines the degree to which the long-term memory $c^{(t-1)}$ at $t-1$ moment is forgotten, and the input gate selects the degree to which the short-term memory is increased, which together determine the long-term memory $c^{(t)}$ of the current moment, as shown in Equation (36).

$$c^{(t)} = g_i * c_1^{(t)} + f^{(t)} * c^{(t-1)} \tag{36}$$

As indicated by the pink dashed box, the combination of $s^{(t-1)}$and $x^{(t)}$is passed through the sigmoid layer to obtain the output gate's activation vector $o^{(t)}$, as shown in Equation (37). Combined with the tanh layer's processing of $c^{(t)}$, the output gate's activation vector $s^{(t)}$, as shown in Equation (38), is obtained.

$$o^{(t)} = \sigma(W_o s^{(t-1)} + U_o x^{(t)} + b_o]) \tag{37}$$

$$s^{(t)} = o^{(t)} * tanh(c^{(t)}) \tag{38}$$

where W_* and U_* denote the weight matrices, and b_* denotes the bias vector parameter.

4. Experimental Results

This section will mainly present the performance of the virtual GPS model and its correction of the navigation system in combination with the real GPS position.

To validate the proposed algorithm, we conducted a series of carefully controlled experiments. The experimental data were collected from the Sailfish AUV, shown in Figure 2, under optimal sea conditions. The datasets used for this validation are 'NavD' and 'VGD,' as described in Section 3.2.1. The 'NavD' dataset, with a time interval of 0.1 s, serves as the foundational dataset from which the 'VGD' dataset was derived. 'VGD' was constructed by subsampling 'NavD' at 1-second intervals to match the GPS update frequency.

The sensor data incorporated in these experiments include 3D attitude, 3D acceleration, and 3D angular velocity from AHRS, 3D velocity from DVL, latitude and longitude from GPS, and depth measurements from IPS. The sensors used in these experiments are detailed in Table 1, which lists their models and key specifications. Furthermore, the primary parameters employed in the experiments are summarized in Table 2.

Table 2. Primary parameters employed in the experiments.

Parameter Name	Symbol	Value	Description
Time Step	δt	0.1	Time interval for each time step.
Time Interval	m	10	Number of time steps in one GPS update cycle
Initial State Vector	X_0	$0_{9\times1}$	The initial 3D position, velocity, and attitude in navigation coordinate system.
Initial State Covariance	P_0	$\begin{bmatrix} 0.1 * I_3, 0_{3\times6} \\ 0_{3\times3}, 0.01 * I_3, 0_{3\times3} \\ 0_{3\times6}, 0.001 * I_3 \end{bmatrix}$	Covariance matrix of initial state error.
Jacobian matrix 1	H_{1k}	$[0_{7\times2}, I_7]$	Jacobian matrix of measurement function $h_1(\cdot)$
Jacobian matrix 2	H_{2k}	I_9	Jacobian matrix of measurement function $h_2(\cdot)$

4.1. The Results of Virtual GPS Model

The virtual GPS model provides position information for the AUV underwater without GPS assistance to improve the navigation system accuracy. Figure 5 shows the result of the virtual GPS model. The northward and eastward displacement of GPS is taken as the true value, and that of the virtual GPS model is compared in terms of displacement waveform, root mean square error (RMSE), and single-step error.

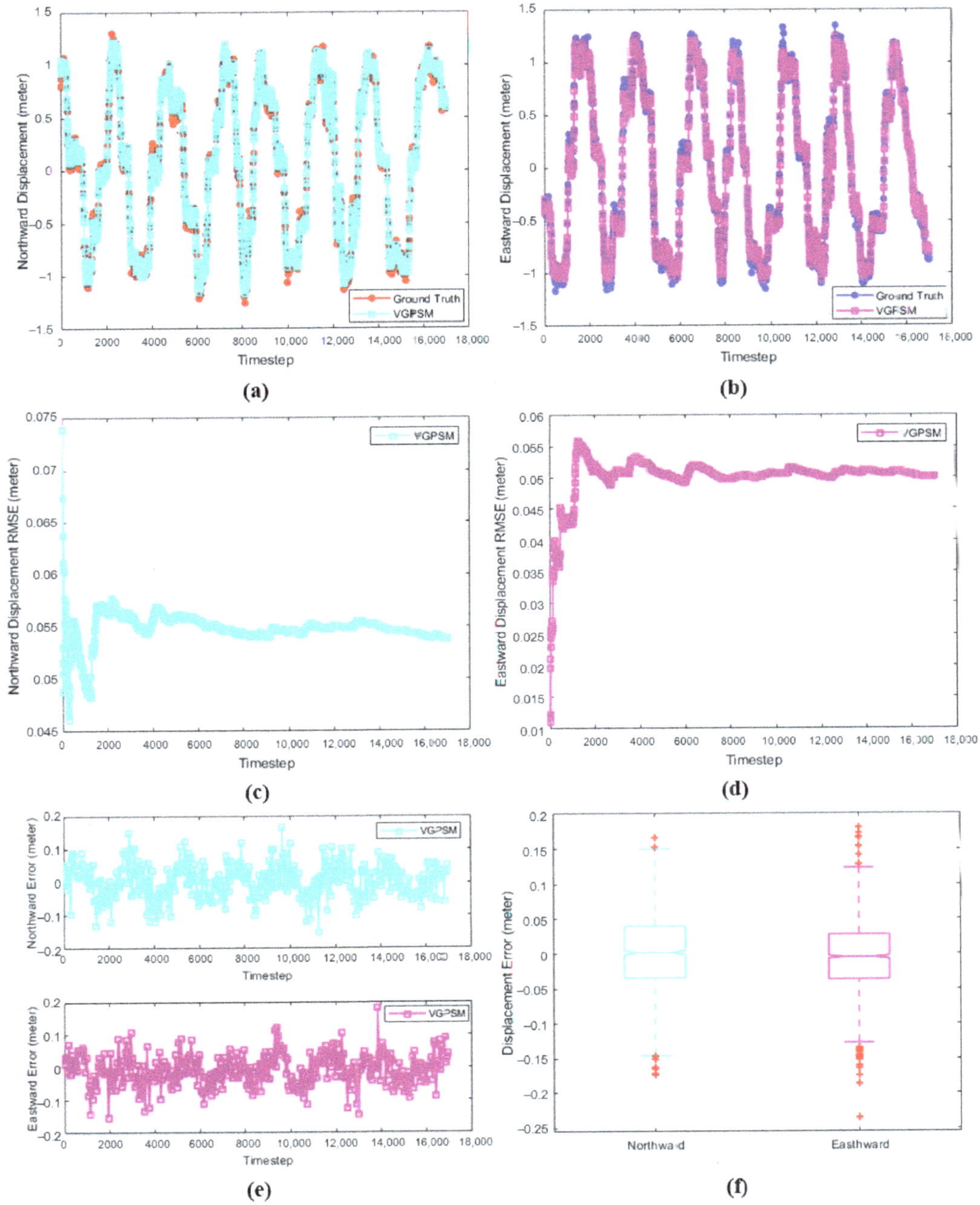

Figure 5. The result of virtual GPS model (**a**) Northward Displacement. (**b**) Eastward Displacement. (**c**) RMSE of Northward Displacement. (**d**) RMSE of Northward Displacement. (**e**,**f**) Error of Northward and Eastward Displacement.

In Figure 5, the red and blue lines represent the true values of northward and eastward displacement, respectively, and the cyan and magenta lines represent the northward and eastward displacement of the virtual GPS model, respectively. (a) and (b) represent the northward and eastward displacement comparison between the virtual GPS model and the GPS truth. (c) and (d) represent the northward and eastward root mean square error between the virtual GPS model and the ground truth, respectively. (e) and (f) compare the error between the virtual GPS model and the truth in two ways, where (e) shows the error at each step, and (f) shows the statistics of all errors.

In order to analyze the performance of the virtual GPS model specifically, we formed a table, as shown in Table 3, which mainly analyzes the performance of the virtual GPS model in terms of the mean, standard deviation, and median of the errors, as well as the root mean square error.

Table 3. Numerical performance of virtual GPS model.

Type	| Error | -M (m)	| Error | -S (m)	Error-M (m)	Error-S (m)	Error-Med (m)	RMSE (m)
Northward Displacement	0.0433	0.032	0.0019	0.0538	0.0021	0.0538
Eastward Displacement	0.0393	0.0312	−0.0059	0.0498	0.0046	0.0502

As shown in Figure 5e, since there are positive and negative errors, we conducted two kinds of analyses on the mean and standard deviation of the errors. One is to take the absolute value of all errors and then solve for the mean (| Error-M |), as shown in Equation (39), and the standard deviation (| Error | -S), as shown in Equation (40). The other is to directly solve for the mean of the errors (Error-M), as shown in Equation (41), and the standard deviation (Error-S), as shown in Equation (42).

$$|Error|_M = \frac{1}{T}\sum_{m=1}^{T} |x_m - \hat{x}_m| \tag{39}$$

$$|Error|_S = \sqrt{\frac{1}{T}\sum_{m=1}^{T} (|x_m - \hat{x}_m| - |Error|_M)^2} \tag{40}$$

$$Error_M = \frac{1}{T}\sum_{m=1}^{T} (x_m - \hat{x}_m) \tag{41}$$

$$Error_S = \sqrt{\frac{1}{T}\sum_{m=1}^{T} (x_m - \hat{x}_m - Error_M)^2} \tag{42}$$

The calculation of RMSE is shown in Equation (43).

$$RMSE = \sqrt{\frac{1}{T}\sum_{m=1}^{T} (x_m - \hat{x}_m)^2} \tag{43}$$

where x_m denotes the true value of northward or eastward displacement at time step m, $\hat{x}_m$ denotes the predicted value of northward or eastward displacement output using the virtual GPS model at time step m, and T denotes the total time steps.

4.2. The Results of Proposed Algorithm

To evaluate the effectiveness of the proposed algorithm, we have meticulously designed three experimental scenarios.

Scenario 1, as depicted in Figure 6, provides continuous GPS assistance throughout the entire operation. In this scenario, both EKF-GPS and the proposed algorithm receive position updates directly from the GPS, while EKF-VGPSM utilizes a virtual velocity model for position correction.

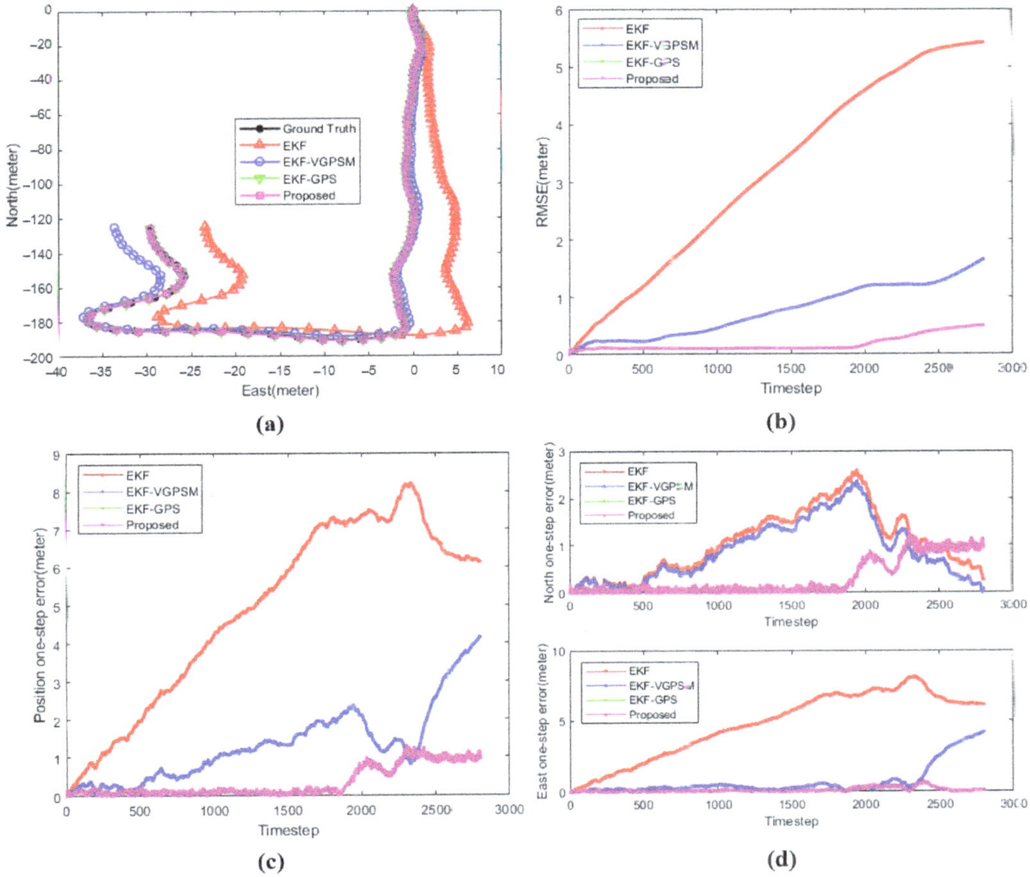

Figure 6. The comparison between proposed method and other algorithms in Scenario 1. (**a**) AUV trajectory. (**b**) Position RMSE. (**c**) Position error. (**d**) Northward and Eastward error.

Scenario 2, illustrated in Figure 7, involves GPS assistance only during the first third of the operation, followed by a period without GPS support during the middle third. Specifically, in this scenario, EKF-GPS and the proposed algorithm are aided by GPS during the initial phase. During the subsequent two-thirds of the operation, the proposed algorithm relies on a virtual GPS model for position updates, while EKF-VGPSM depends entirely on the virtual GPS model for the entire duration.

Scenario 3, shown in Figure 8, operates without any GPS assistance for the entire duration. In this case, EKF-GPS behaves identically to a standard EKF without GPS, while the proposed algorithm functions equivalently to EKF-VGPSM, both relying solely on the virtual GPS model for positional updates.

These three scenarios are designed to comprehensively assess the performance of the proposed algorithm under varying conditions, particularly focusing on its robustness and accuracy in the absence of real-time GPS data, which is critical for underwater navigation systems.

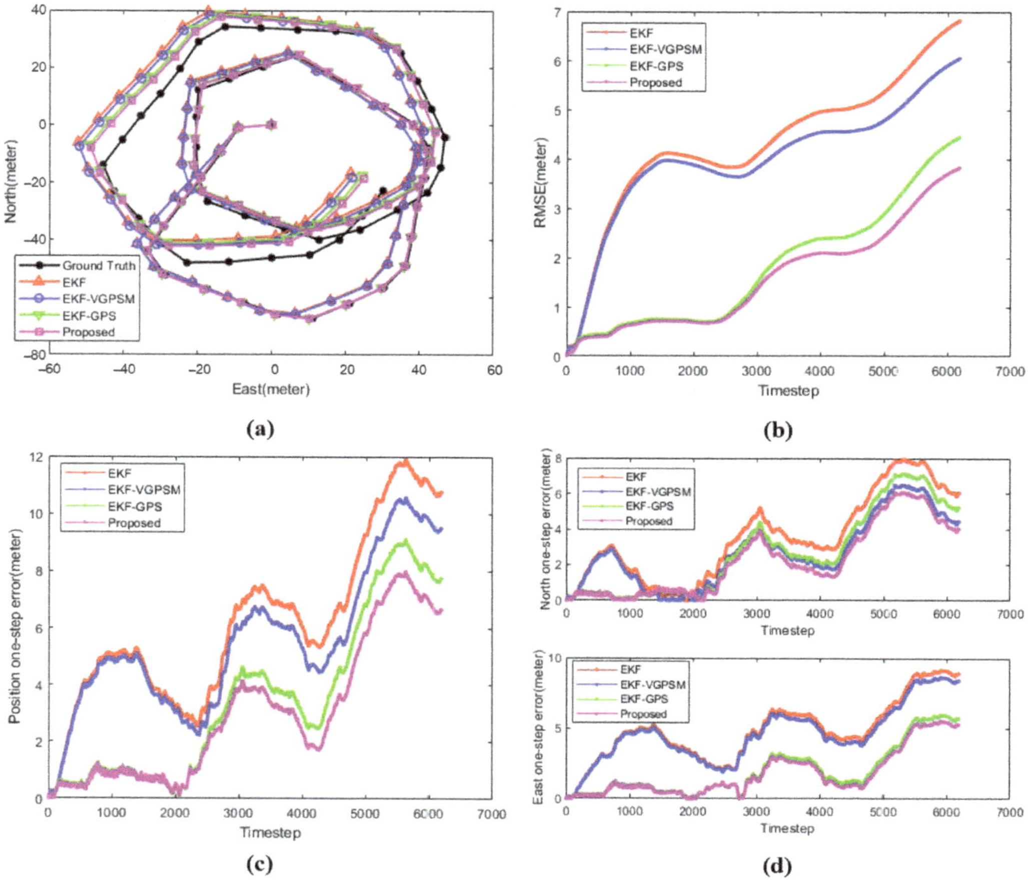

Figure 7. The comparison between proposed method and other algorithms in Scenario 2. (**a**) AUV trajectory. (**b**) Position RMSE. (**c**) Position error. (**d**) Northward and Eastward error.

In Figures 6–8, the black line represents the ground truth, the red line denotes the trajectory estimated using the standard extended Kalman filter (EKF), the blue line illustrates the trajectory estimated using the EKF augmented with the virtual GPS model (VGPSM), the green line depicts the trajectory estimated using the EKF with GPS positional assistance (EKF-GPS), and the magenta line represents the trajectory generated using the proposed algorithm. Subfigure (a) shows a trajectory comparison among the aforementioned algorithms against the ground truth. Subfigure (b) compares the root mean square error (RMSE) of the different methods. Subfigure (c) contrasts the single-step error across the algorithms, while Subfigure (d) breaks down the single-step error into the northward and eastward components.

To enhance the clarity of the results presented in Figures 6–8, we summarize the key findings in Table 4. This table primarily includes the trajectory length, the root mean square error (RMSE), the endpoint error, the northward error (Error-N), and the eastward error (Error-E) at the endpoint, and the overall navigation accuracy for each scenario. The RMSE represents the root mean square of the displacement errors, calculated as shown in Equation (44). The displacement error at time step i is determined according to Equation (45), and the navigation accuracy is computed using Equation (46). Additionally, the optimal values achieved by each algorithm across the three scenarios are highlighted in bold to facilitate a clearer comparison of the strengths and weaknesses of each method.

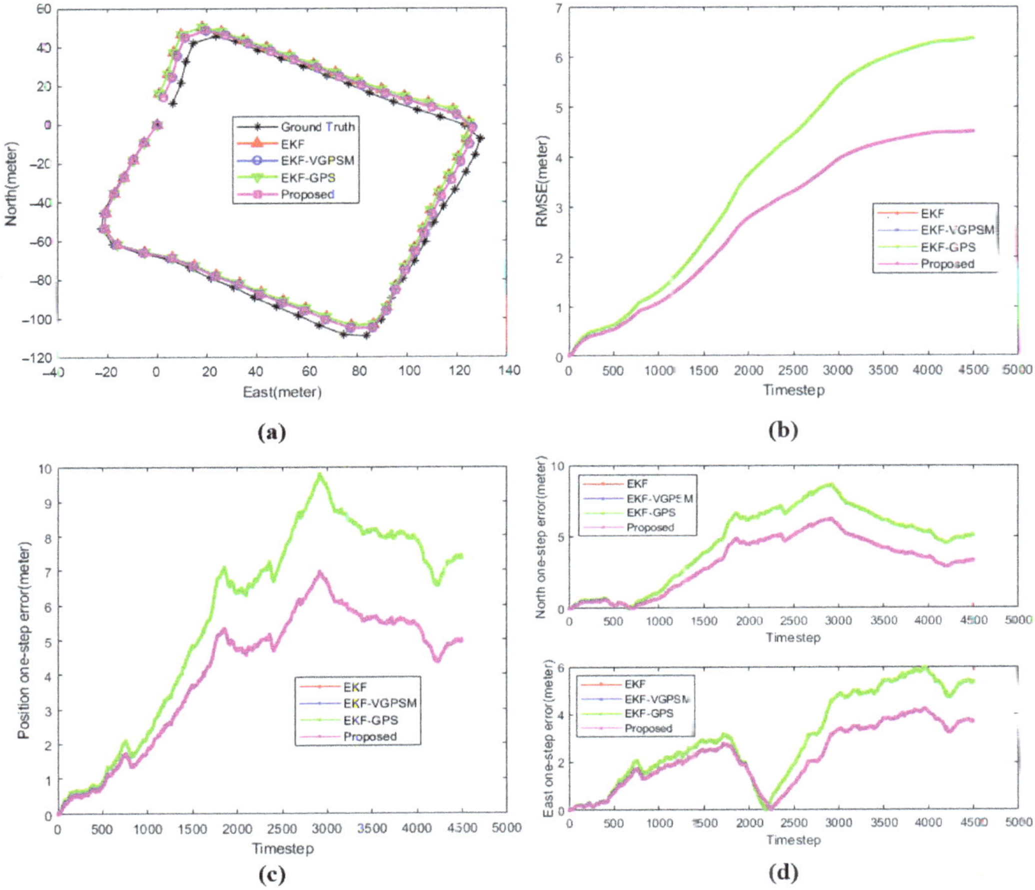

Figure 8. The comparison between proposed method and other algorithms in Scenario 3. (**a**) AUV trajectory. (**b**) Position RMSE. (**c**) Position error. (**d**) Northward and Eastward error.

Table 4. Numerical performance of different algorithms for AUV navigation.

Data	Distance (m)	Algorithm	RMSE (m)	Error (m)	Error-N (m)	Error-E (m)	Accuracy
Scenario 1	287.33	**EKF**	5.42	6.14	0.27	6.13	0.01886
		EKF-VGPSM	1.65	4.16	**0.07**	4.15	0.00574
		EKF-GPS	**0.5**	**0.95**	0.94	**0.13**	**0.00174**
		Proposed	**0.5**	**0.95**	0.94	**0.13**	**0.00174**
Scenario 2	641.48	**EKF**	6.83	10.75	6.04	8.9	0.01065
		EKF-VGPSM	6.06	9.5	4.45	8.39	0.00945
		EKF-GPS	4.45	7.76	5.22	5.74	0.00694
		Proposed	**3.8**	**6.67**	**4.03**	**5.31**	**0.00592**
Scenario 3	457	**EKF**	6.37	7.35	5.06	5.33	0.01394
		EKF-VGPSM	**4.51**	**4.94**	**3.31**	**3.67**	**0.00987**
		EKF-GPS	6.37	7.35	5.06	5.33	0.01394
		Proposed	**4.51**	**4.94**	**3.31**	**3.67**	**0.00987**

$$RMSE = \sqrt{\frac{1}{T}\sum_{i=1}^{T} Error_i^2} \quad (44)$$

$$Error_i = \sqrt{(x_i - \hat{x}_i)^2 + (y_i - \hat{y}_i)^2} \quad (45)$$

$$Accuracy = \frac{RMSE}{Distance} \quad (46)$$

where x_i and y_i denote the ground truth of the displacement at time step i in the north and east directions, respectively. $\hat{x}_i$ and $\hat{y}_i$ denote the outputs of the virtual GPS model at time step i in the north and east directions, respectively, T denotes the total number of time steps, and $Distance$ denotes the total distance of this set of data.

5. Experimental Analysis

5.1. The Performance Analysis of Virtual GPS Model

In order to verify the effectiveness of the virtual GPS model, we first carried out the comparison between the virtual GPS model and the true GPS data in the northward and eastward displacements, the results of which are shown in Figure 5 and Table 3.

From the displacement comparison in (a) and (b), it can be seen that the northward and eastward displacements output by the virtual GPS model are very close to the true displacement values of GPS. From the comparison of the root mean square error in the sub-figures (c) and (d), it can be seen that the RMSE of northward and eastward displacements is basically stable at around 0.05 m. Combined with Table 3, the root mean square errors of the virtual GPS model in the north and east directions throughout the entire process are 0.0538 m and 0.0502 m, respectively. From the error comparisons in (e) and (f), it can be seen that the single-step error fluctuates around −0.1 m to 0.1 m, and the median error is close to 0 m. Combined with Table 3, it can be seen that the mean and covariance of the errors between the northward displacement and the true value of the virtual GPS model are 0.0019 m and 0.0538 m, respectively, and the mean and covariance of the errors between the virtual GPS model and the true value in northward displacement are 0.0019 m and 0.0538 m, respectively, and those in eastward displacement are −0.0059 m and 0.0498 m, respectively. To show the distribution of the data, the mean of the absolute value of the error is higher than that of the error, and the mean squared error of the absolute value of the error is lower than that of the error.

5.2. The Performance Analysis of Proposed Method

To evaluate the effectiveness of the proposed algorithm, we conducted three experimental scenarios under different conditions. In the first scenario, depicted in Figure 6, both the EKF-GPS and the proposed algorithm operated with continuous GPS assistance throughout the entire process. In this setup, EKF-VGPSM relied on the position data provided by the output of the virtual GPS model. The results clearly show that the navigation accuracy of EKF-VGPSM was significantly better than that of the standard EKF. Moreover, the trajectories produced by both EKF-GPS and the proposed algorithm coincided and were noticeably more accurate than those of EKF-VGPS alone. As summarized in Table 4, the proposed algorithm and EKF-GPS exhibited the smallest root mean square error (RMSE), endpoint error, and eastward error at the endpoint. The proposed algorithm improved navigation accuracy by 90.77 % compared to EKF, primarily due to the continuous GPS support. Additionally, EKF-VGPSM, leveraging the virtual GPS model, achieved a 69.56 % improvement in accuracy over EKF, further validating the effectiveness of the proposed method.

In the second scenario, illustrated in Figure 7, both EKF-GPS and the proposed algorithm were initially assisted by GPS for the first one-third of the experiment. However, during the remaining two-thirds of the time, the EKF-GPS lost its positional assistance, while the proposed algorithm continued to benefit from the virtual GPS model's output. In the first phase, the errors of both EKF-GPS and the proposed algorithm were minimal due to GPS assistance. However, in the second phase, the error of EKF-GPS began to increase as it lost GPS support. In contrast, the proposed algorithm, aided by the virtual GPS model, maintained significantly lower errors and outperformed all other algorithms. According to Table 4, the proposed algorithm demonstrated the smallest RMSE, endpoint error, and errors in both the northward and eastward directions. The navigation accuracy of the proposed algorithm was improved by 37.77 % compared to EKF, showcasing its robustness even with partial GPS assistance.

In the third scenario, shown in Figure 8, no GPS assistance was available. Under these conditions, EKF-GPS lost its available GPS position data, and its navigation results aligned with those of the standard EKF, as indicated by the overlapping red and green trajectories in the figure. Both EKF-VGPSM and the proposed algorithm relied entirely on the virtual GPS model for positional data, resulting in consistent navigation outcomes, as reflected by the overlapping magenta and blue lines. This scenario effectively compares the navigation performance of EKF and EKF-VGPSM. The results clearly indicate that EKF-VGPSM and the proposed algorithm, both supported by the virtual GPS model, achieved significantly better navigation accuracy than the standard EKF and EKF-GPS. Table 4 confirms that the proposed algorithm reduced the root mean square error to 4.51 m compared to 6.37 m for EKF, resulting in a 29.2% improvement in navigation accuracy.

The effectiveness of the proposed method has been validated through a series of experiments under varying conditions. The results demonstrate that the proposed approach effectively reduces cumulative errors, with navigation accuracy improvements ranging from 29.2% to 69.56% compared to the conventional EKF. Additionally, the virtual GPS model's output accurately predicts real GPS displacement changes, with root mean square errors of 0.0538 m northward and 0.0502 m eastward, respectively. These experimental results robustly demonstrate the proposed method's effectiveness in improving navigation accuracy.

5.3. Error Analysis

The primary sources of error stem from sensor inaccuracies and model limitations. AHRS, while essential for orientation data, can introduce drift over time due to inherent biases and noise in gyroscope and accelerometer readings. DVL, used to measure velocity, can suffer from inaccuracies in low-quality bottom lock or turbulent conditions, leading to velocity estimation errors. GPS, although accurate, can introduce errors due to multipath effects or signal loss, especially in challenging underwater environments.

In the virtual GPS model, errors may arise from the integration of these noisy sensor data, where slight inaccuracies in velocity or orientation measurements can accumulate over time, leading to larger discrepancies. The proposed algorithm mitigates some of these errors through its innovative use of deep neural networks, which likely improves the sensor fusion process. However, the residual errors observed, such as the fluctuations in single-step errors and the slight biases in displacement predictions, indicate that the model could still be susceptible to overfitting or insufficient generalization in diverse conditions.

Overall, while the virtual GPS model and the proposed method significantly enhance navigation accuracy, particularly when GPS data are sparse or unavailable, the errors observed underscore the need for continued refinement in sensor fusion algorithms and model robustness to achieve even higher precision in underwater navigation.

5.4. Computational Complexity Analysis

The computational complexity of the extended Kalman filter primarily hinges on the matrix operations performed during the prediction and correction steps. Let n denote the dimension of the state vector, and o the dimension of the observation vector. The prediction step involves matrix multiplication and addition, leading to a complexity of $O(n^2)$. The correction step requires matrix inversion and multiplication, with a complexity of $O(n^3)$. Therefore, the overall complexity for one EKF iteration is $O(n^3)$.

The virtual GPS model, which integrates Bi-LSTM and LSTM networks, has its computational complexity determined by the number of LSTM units, the sequence length, and the structure of the fully connected layers. Let d represent the input feature dimension, T the sequence length, u the number of LSTM units, and p and q the number of units in the fully connected layers. Given the bidirectional nature of Bi-LSTM, the complexity for this layer is $O(2mdu^2)$. For the LSTM layer, processing the sequence unidirectionally yields a complexity of $O(mdu^2)$. The fully connected layer computes the output from the LSTM units to the final regression layer, with a complexity of $O(up + pq + 2q)$. Thus, the total complexity for processing a sequence through VGPSM is $O(mdu^2)$, assuming the fully connected layers and dropout do not significantly affect the complexity.

In practical AUV navigation applications, real-time performance is of paramount importance. The computational complexity of the EKF under standard conditions is $O(n^3)$. When invoking the VGPSM every m time step, the complexity increases by $O(mdu^2)$. While the introduction of VGPSM enhances navigational accuracy, it simultaneously imposes additional demands on computational resources. However, with the support of modern hardware, this level of computational complexity remains within an acceptable range. Specifically, by employing extended time steps m and leveraging efficient parallel computing frameworks, the method can meet the demands of most real-time AUV navigation tasks. Consequently, this approach successfully maintains high accuracy while still offering a significant real-time advantage.

6. Conclusions

To address the issue of increasing navigation errors in low-cost AUV without assisted positioning underwater, this paper introduces a virtual GPS model (VGPSM) based on deep sequence learning. By integrating VGPSM with the extended Kalman filter (EKF), the proposed approach provides a high-precision navigation solution for AUVs. The model leverages time-series data from sensors such as AHRS and DVL during surface navigation. It learns the relationship between these sensor inputs and GPS data using a hybrid architecture of LSTM and Bi-LSTM networks, which are particularly suited for processing and predicting time-series data. The VGPSM generates virtual GPS displacements updated at the same frequency as real GPS data. During underwater navigation, these virtual GPS displacements are utilized as measurements to assist the EKF in state estimation, significantly improving the accuracy and robustness of the navigation system.

However, this study presents certain limitations that warrant consideration. Firstly, the navigation datasets used for training and validation were obtained from a single AUV operating under favorable sea conditions. As a result, the generalizability of the model across different AUV platforms and in more challenging or variable marine environments remains uncertain. Additionally, the model's computational efficiency requires further optimization to ensure its viability for deployment in real-world scenarios, where computational resources may be limited, and real-time processing demands are more stringent.

Future work will focus on addressing these limitations to enhance the robustness and applicability of the proposed method. Specifically, we aim to diversify the training datasets by incorporating data from multiple AUVs operating under varied environmental conditions, thereby improving the model's adaptability and performance across diverse scenarios. Furthermore, we will explore advanced optimization techniques to reduce computational overhead, enabling the system to operate more efficiently across a broader range of hardware platforms.

Author Contributions: Conceptualization, P.-F.L.; methodology, P.-F.L.; software, P.-F.L.; validation, P.-F.L. and L.-X.X.; formal analysis, P.-F.L.; investigation, P.-F.L. and Z.-C.H.; resources, P.-F.L.; data curation, P.-F.L.; writing—original draft preparation, P.-F.L.; writing—review and editing, P.-F.L. and J.-Y.L.; visualization, P.-F.L. and Z.-C.H.; supervision, P.-F.L. and L.-X.X.; project administration, P.-F.L.

and L.-X.X.; funding acquisition, P.-F.L. and L.-X.X. All authors have read and agreed to the published version of the manuscript.

Funding: This research was funded by Natural Science Research of Jiangsu Higher Education Institutions of China (22KJB570003), Jiangsu Marine Technology Innovation Center (1012992301), and National Key Research and Development Program of China (2022YFC2806600, 2022YFC2806604).

Data Availability Statement: Data set available on request from the authors.

Conflicts of Interest: The authors declare no conflicts of interest.

References

1. Zhou, J.; Si, Y.; Chen, Y. A review of subsea AUV technology. *J. Mar. Sci. Eng.* **2023**, *11*, 1119. [CrossRef]
2. Leonard, J.J.; Bahr, A. Autonomous underwater vehicle navigation. In *Handbook of Ocean Engineering*; Springer: Berlin/Heidelberg, Germany, 2016; pp. 341–358.
3. Stutters, L.; Liu, H.; Tiltman, C.; Brown, D.J. Navigation technologies for autonomous underwater vehicles. *IEEE Trans. Syst. Man, Cybern. Part C Appl. Rev.* **2008**, *38*, 581–589. [CrossRef]
4. Zhang, B.; Ji, D.; Liu, S.; Zhu, X.; Xu, W. Autonomous Underwater Vehicle navigation: A review. *Ocean Eng.* **2023**, *273*, 113861. [CrossRef]
5. Maurelli, F.; Krupiński, S.; Xiang, X.; Petillot, Y. AUV localisation: A review of passive and active techniques. *Int. J. Intell. Robot. Appl.* **2022**, *6*, 246–269. [CrossRef]
6. El-Sheimy, N.; Youssef, A. Inertial sensors technologies for navigation applications: State of the art and future trends. *Satell Navig.* **2020**, *1*, 2. [CrossRef]
7. Bian, H.; Li, A.; Ma, H.; Wang, R. Velocity Measurement. In *Essentials of Navigation: A Guide for Marine Navigation*; Springer: Berlin/Heidelberg, Germany, 2024; pp. 149–168.
8. Zhang, L.; Wu, K.; Zhou, G.; Lan, Z. Design, Preparation, and Experiment of the Titanium Alloy Thin Film Pressure Sensor for Ocean Depth Measurements. *IEEE Sens. J.* **2024**, *24*, 14042–14049. [CrossRef]
9. Kussat, N.H.; Chadwell, C.D.; Zimmerman, R. Absolute positioning of an autonomous underwater vehicle using GPS and acoustic measurements. *IEEE J. Ocean. Eng.* **2005**, *30*, 153–164. [CrossRef]
10. Taraldsen, G.; Reinen, T.A.; Berg, T. The underwater GPS problem. In Proceedings of the Oceans 2011 IEEE, Santander, Spain, 6–9 June 2011; IEEE: Piscataway, NJ, USA, 2011; pp. 1–8.
11. Zhang, L.; Hsu, L.T.; Zhang, T. A novel INS/USBL integrated navigation scheme via factor graph optimization. *IEEE Trans. Veh. Technol.* **2022**, *71*, 9239–9249. [CrossRef]
12. Brossard, M.; Barrau, A.; Bonnabel, S. AI-IMU dead-reckoning. *IEEE Trans. Intell. Veh.* **2020**, *5*, 585–595. [CrossRef]
13. Konrad, T.; Gehrt, J.J.; Lin, J.; Zweigel, R.; Abel, D. Advanced state estimation for navigation of automated vehicles. *Annu. Rev. Control* **2018**, *46*, 181–195. [CrossRef]
14. Potokar, E.R.; Norman, K.; Mangelson, J.G. Invariant extended kalman filtering for underwater navigation. *IEEE Robot. Autom. Lett.* **2021**, *6*, 5792–5799. [CrossRef]
15. Mahesh, B. Machine learning algorithms-a review. *Int. J. Sci. Res.* **2020**, *9*, 381–386. [CrossRef]
16. Ma, D.; Chen, X.; Ma, W.; Zheng, H.; Qu, F. Neural network model-based reinforcement learning control for auv 3-d path following. *IEEE Trans. Intell. Veh.* **2023**, *9*, 893–904. [CrossRef]
17. LeCun, Y.; Bengio, Y.; Hinton, G. Deep learning. *Nature* **2015**, *521*, 436–444. [CrossRef] [PubMed]
18. Zhao, Z.; Chen, W.; Wu, X.; Chen, P.C.; Liu, J. LSTM network: A deep learning approach for short-term traffic forecast. *IET Intell. Transp. Syst.* **2017**, *11*, 68–75. [CrossRef]
19. Hou, X.; Wang, J.; Bai, T.; Deng, Y.; Ren, Y.; Hanzo, L. Environment-aware AUV trajectory design and resource management for multi-tier underwater computing. *IEEE J. Sel. Areas Commun.* **2022**, *41*, 474–490. [CrossRef]
20. Lv, P.F.; He, B.; Guo, J.; Shen, Y.; Yan, T.H.; Sha, Q.X. Underwater navigation methodology based on intelligent velocity model for standard AUV. *Ocean Eng.* **2020**, *202*, 107073. [CrossRef]
21. Vandavasi, B.N.J.; Gidugu, A.R.; Venkataraman, H. Deep Learning Aided Magnetostatic Fields Based Real-Time Pose Estimation of AUV for Homing Applications. *IEEE Sens. Lett.* **2023**, *7*, 1–4. [CrossRef]
22. Cohen, N.; Klein, I. BeamsNet: A data-driven approach enhancing Doppler velocity log measurements for autonomous underwater vehicle navigation. *Eng. Appl. Artif. Intell.* **2022**, *114*, 105216. [CrossRef]
23. Qiao, Y.; Yin, J.; Wang, W.; Duarte, F.; Yang, J.; Ratti, C. Survey of deep learning for autonomous surface vehicles in marine environments. *IEEE Trans. Intell. Transp. Syst.* **2023**, *24*, 3678–3701. [CrossRef]
24. He, Q.; Yu, H.; Fang, Y. Deep Learning-Based Inertial Navigation Technology for Autonomous Underwater Vehicle Long-Distance Navigation—A Review. *Gyroscopy Navig.* **2023**, *14*, 267–275.
25. Xie, Y.X.; Liu, J.; Hu, C.Q.; Cui, J.H.; Xu, H. AUV dead-reckoning navigation based on neural network using a single accelerometer. In Proceedings of the 11th International Conference on Underwater Networks & Systems, Shanghai, China, 24–26 October 2016; pp. 1–5

26. Huang, H.; Chen, X.; Zhang, B.; Wang, J. High accuracy navigation information estimation for inertial system using the multi-model EKF fusing adams explicit formula applied to underwater gliders. *ISA Trans.* **2017**, *66*, 414–424. [CrossRef] [PubMed]
27. Li, W.; Chen, M.; Zhang, C.; Zhang, L.; Chen, R. A Novel Neural Network-Based SINS/DVL Integrated Navigation Approach to Deal with DVL Malfunction for Underwater Vehicles. *Math. Probl. Eng.* **2020**, *2020*, 2891572. [CrossRef]
28. Sabet, M.T.; Daniali, H.M.; Fathi, A.; Alizadeh, E. A low-cost dead reckoning navigation system for an AUV using a robust AHRS: Design and experimental analysis. *IEEE J. Ocean. Eng.* **2017**, *43*, 927–939. [CrossRef]
29. Taud, H.; Mas, J. Multilayer Perceptron (MLP). In *Geomatic Approaches for Modeling Land Change Scenarios*; Camacho Olmedo, M.T., Paegelow, M., Mas, J.F., Escobar, F., Eds.; Springer International Publishing: Cham, Switzerland, 2018; pp. 451–455.
30. Gonzalez, J.; Yu, W. Non-linear system modeling using LSTM neural networks. *IFAC-PapersOnLine* **2018**, *51*, 485–489. [CrossRef]
31. Luo, Z.; Shi, D.; Ji, J.; Shen, X.; Gan, W.S. Real-time implementation and explainable AI analysis of delayless CNN-based selective fixed-filter active noise control. *Mech. Syst. Signal Process.* **2024**, *214*, 111364. [CrossRef]
32. Xia, H.; Zhu, R.; Yuan, H.; Song, C. Rapid quantitative analysis of cotton-polyester blended fabrics using near-infrared spectroscopy combined with CNN-LSTM. *Microchem. J.* **2024**, *200*, 110391. [CrossRef]
33. Topini, E.; Fanelli, F.; Topini, A.; Pebody, M.; Ridolfi, A.; Phillips, A.B.; Allotta, B. An experimental comparison of Deep Learning strategies for AUV navigation in DVL-denied environments. *Ocean Eng.* **2023**, *274*, 114034. [CrossRef]
34. Prieto, A.; Prieto, B.; Ortigosa, E.M.; Ros, E.; Pelayo, F.; Ortega, J.; Rojas, I. Neural networks: An overview of early research, current frameworks and new challenges. *Neurocomputing* **2016**, *214*, 242–268. [CrossRef]
35. Lv, P.F.; He, B.; Guo, J. Position correction model based on gated hybrid RNN for AUV navigation. *IEEE Trans. Veh. Technol.* **2021**, *70*, 5648–5657. [CrossRef]
36. Lv, P.; Guo, J.; Song, Y.; Sha, Q.; Jiang, J.; Mu, X.; Yan, T.; He, B. Autonomous navigation based on iSAM and GPS filter for AUV. In Proceedings of the 2017 IEEE Underwater Technology (UT), Busan, Republic of Korea, 21–24 February 2017; pp. 1–4.
37. Guo, J.; Li, D.; He, B. Intelligent collaborative navigation and control for AUV tracking. *IEEE Trans. Ind. Inform.* **2020**, *17*, 1732–1741. [CrossRef]
38. Mu, X.; He, B.; Zhang, X.; Song, Y.; Shen, Y.; Feng, C. End-to-end navigation for autonomous underwater vehicle with hybrid recurrent neural networks. *Ocean Eng.* **2019**, *194*, 106602. [CrossRef]
39. Staudemeyer, R.C.; Morris, E.R. Understanding LSTM–a tutorial into long short-term memory recurrent neural networks. *arXiv* **2019**, arXiv:1909.09586.
40. Greff, K.; Srivastava, R.K.; Koutník, J.; Steunebrink, B.R.; Schmidhuber, J. LSTM: A search space odyssey. *IEEE Trans. Neural Networks Learn. Syst.* **2016**, *28*, 2222–2232. [CrossRef] [PubMed]

drones

MDPI

Article

A Framework for Autonomous UAV Navigation Based on Monocular Depth Estimation

Jonas Gaigalas *, Linas Perkauskas, Henrikas Gricius, Tomas Kanapickas and Andrius Kriščiūnas

Department of Applied Informatics, Faculty of Informatics, Kaunas University of Technology, Studentu St. 50, LT-51368 Kaunas, Lithuania

* Correspondence: jonas.gaigalas@ktu.edu

Abstract: UAVs are vastly used in practical applications such as reconnaissance and search and rescue or other missions which typically require experienced operators. Autonomous drone navigation could aid in situations where the environment is unknown, GPS or radio signals are unavailable, and there are no existing 3D models to preplan a trajectory. Traditional navigation methods employ multiple sensors: LiDAR, sonar, inertial measurement units (IMUs), and cameras. This increases the weight and cost of such drones. This work focuses on autonomous drone navigation from point A to point B using visual information obtained from a monocular camera in a simulator. The solution utilizes a depth image estimation model to create an occupancy grid map of the surrounding area and uses an A* path planning algorithm to find optimal paths to end goals while navigating around the obstacles. The simulation is conducted using AirSim in Unreal Engine. With this work, we propose a framework and scenarios in three different open-source virtual environments, varying in complexity, to test and compare autonomous UAV navigation methods based on vision. In this study, fine-tuned models using synthetic RGB and depth image data were used for each environment, demonstrating a noticeable improvement in depth estimation accuracy, with reductions in Mean Absolute Percentage Error (MAPE) from 120.45% to 33.41% in AirSimNH, from 70.09% to 8.04% in Blocks, and from 121.94% to 32.86% in MSBuild2018. While the proposed UAV autonomous navigation framework utilizing depth images directly from AirSim achieves 38.89%, 87.78%, and 13.33% success rates of reaching goals in AirSimNH, Blocks, and MSBuild2018 environments, respectively, the method with pre-trained depth estimation models fails to reach any end points of the scenarios. The fine-tuned depth estimation models enhance performance, increasing the number of reached goals by 3.33% for AirSimNH and 72.22% for Blocks. These findings highlight the benefits of adapting vision-based models to specific environments, improving UAV autonomy in visually guided navigation tasks.

Keywords: UAV; control; vision-based navigation; path planning; collision avoidance; simulation; deep learning; monocular depth estimation

1. Introduction

Unmanned aerial vehicles (UAVs) have become an active object of recent research. The quad-rotor drones are especially appealing due to their lightweight frame, sensor modularity, and availability of other cost-effective off-the-shelf parts. Essential topics such as search and rescue, remote sensing, precision agriculture, goods delivery, and other civil applications have seen innovation utilizing UAVs [1]. However, the human operation of such vehicles can become a problem in some scenarios, for example, where the GPS and radio signals are unavailable, and it is impossible to preplan a trajectory due to non-existing 3D maps or landmark references. Autonomous working drones could be used to perform such missions. This way, the signal interference would not be a threat, as all the computing and control processes would be internal.

The autonomous navigation of a vehicle requires robot perception and localization, motion planning, and the execution of commands with controllers. Usually, for this complicated system to be robust, full-navigation solutions employ the fusion of multiple sensors. The most common of these are light detection and ranging (LiDAR) scanners, multiple cameras, and acoustic or radar sensors accompanied by inertial measurement units (IMUs) [2]. These solutions can be quite expensive in terms of both the price and computational cost. Reliable LiDAR scanners typically range in price from a few thousand to tens of thousands of dollars. In contrast, simple RGB cameras and the necessary system hardware offer a more affordable alternative. Developing a system that achieves results comparable or superior to LiDAR while being cost-effective would provide significant advantages in various systems.

A solution for these problems is to use a monocular camera as the primary sensor for the perception and localization of the drone. RGB cameras are relatively cheap and lightweight, bringing the cost of such a system down. Also, the captured images can provide more semantic information to aid the operations. However, using a simple RGB camera comes with its challenges. The algorithms involved in visual localization are computationally demanding and prone to problems arising from motion blur, variation in illumination, and inherent scale drift [3]. Many researchers opt into RGB-D cameras for their setups to gather additional depth information to counter some of these problems. However, these sensors are relatively expensive in comparison to simple RGB cameras.

In recent years, research has brought attention to various depth estimation methods [4]. By utilizing machine learning models, it is now possible to achieve accurate depth images from a single RGB image. This enables the use of algorithms traditionally only available in setups containing RGB-D cameras, thus expanding the capabilities of monocular RGB camera drone systems, albeit with increased computational cost.

With the rise in UAVs (and robotics in general), new tools for the simulation of such systems have seen the light as software-in-the-loop (SiL) testing is an integral part of building a robust autonomous navigation solution. However, autonomous UAV simulation research mainly consists of indoor environments where robots navigate corridors, areas of sparsely distributed box-shaped objects or smaller, and specific segments of areas [5–7]. Also, most researchers only use a few select tools in their papers, and, most of the time, do not provide the environments or the exact coordinates for the start and end points [8]. All of this does not allow for reasonable comparison between the research or real-world conditions.

This study aims to develop an autonomous navigation system for UAVs that relies on visual information acquired with a monocular RGB camera in simulated environments. By implementing a depth estimation model, an occupancy grid map is generated from point clouds. This allows the drone to dynamically plan a path for obstacle avoidance during the navigation from point A to point B. In this study, the results of simulated flights in different scenarios utilizing estimated depth images are compared with results using depth images from the simulation software AirSim [9]. Different voxel sizes are compared to check their effect on navigation performance during flight. In addition, each flight scenario's exact environments and start/end points are provided for future comparisons between autonomous navigation solutions in AirSim platform.

2. Related Work

Building a UAV navigation system that fully functions without human input is no easy feat, as many components must work in coordination. Therefore, only about 16% of works in the field of UAV navigation—particularly in scenarios without GNSS (Global Navigation Satellite System) availability— have proposed complete navigation solutions, while the rest of the research focused on individual parts of UAV navigation like localization, estimation of various inertial parameters of flight, mapping of surroundings, and path planning [2].

2.1. Localization

Localization is a fundamental aspect of UAV navigation, providing the necessary information by estimating a vehicle's position and orientation in its operating environment. It is essential in other robotics applications and is also the most researched navigation area [2].

Autonomous navigation solutions utilizing vision sensors can be broken down into two main categories: mapless and map-based [8,10,11]. In mapless systems, UAVs do not need prior knowledge of the area they fly in, and no maps are built during the navigation process. Solutions can be categorized as map-based when UAVs are either provided with a map of the area they need to navigate or they simultaneously build one whilst exploring the unknown territory.

Mapless systems rely on real-time sensor data to estimate the motion and position of the vehicle without prior knowledge of the environment. Such vision-based systems extract visible features from the environment to make navigation decisions. The most prominent methods in mapless solutions are optical flow (OF) and feature tracking [11]. In recent years, researchers have also shown systems that use various machine-learning approaches [8].

The optical flow technique compares pixels between consecutive frames and estimates the motion between the camera and the scene. Two main categories of optical flow include global [12] and local [13]. In the case of global OF, the neighboring pixels of a point in the image and the smoothness of their movements are considered. In the local approach, all the pixels in the image are compared, and their differences are measured. The earliest application of optical flow to UAV navigation was based on a stereo vision setup imitating a bee's flight, where the UAV estimates the distance to walls and obstacles from two cameras and traverses through a corridor accordingly [14]. Throughout the years, navigation methods that utilize optical flow have improved. For example, researchers showed that using inverse optical flow techniques with a single forward-facing camera can overcome the previously existing limitations of balance strategy when UAVs face an obstacle directly in front [15]. This enables vehicles to navigate in urban canyon settings.

The feature tracking method extracts points or patterns from images visible in the environment and follows them over time to estimate their motion relative to the camera. This technique can track invariant features such as corners, lines, or other points that could be detected in different images of the same area under changed conditions like rotation, illumination, and scale [16]. Well-known feature point detectors and descriptor extraction algorithms include Harris corners [17], features from accelerated segment test (FAST) [18], scale-invariant feature transform (SIFT) [19], binary robust independent elementary features (BRIEF) [20], speeded up robust features (SURF) [21], and Oriented FAST and Rotated BRIEF (ORB) [22]. In recent years, feature descriptors have become more invariant to illumination and scale changes and more robust than OF methods. However, due to the computational overhead that feature detection and matching brings, the feature-based systems compute sparser environment representation in points. Furthermore, in textureless (feature-poor) environments, tracking performance can be poor or even lost [3].

Map-based systems rely on sensor data and prior knowledge about the environment, such as known or partially known maps or spaces. It also includes map-building strategies that can reconstruct the map of the area in which drones navigate dynamically [8].

Prior knowledge of surroundings can be specified as landmarks such as waypoints made up of roof positions that can be matched from satellite images during navigation or specialized patterns such as QR or other ground patterns that can be detected with the help of machine learning and feature-based algorithms [8]. These landmarks can serve as access points to guide the drone, show where obstacles are located, or help track targets. Moreover, the knowledge can come in the form of maps of the environment. Two main types of maps are used in UAV navigation: octree and occupancy grid maps [11]. Volumetric 3D environment models can be generated utilizing sensor data and represented as hierarchical tree structures based on octrees. An efficient open-source framework used by many robotic systems is OctoMap [23]. The occupancy grid method represents the environment as a grid of cells where each cell is likely occupied. An example of such a mapping technique is shown in [24]. Information about obstacles and free space is stored in multi-volume occupancy grids, which allows them to incrementally fuse new positive or negative sensor information, enabling correction for errors and a more robust 3D mapping of the environment.

These methods can also be used in a similar map-building approach, which dominates the full-navigation solutions and is called simultaneous localization and mapping (SLAM) [2]. This technique allows UAVs to generate a map and simultaneously estimate their position while navigating an unknown environment [25]. Methods like feature-based ORB-SLAM [26], MonoSLAM [27], or direct LSD-SLAM [28] can be utilized in monocular vision-based setups.

2.2. Monocular Depth Estimation

Monocular depth estimation is the process of assessing depth from singular RGB images and inferring depth information. Researchers try to differentiate the types of depth estimation methods into two categories: single-shot and multi-frame [4].

Single-shot depth estimation drops the frames and estimates the depth from any taken image without a broader context about the environment. These methods often use deep learning models trained on large datasets to extract depth cues from visual patterns such as texture, shading, and perspective. Depth estimation models like Depth Anything [29] provide a way to generalize depth predictions for small and large details by incorporating multi-scale attention mechanisms. Another prominent approach, Monodepth2 [30], achieves accurate depth predictions by combining self-supervised learning with encoder–decoder architectures, allowing it to adapt to various domains and datasets without requiring extensive ground truth depth annotations.

Multi-frame depth estimation leverages temporal information from consecutive frames to infer depth, exploiting motion cues such as parallax and changes in camera position. These methods often integrate principles from structure-from-motion (SfM) or visual-inertial odometry (VIO). By analyzing object shifts across frames, multi-frame methods achieve higher depth accuracy and reconstruct 3D structures more effectively. Recent advancements, such as the Multi-Frame Depth Transformer (MFDT), demonstrate the effectiveness of transformer-based architectures and how depth estimation can be improved by capturing temporal relationships in scene geometry and motion dynamics [31].

2.3. Path Planning

Path planning is a major navigation system component that aims to swiftly and accurately navigate a sufficiently complex environment. One type of path-planning algorithm that can be used for online UAV navigation is called graph-based path planning. These types of algorithms work by representing the environment as a graph. In such a graph, nodes represent discrete navigable locations in the environment, and edges represent possible paths between these locations. Graph-based search algorithms aim to find the optimal path from the start to the goal node on a pre-defined graph. These methods are highly effective in structured environments with known maps and can efficiently handle static obstacles. Nevertheless, these algorithms can still be used for online path planning. However, such use introduces the need for recurrent replanning. Examples of viable graph-based search algorithms are Dijkstra's Algorithm [32], A* [33], Theta* and Lazy Theta* [34], and Jump Point Search (JPS) [35]. Other graph-based search algorithms are designed to work in dynamic/unknown environments, e.g., D* [36] or D* lite [37].

Another widely used path planning algorithm type is sampling-based algorithms. Sampling approaches create a graph representation of the environment by sampling points and connecting them to form a possible path. These methods are especially useful in unknown or partially known, high-dimensionality environments as these algorithms focus on finding any valid solution instead of the globally optimal one. Examples of such algorithms are Rapidly Exploring Random Trees (RRT) [38], Probabilistic Roadmap (PRM) [39], and Voronoi Diagrams [40].

2.4. Autonomous UAV Navigation

Complete solutions for autonomous UAV navigation combine all the parts into one system. Usually, such systems employ numerous different sensors for more robust identification and mapping of the environment to avoid obstacles in the UAV's path [2].

LiDAR scans environments with laser rangefinders to provide accurate distance and position measurements for single points. These sensors are usually used as complimentary devices to provide depth information as they cannot extract texture or color. Techniques like 2D or 3D LiDAR SLAM are utilized in such solutions [5,6,41–44].

RGB-D can also provide depth-sensing capabilities. Devices like Microsoft Kinect, Intel RealSense, or Asus Xtion retrieve red, green, and blue alongside depth data for each pixel. This is usually performed by utilizing infrared projectors and receivers. The depth information combined with SLAM techniques can result in highly detailed 3D maps [7,45–47]. Stereo camera setups using multiple cameras can also generate depth information from the disparity between captured images [48–50]. Numerous approaches also make use of monocular cameras [25,51–53].

3. Materials and Methods

3.1. Simulation Environment

This study uses the AirSim simulator to conduct autonomous drone flights. There are ten default environments built with Unreal Engine and provided by AirSim. Each environment is based on an outdoor setting with varying numbers of objects and complexity. The simulator can extract depth, RGB, and segmentation images from the drone camera's point of view. The AbandonedPark environment has a moving Ferris wheel, the CityEnviron has moving cars and pedestrians implemented, and TrapCam has simulated moving deer and vegetation generation, which can be tuned with different parameters. For experimentation, three of the available environments (AirSimNH, Blocks, and MSBuild2018) were chosen. Overview screenshots from the simulator in these environments can be seen in Figure 1.

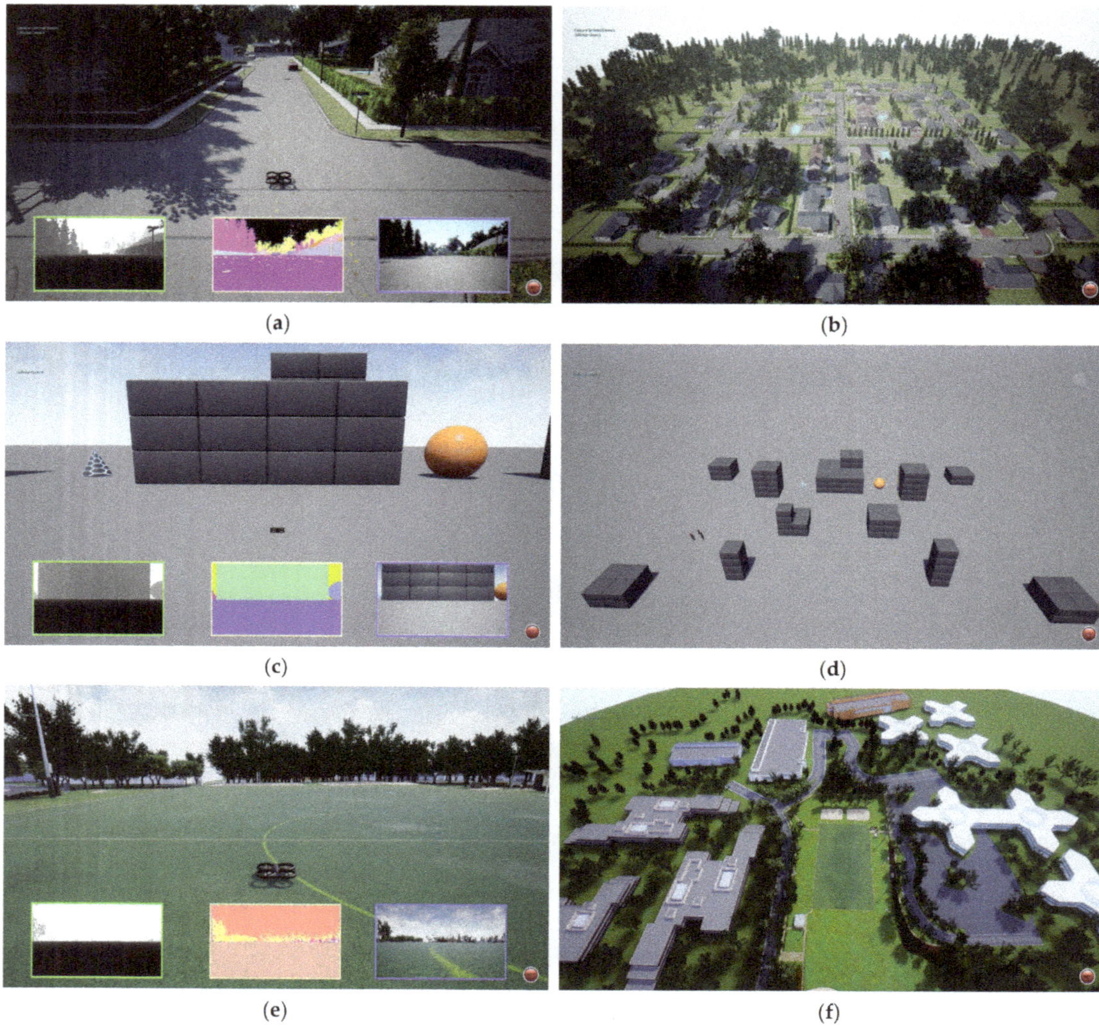

Figure 1. Snapshots from different AirSim environments that show depth, segmentation, and RGB images from the drone's front-facing camera and the environment from higher up: AirSimNH (**a**,**b**); Blocks (**c**,**d**); and MSBuild2018 (**e**,**f**).

The drone was setup to acquire 518 × 518 pixel-sized RGB and depth images with a 60-degree field of view (FOV). During navigation, the drone keeps distance from obstacles in a 3 × 3 voxel grid, as shown as a blue grid in Figure 2. The size of the voxel V can be changed, which in turn scales this grid, allowing the drone to traverse in either narrower or broader spaces. Other drone settings, such as flying speed or other flight parameters, can be configured in the simulator; however, simulations in this study scope were conducted to explore the effects of varying voxel size; therefore, the other parameters were set to default values such as flying velocity of 2 m/s, an occupancy map size of 1800 × 1800 × 100 voxels, a maximum of 200 iterations to find paths, and a maximum of 20 consecutive steps with collisions registered to quit the simulation.

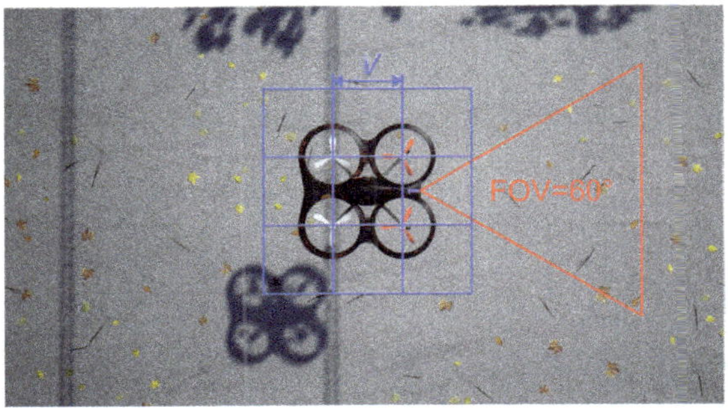

Figure 2. Snapshot of the drone inside the AirSim environment. The blue grid shows the scalable boundary of the drone in voxels of size V. The red area depicts the drone's field of view.

All the simulations were performed on a personal computer. The environments are computationally heavy on graphics; therefore, a dedicated graphics card (GPU) had to be installed in the computer. All the system's specifications are noted in Table 1.

Table 1. Specifications of the personal computer used to run flight simulations in this work.

GPU	CPU	RAM
EVGA GeForce RTX 3070 Ti; 8 GB GDDR6X @ 1575 MHz	AMD Ryzen 7 5800X; 8 cores @ 3.8 GHz	Kingston Fury: 2 × 32 GB @ 3200 MHz

Three scenarios were chosen for each environment to test the autonomous drone navigation system. Start and end coordinates alongside the Euclidean distance of each proposed scenario are in Table 2. Here, the last column notes where each scenario is depicted in Figure 3 where yellow circles and triangles depict the start and end positions of different scenarios in each environment, respectively. Red lines show the paths that a drone could hypothetically take to navigate through each scenario.

Table 2. Start and end coordinates of different scenarios in chosen environments.

Environment	Scenario	Start (X, Y, Z)	End, (X, Y, Z)	Distance	Snapshot
AirSimNH	1	0, 0, 0	−130, −115, 3	173.59	a
	2	0, 0, 0	130, 130, 3	183.87	b
	3	0, 0, 0	−115, 130, 3	173.59	c
Blocks	1	0, 0, 0	70, 150, 3	165.56	d
	2	0, 0, 0	70, −135, 3	152.10	e
	3	0, 0, 0	−105, −150, 3	183.12	f
MSBuilding2018	1	0, 0, 0	30, 260, 3	261.74	g
	2	0, 0, 0	30, −280, 3	281.62	h
	3	0, 0, 0	−400, 30, 3	401.14	i

As the AirSimNH environment (Figure 3a–c) has houses and their backyards separated by roads, all three proposed flight paths start at the intersection of these roads. The end points are located at the corners of the neighborhood. These paths force the drone to navigate complex environments containing houses, trees, fences, and other objects typically found in such backyards. It is worth noting that deer and raccoons are also moving in this environment. Also, along the outskirts of the environment, there is a densely populated forest, which is well suited for obstacle avoidance experiments.

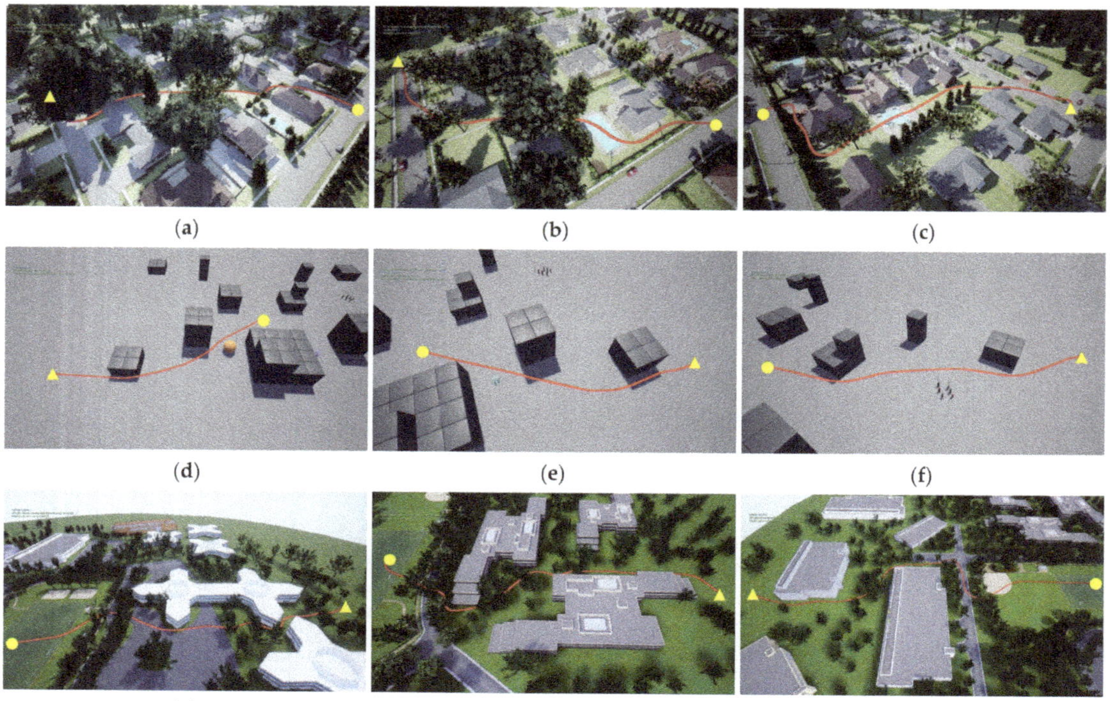

(a) (b) (c) (d) (e) (f) (g) (h) (i)

Figure 3. Snapshots of different scenarios in different environments. Yellow circles show start positions, triangles—set goal coordinates. Red lines depict hypothetical paths for simulated flights. Scenarios no.1–3 in different AirSim environments: (**a–c**) AirSimNH; (**d–f**) Blocks; and (**g–i**) MSBuild2018.

For tests in less complex settings, the Blocks environment (Figure 3d–f) was chosen. Relatively big-sized blocks are scattered sparsely in this environment alongside some other basic shapes like balls, cones, or cylinders. The end point coordinates are located at the corners of the map. To reach denoted positions, the drone would need to evade simple-shaped boxes and other objects.

MSBuild2018 is one of the AirSim provided environments (Figure 3g–i) that is composed of an open field, roads, and building complexes densely surrounded by trees and other objects. In proposed scenarios, the drone starts in the open field. To reach the goals, the drone must navigate through thick vegetation along the roads and between the buildings, making the task more complex.

3.2. System Architecture

The proposed architecture for an autonomous monocular drone navigation system is depicted in Figure 4. The system is divided into a simulation environment and a drone navigation system called a "Monocular drone". For the simulation environment, Airsim v1.8.1 was used. The navigation system runs in a containerized Docker environment running Ubuntu 20.04 guest. The ROS noetic v1.17.0 framework was used for communication between these two parts.

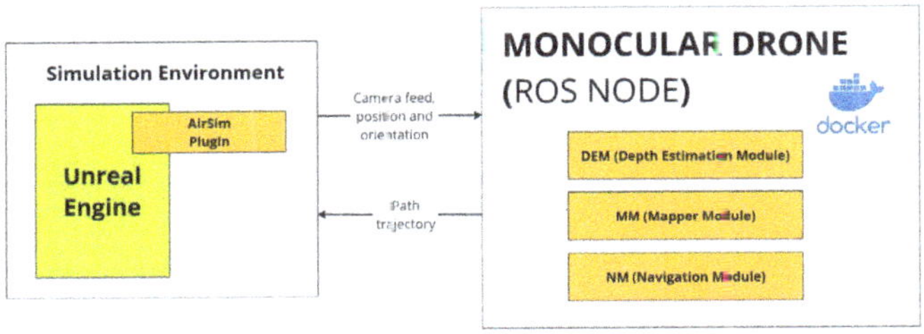

Figure 4. Diagram of the autonomous monocular drone navigation system.

The navigation system was further divided into three modules:

- Depth estimation module (DEM). This module is responsible for estimating depth images from the RGB camera feed provided by the simulation environment. For this task, the Depth Anything V2 model was utilized.
- Mapper module (MM). The purpose of MM is to build and iteratively update the occupancy map-based 3D environment using depth images supplied by DEM and camera position and orientation retrieved from the simulation environment.
- Navigation module (NM). This module finds viable path trajectories to specified points in an everchanging mapped 3D environment. Path-finding logic is based on the A* algorithm. The output of this module is fed back to the simulation environment.

3.3. Depth Estimation

3.3.1. Preparing Synthetic Data for Depth Estimation Models

The inherent scale-drift problem occurred using the pretrained single-shot monocular depth estimation model. However, a hypothesis was proposed that the errors in depth estimation come from insufficient synthetic images and that the results could be improved by finetuning the model.

The models were finetuned with the simulation environment data to obtain more accurate depth images from RGB images. A dataset was generated for each environment to improve the depth estimation accuracy, as the pre-trained model struggled to adjust to environment lighting conditions, textures, objects, and other scene features. A script was developed to generate datasets to fine-tune the model. The script in Algorithm 1 used parameters such as the following:

- Number of coordinates that define how many points will be generated;
- Map boundaries specifying the range in which the drone can be placed;
- Height list specifying predefined height (1, 3, 5, 7, and 9 m) for generated coordinates along the *z*-axis;
- Camera IDs, representing cameras used to capture depth and RGB images. Cameras were positioned at 0, 45, 90, 135, 180, 225, 270, and 315 degrees, allowing images to be captured from all sides for a specific coordinate.

For each different height, a set of coordinates was generated, and at these coordinates, all cameras captured depth and RGB images, ensuring that various viewpoints and heights would be captured. The visualization of all the angles of cameras and examples of acquired images are depicted in Figure 5.

Algorithm 1. Data generation script

Input: Number of coordinates *num_coord*, map boundaries *x_start*, *x_end*, *y_start*, *y_end*, Height list *z_heights*,
Constants: Camera id list *camera_ids*
Output: Saved images.

Generate random coordinates:
Initialize coordinate list *coord_list*
For z **to** *z_heights*:
  For *i* **to** *num_coord*:
    add the generated coordinate to coordinate list *coord_list.add(random(x_start, x_end), random(y_start, y_end), z)*
  End For
End For
Start and stabilize the drone in environment
For *coordinate* **in** *coord_list:*
  Spawn and hover drone in *coordinate*
  For a *camera* **in** *camera_ids:*
    Save the Depth and RGB image of the *camera*
  End For
End For

Figure 5. Generation of RGB and depth images from AirSim for finetuning dataset in AirSimNH environment. On the (**right**), a snapshot of a drone setup in randomly generated coordinates, where red arrows depict the angles at which the cameras are pointing. On the (**left**) are examples of acquired pairs of RGB and depth images from front and back-facing cameras.

3.3.2. Process for Fine-Tuning Depth Estimation Model

Specialized depth estimation models were fine-tuned for each environment. Training images had a resolution of 518 × 518 pixels due to the chosen model's architectural design.

The prepared dataset was split into training and validation sets (approximately 80% for training and 20% for validation). Training images included absolute depth annotations allowing the model to infer statistically accurate metric depths without requiring an explicit scale correction factor. Unlike event cameras, which rely on dynamic scene changes to estimate depth, our method infers depth purely from static image inputs, making it more suitable for structured scene analysis. The chosen architecture is the Depth Anything V2 [29] model with a Vision Transformer (ViT-S) encoder. The model using the (ViT-S) encoder is the smallest but one of the fastest and least resource-intensive among released models. The fine-tuning processes involved adapting the network to the specific domain. The dataset was preprocessed by normalizing pixel values and adjusting them to fit the model architecture. The training was performed using the BerHu loss function and AdamW optimizer with a learning rate scheduler which helped reduce the learning rate reduction over time and improve convergence.

Additionally, models were trained without references to objects in a simulated environment such as object coordinates, object classes, or object mapping. The fine-tunned model only relies on the RGB, and the depth images extracted from the AirSim simulation environment, which means that the model only learns from the visuals of the environment.

The BerHu loss function has been chosen to evaluate performance because it combines $L1$ and $L2$ terms. The loss function behaves similarly to the standard $L1$ measure for minor prediction errors. It is more robust to outliers due to its constant gradient and avoidance of massive updates. As the mistakes grow, they transition smoothly to the $L2$ penalty that prevents gradients from exploding while assigning proportionately higher penalties to substantial discrepancies. Such a loss definition is particularly effective in handling large-depth errors while remaining sensitive to more minor errors. This approach is motivated by prior work on depth estimation, where the BerHu loss has been shown to improve the robustness of deep learning models [54].

$$l_i = \begin{cases} \delta_i, & if\ \delta_i \le \tau, \\ \frac{\delta_i^2+\tau^1}{2\tau}, & if\ \delta_i > \tau. \end{cases} \quad (1)$$

The overall loss is then the average over all the pixels:

$$L_{BerHu}(y, \acute{y}) = \frac{1}{N}\sum_{i=1}^{N} l_i \quad (2)$$

This can be expressed mathematically as a piecewise function where y is the ground-truth depth, $\acute{y}$ is the predicted depth. τ is the threshold (in this work, the threshold $\tau = 0.2$), and N is the total number of pixels in the batch. Define $\delta_i = |y_i - \acute{y}_i|$.

3.4. Occupancy Grid Map

A probability-based occupancy map was used for environment mapping. The mapping process consists of iteration, where each iteration is divided into three stages: in the first stage, the mapping subsystem takes $W \times H$ sized normalized depth images $dimg$ received from the depth estimation model as input. Here, $dimg$ is a single-channel image, where pixel intensity is directly linearly proportional to the distance from the point of observation; the input image is then downscaled by filtering pixels with a horizontal and vertical stride of 2., i.e. each second pixel in each second row is used; the resulting image is then converted to a point cloud by applying Algorithm 2.

Algorithm 2. Depth Image to Point Cloud Conversion

Input: Depth image *dimg*, Field of view *fov*, Image dimensions *(H, W)*, Stride *stride*, Minimum depth *dmin*, Maximum depth *dmax*.
Output: Point cloud *pcd* in ENU coordinate system.

Initialize empty point cloud *pcd*
Compute horizontal FOV in radians $hfov_{rad} = fov \cdot \pi/180.0$
Calculate focal lengths $fx = fy = (W/2.0)/tan(hfov_{rad}/2.0)$
Set principal point coordinates $cx = W/2.0$, $cy = H/2.0$
For $v = 0$ **to** H **step** *stride*:
    For $u = 0$ **to** W **step** *stride*:
        Retrieve depth value $z = dimg[v,u]$
        If $dmin \leq z < dmax$ **then**:
            Compute $x = (u - cx) \cdot z/fx$
            Compute $y = (v - cy) \cdot z/fy$
            Rotate and add point [z,−x,−y] to *pcd* ENU point
        End If
    End For
End For
Return point cloud *pcd*

This conversion discards points that are too close or too far away, then rotates 90 degrees along the pitch axis and converts them to use the ENU (East–North–Up) world reference system. This results in a point cloud from the drone's perspective. This point cloud has to be converted to a global perspective. It is performed by rotating the whole point vector by camera orientation quaternion, which was already converted from NED (North–East–Down) to the ENU world reference system. Then, it follows the translation to the camera position. The position was also converted from NED to ENU world reference system. The resulting point cloud must then be discretized. The point cloud is scaled by performing an element-wise division by the selected V value and rounding each element to the nearest integer. This results in the voxel indices array O of currently visible obstacles.

The second stage of mapping is known as empty space detection. This is crucial for cleaning up obstacle data that is no longer valid due to introduced depth estimation positioning errors and environmental changes. The known empty space voxel indices array E is collected by applying Bresenham [55] 3D ray-casting algorithm from camera voxel to each obstacle voxel from O.

The final stage combines previously collected obstacle data O and empty space data E into the global map. In our case, the global map is a continuous 3D byte array with default values set to the chosen *occup_unkn* value representing unknown space. The known space voxel values fall into the occupancy intensity interval [*occup_min; occup_max*]. A lower value means lower obstacle certainty and higher values mean vice versa. In our study, intensity values are capped in [0; 12] intervals. One specific value in this interval is chosen as the occupancy threshold value. Reaching it would mean that the voxel can be treated as occupied. Modification of occupancy intensity values starts by changing the voxel value from *occup_unkn* to *occup_min* when the voxel index first occurs in either O or E. Then, values at E are decreased by *ray_miss* constant (1 was used in this work). Finally, values at O are increased by the *ray_hit* constant (8 was used in this work).

A series of these iterations results in one global dynamic probability-based occupancy map that can be used for online navigation. The result of such a mapping process is visualized in Figure 6.

(a) (b)

Figure 6. Snapshot of AirSimNH environment mapping result using depth images acquired from AirSim. (**a**) RGB image; (**b**) occupancy grid map visualized in rviz. Different color blocks represent known occupied space: blue means higher occupancy certainty; red means lower.

3.5. Path Planning and Collision Avoidance

The path planning and collision avoidance in the occupancy map are implemented using the A* search algorithm [56] limited to N_{max} number of iterations. In order for A* to be able to find a path point, set p to a given destination point G in an unknown and/or dynamic environment 3D grid $\mathbb{Z}^3$, perform an optimistic search in the unobserved space, and replanning actions, when certain criteria are met, have to be added. However, these additions do not come without challenges.

As any point in space, including the drone's exact position, is abstracted by a single voxel, it is hard to estimate the precise distance to obstacles., i.e., 2 points approximated by neighboring voxels could be vanishingly close to each other or as far as two V values apart. This increases the probability of a collision with an environment. To counteract that, path planning is only allowed in voxels that are unknown or not occupied and have 26 neighboring voxels that are also unknown or not occupied. In other words, path p is valid if set of voxels E does not contain know occupied voxels. This increases the distance to obstacles by a single V value and reduces the number of collision scenarios.

$$E = P \cup \left\{ (x', y', z') \in \mathbb{Z}^3 \mid \exists (x, y, z) \in P, \mid x - x' \mid \leq 1, \mid y - y' \mid \leq 1, \mid z - z' \mid \leq 1 \right\} \quad (3)$$

As we allow path planning in known space, even static obstacles will inevitably appear in the middle of the planned path, requiring path adjustments. As mentioned, these adjustments, in our case, are made in the form of replanning, which itself is a repeated path planning action from current point C to previous goal point G.

Replanning is triggered after one of these conditions are met:

1. A new obstacle is detected inside set E;
2. Replanning is also performed if the drone drifts away from the planned path by a specified distance d., i.e., if $\mid p \in P \mid \| p - C \| \leq d \mid = 0$;
3. Intermediate goal point I is reached. Point I is a voxel with best known distance heuristic to goal G after failure to find a complete path to G in N_{max} iterations.

4. Results

4.1. Fine-Tunning of Depth Estimation Models

Three datasets were generated for the Blocks, MSBuild2018, and AirsimNH environments using the previously provided algorithm. Each dataset consisted of 50 thousand pairs of RGB and depth images. Additionally, the dataset was generated in subsets of

10 thousand image pairs. Each time a subset generation was completed, a new depth estimation model was trained alongside the previously generated images and the new subset. In simulations, it was observed that as the number of images increased, the performance of the fine-tuned depth estimation model improved. The drone demonstrated improved capabilities in planning the path towards the goal coordinate as more data were introduced to the depth estimation model. Moreover, while inspecting the dataset, some outlier coordinates were observed, such as locations inside buildings that lacked a complete interior and were below the environment's ground. Furthermore, some coordinates were generated inside buildings with complete interiors. However, these coordinates could also be considered outliers because the drone is not expected to navigate inside buildings. These outliers likely harmed the quality of the depth estimation model.

Nevertheless, despite probable outliers, these images were used to finetune models. After each training epoch, the models were evaluated and compared against collected AirSim depth images to quantify performance (Figure 7). The depth estimation models were trained using the following hyperparameters: the learning rate of 1×10^{-4}, a batch size of 8, the AdamW optimizer, and 15 epochs. All three datasets show a steady loss reduction across early epochs, but no noticeable improvements are seen after the 11th epoch. The model trained on the Blocks environment retained the lowest loss compared to other environments. This is likely due to fewer depth ambiguities and more consistent scene structure in Blocks, while AirSimNH and MSBuild2018 had more complex scenes.

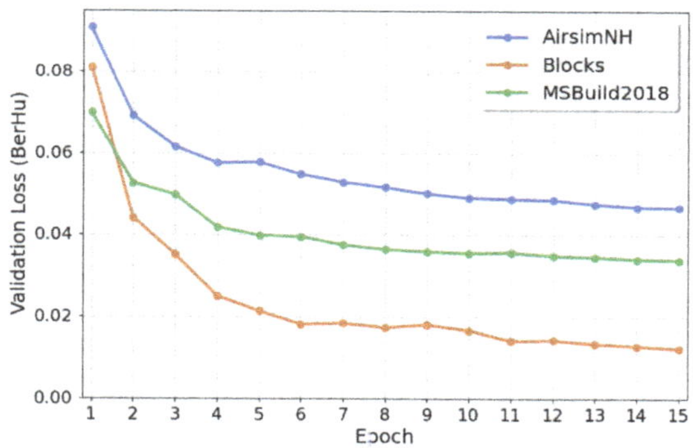

Figure 7. BerHu loss was experienced during training on the depth estimation model. Different lines for different environments: blue for AirsimNH, orange for Blocks, and green for MSBuild2018.

After conducting qualitative inspections and examining whether the predicted depth maps accurately captured major scene structures (e.g., walls, objects, and terrain features), it can be seen that the results are quite noticeable. From the evaluation in Figure 8, it is visually noticeable that the base pretrained model on the AirSimNH environment tends to estimate objects a lot further in comparison to ground truth images from simulation. In contrast, the fine-tuned model captures objects better with up-close details, higher mapping accuracy, and a more diverse depth range. A high difference in MAPE can be observed in Table 3 between models. Depth image discrepancies between pretrained and fine-tuned models can lead to the belief that models should be specialized for each environment. Since higher error discrepancies become more apparent in mismatched domains, the likelihood of crashes or collisions increases significantly, as the model's inaccurate depth predictions may fail to capture obstacles and distances correctly.

(a) (b) (c)

Figure 8. Visualizations comparing (**a**) RGB input; (**b**) pre-trained Depth Anything V2 small model's depth image output; and (**c**) fine-tuned version's depth image. The image was captured in the AirSimNH environment.

Table 3. Mean Absolute Percentage Error (MAPE) comparison between pre-trained and fine-tuned models across different environments.

Environment	Pre-Trained Model	Finetuned Model
AirSimNH	120.45%	33.41%
Blocks	70.09%	8.04%
MSBuild2018	121.94%	32.86%

4.2. Autonomous Navigation Simulation Results

4.2.1. General Experimentation Results

During the experimentation phase of this work, the navigation system was tested in three different environments. In total, 810 flights were conducted: 10 for each scenario and voxel size V [0.5, 1.0, 2.0] pair for each environment. These autonomous flights were performed using depth images acquired directly from AirSim, depth images estimated with the pre-trained Depth Anything V2 model, and depth images estimated by models finetuned for each environment.

The resulting metrics of the flights conducted during the experimentation are shown in Table 4. Here, the results from all the scenarios and voxel sizes used in each environment are added for each depth image acquisition method. The third column shows the number of flights that reached the goal position successfully within 5 m. The fourth column shows how close to the goal all the failed flights ended on average.

As can be seen, when depth images were estimated by a pre-trained model, not a single simulation successfully reached the goal. The paths flown during these simulations are depicted in Appendix A, Figure A1. Most of the flights ended prematurely when the drone would become stuck in the ground or other obstacles and could not move further because of the simulator's physics. This is mainly because the used model does not generalize well for these digital environments and also suffers from scale-drift issues or problems where closer objects occupying more of the view are estimated further away or vice versa.

Table 4. Results from autonomous flights simulated in different AirSim environments with varying methods of depth image acquisition.

Environment	Method	Reached Goal	Distance to Goal
AirSimNH	Depth Images from AirSim	35 of 90	117.40 (55 unsuccessful)
	Estimated Depth Images (finetuned)	3 of 90	135.28 (87 unsuccessful)
	Estimated Depth Images (pre-trained)	0 of 90	171.20 (90 unsuccessful)
Blocks	Depth Images from AirSim	79 of 90	67.75 (11 unsuccessful)
	Estimated Depth Images (finetuned)	65 of 90	95.95 (25 unsuccessful)
	Estimated Depth Images (pre-trained)	0 of 90	152.34 (90 unsuccessful)
MSBuild2018	Depth Images from AirSim	12 of 90	187.80 (78 unsuccessful)
	Estimated Depth Images (finetuned)	0 of 90	230.50 (90 unsuccessful)
	Estimated Depth Images (pre-trained)	0 of 90	311.25 (90 unsuccessful)

The finetuning of the models improved the number of successful navigations by 72.22% and 3.33% in Blocks and AirSimNH environments, respectively. However, using this method, the drone still did not manage to reach a single goal in MSBuild2018 simulations. Utilization of finetuned models for depth estimation also helped to get the drone closer to the goal in flights where the simulation ended prematurely. The average distance to goals improved by 36, 56, and 81 m in AirSimNH, Blocks, and MSBuild2018 environments.

Across all environments, the processed frames per second (FPS) range from approximately 2.7 to 4.35, depending on the voxel size and complexity of the environment. The full list of averaged frames per second can be seen in Appendix A, Table A1. All of the values indicate that the drone travels ~0.46 m per frame in the highest FPS scenario, while in the lowest FPS scenario, it travels ~0.74 m per frame. These measurements show that the drone replans its path after covering these distances.

Increasing FPS is critical when solving problems like this because if the drone does not receive new depth images that indicate an obstruction before a collision, it cannot replan its path before the collision with an object. For this reason, a higher FPS indicates improved drone performance, faster responses to obstacles, and more frequent optimal path re-evaluations.

Furthermore, FPS measurements show that the lowest performance was observed in the AirsimNH environment, while the best was in the Blocks environment. Visually, Blocks is the least detailed environment, and AirsimNH is the most thorough. It indicates that either environment complexity or graphical demand impacts the image processing rate. Due to limited data, it is not possible to determine which factor plays a more significant role.

When voxel size is increased, the image processing rate improves due to reduced computational requirements for path planning as larger voxel sizes simplify the complexity of the environment—additionally, safety margins increase, leading to a reduced risk of collision. However, when V is decreased, smaller objects and narrower paths may go unnoticed or overestimated, leading to suboptimal paths, blocked narrow passages, or even collisions. While lower voxel sizes may provide a more precise path, it also increases execution costs, reduces image processing rate, and decreases reaction times, leading to more challenging path planning.

Lastly, testing cases utilizing AirSim depth images demonstrate the lowest FPS values, primarily due to additional preprocessing steps such as depth image normalizations and format conversion. Additionally, the finetuned depth estimation model displays lower FPS than a pre-trained model, which is expected, as fine-tuning increases the model complexity of the hidden layers, leading to increased processing times.

4.2.2. Acquired AirSim Depth Images vs. Finetuned Depth Estimation

The averaged results for each pair of environment and depth image acquisition methods are in Tables 5–7. Here, the columns denote the voxel size in meters discretized from point cloud data, the scenario number specifying which coordinates in Table 2 were used for the end goal, the number of flights that reached the goal within 5 m, the average number of steps that collided with objects in the area, the average calculated Euclidean flight distance in meters, the average time of flight in seconds, and average calculated distance in meters to goal coordinates from the last step before premature termination of simulation such as collision for 20 consecutive steps (the drone is stuck in an obstacle and cannot move) or failure of 200 consecutive iterations to find a new path. These constraints ensure that the drone does not keep hovering in place for an infinite time during experiments.

The snapshots in Figures 9–11 depict all the paths the drone took to reach the end positions from the start in different environments for the AirSim acquired and estimated depth image usage. Here, red lines are the flights with voxel size V of 0.5 m; green lines show paths using 1.0 m; blue lines are 2.0 m.

Table 5. Autonomous flight results comparison between AirSim and estimated depth image usage in AirSimNH environment. The results are calculated from ten flights for each voxel size and scenario pair. The best results for each scenario are bolded.

		AirSim Depth Images					Estimated Depth Images				
Voxel Size	**Scenario**	**Reached Goal**	**Collisions**	**Distance**	**Time**	**Distance to Goal**	**Reached Goal**	**Collisions**	**Distance**	**Time**	**Distance to Goal**
0.5	1	3	12.00	**226.75**	**136.42**	129.56	0	-	-	-	120.26
	2	3	0.67	**228.47**	**132.29**	125.93	3	5.00	282.79	176.08	126.85
	3	0	-	-	-	142.63	0	-	-	-	**128.47**
1.0	1	4	3.00	345.22	212.07	**41.13**	0	-	-	-	124.08
	2	7	0.43	288.95	174.66	**78.63**	0	-	-	-	133.49
	3	3	**4.67**	**439.83**	**271.87**	131.08	0	-	-	-	147.58
2.0	1	**9**	**0.22**	346.49	225.01	72.87	0	-	-	-	130.44
	2	6	**0**	348.58	224.92	172.17	0	-	-	-	140.68
	3	0	-	-	-	162.58	0	-	-	-	165.69

Table 6. Autonomous flight results comparison between AirSim and estimated depth image usage in the Blocks environment. The results are calculated from ten flights for each voxel size and scenario pair. The best results for each scenario are bolded.

		AirSim Depth Images					Estimated Depth Images				
Voxel Size	**Scenario**	**Reached Goal**	**Collisions**	**Distance**	**Time**	**Distance to Goal**	**Reached Goal**	**Collisions**	**Distance**	**Time**	**Distance to Goal**
0.5	1	7	0.71	**178.51**	**95.62**	122.58	**10**	**0**	196.28	112.91	-
	2	**10**	**0**	**160.09**	**83.25**	-	9	**0**	160.16	88.51	106.87
	3	**8**	2.63	221.09	**122.61**	129.52	3	**0.63**	222.76	128.11	90.84
1.0	1	**9**	**0**	179.57	97.30	124.76	7	**0**	207.58	118.11	129.32
	2	**10**	**0**	160.64	85.98	-	7	**0**	161.32	92.70	102.68
	3	7	0.71	226.13	125.64	109.51	5	**0**	226.93	133.94	**58.17**
2.0	1	8	0.25	188.67	118.40	123.39	4	**0**	350.40	217.18	**119.89**
	2	**10**	**0**	166.80	106.18	-	9	**0**	186.23	119.91	**96.14**
	3	**10**	**0**	**209.12**	131.73	-	9	**0**	235.37	153.67	159.64

Table 7. The autonomous flight result comparison between AirSim and estimated depth image usage in the MSBuild2018 environment. The results are calculated from ten flights for each voxel size and scenario pair. The best results for each scenario are bolded.

		AirSim Depth Images					Estimated Depth Images				
Voxel Size	Scenario	Reached Goal	Collisions	Distance	Time	Distance to Goal	Reached Goal	Collisions	Distance	Time	Distance to Goal
0.5	1	**6**	**2.33**	376.99	**217.90**	**53.96**	0	-	-	-	104.55
	2	**4**	**25.00**	**472.27**	**283.29**	176.02	0	-	-	-	222.09
	3	0	-	-	-	**218.00**	0	-	-	-	254.27
1.0	1	1	43.00	376.29	224.41	174.06	0	-	-	-	216.62
	2	0	-	-	-	**160.91**	0	-	-	-	246.09
	3	0	-	-	-	234.73	0	-	-	-	268.21
2.0	1	4	14.25	**375.22**	220.60	158.99	0	-	-	-	227.49
	2	0	-	-	-	236.57	0	-	-	-	247.53
	3	0	-	-	-	276.91	0	-	-	-	287.67

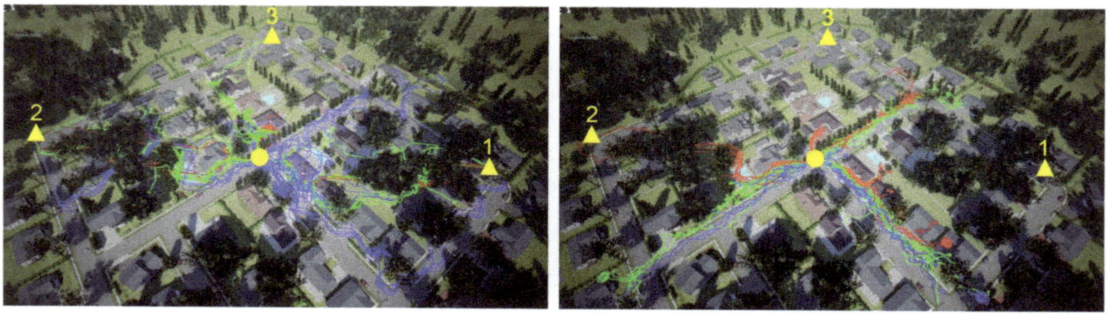

(a) (b)

Figure 9. Comparison of flight paths during experimentation in the AirSimNH environment between (**a**) AirSim depth image and (**b**) estimated depth image usage. Different path colors mean different occupancy grid voxel size: red lines 0.5, green lines 1.0, blue lines 2.0. Yellow circles show start positions, triangles—goals. Scenarios are numbered accordingly.

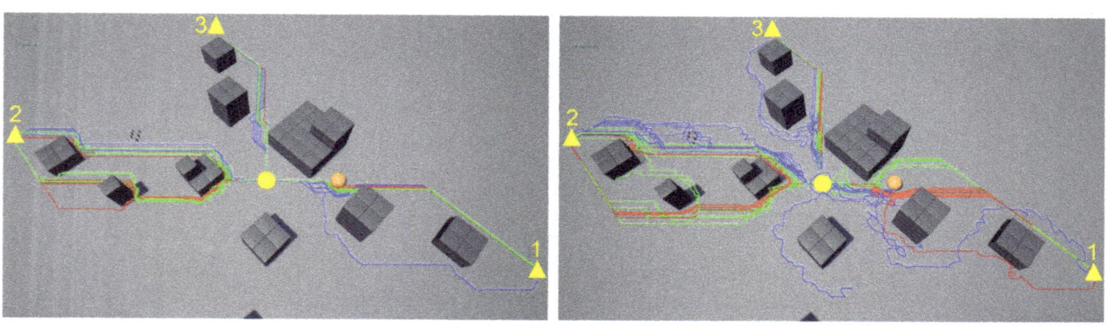

(a) (b)

Figure 10. Comparison of flight paths during experimentation in Blocks environment between (**a**) AirSim depth image and (**b**) estimated depth image usage. Different path colors mean different occupancy grid voxel sizes: red lines 0.5, green lines 1.0, blue lines 2.0. Yellow circles show start positions, triangles—goals. Scenarios are numbered accordingly.

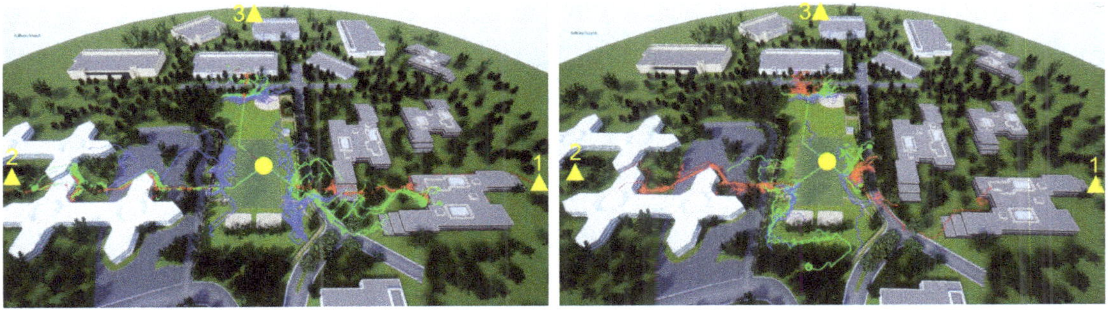

(a) (b)

Figure 11. Comparison of flight paths during experimentation in MSBuild2018 environment between (**a**) AirSim depth image and (**b**) estimated depth image usage. Different path colors mean different occupancy grid voxel sizes: red lines 0.5, green lines 1.0, blue lines 2.0. Yellow circles show start positions, triangles—goals. Scenarios are numbered accordingly.

A clear difference in navigation routes can be seen between AirSim depth image and estimated depth image use when comparing flights in complex AirSimNH environment (Figure 9). Even with bigger voxel sizes, the drone can navigate between houses, trees, and other objects using depth acquired from AirSim (a). However, the drone would often fly too close to obstacles and become stuck when the voxel sizes are smaller. On the other hand, the estimated depth images are still too noisy and allow errors to accumulate in the occupancy grid map of the environment as can be seen in Appendix B, where Figure A2 depicts differences in occupancy grids (a for this environment, scenario no.2). This leaves the drone frantically looking around as the path planning algorithm recalculates to navigate in some small opening, only to close that opening with more voxels discretized from noisy depth data.

The best results for each scenario for the AirSimNH environment are bolded in Table 5. The third scenario was the toughest to navigate, even for a drone acquiring depth images directly from AirSim. Only three flights ended successfully in goal coordinates, all with 1.0 m for voxel size. Scenarios no.1 and no.2 both had sixteen flights reaching the set goals when using AirSim depth images. The most successful flights were recorded using $V = 2.0$ for the first scenario and $V = 1.0$ for the second. However, for the successful flights, the lowest distance and time traveled were reached when $V = 0.5$.

The simpler Blocks environment proved that the depth estimation methodology can be successfully utilized in such conditions. However, the drone navigates more chaotically (Figure 10b) and travels a greater distance. Unlike when using depth data from AirSim, the errors accumulated in estimated depth images steered the drone further from obstacles. This helped the former to outperform the latter in the first scenario with the orange ball, as noted in Table 6. In the case of successfully finished flights, the collision count was generally lower for estimation methodology. As previously stated, the occupancy grid map examples for this environment scenario no.1 can be seen in Appendix B, Figure A2b.

All the flights reached goals set in the second scenario when utilizing AirSim depth images while the least distance and time was traveled with $V = 0.5$. The most successful simulations and least average distance traveled in scenario no.3 were recorded with the largest voxel size. All the flights with a $V = 0.5$ setting reached the goal in the first scenario when depth images were estimated with the finetuned model; however, the distance and time traveled were still better with depth images from AirSim. Out of the failed attempts, the lower distance to the goal was calculated using the estimated depth image method.

The most difficult environment proved to be MSBuild2018. It starts in an open field surrounded by dense tree lines from all directions. There are some openings to the roads and parking lots, but because the path planner starts by giving the most optimal routes, the drone tries to steer between the narrow openings of tree lines when $V = 0.5$ (red lines in Figure 11). As the voxel size increases, the drone can no longer see the gaps in the occupancy map, so it tries to go higher—above the trees and buildings. As the depression angle is low from step to step, the drone is not able to build an occupancy grid map of further area, the planner fails to find the next optimal route, and the simulation finishes prematurely. This happens in flights with AirSim acquired and estimated depth images, as seen from comparison snapshots.

Using the depth images from AirSim, the first scenario was successfully navigated only 11 times out of 30, and only one other time in scenario no. 2 (Table 7). The depth image estimation technique failed to reach any goal. It also did not help to reach fewer distance to goals even in failed attempts. All of this shows that the navigation algorithm needs some improvements to navigate such complex environments.

5. Discussion

One of the primary hurdles that were faced in the implementation of the navigation system was poor obstacle avoidance due to the drone constantly following the walls of the obstacles while the drone's inertia, i.e., external forces, was not being accounted for. This behavior was observed not only with depth estimation but also when utilizing depth images acquired with AirSim. To improve this, instead of an algorithm that searches in neighboring nodes, i.e., A*, we could use path-finding algorithms that search in all directly observable nodes. A good potential candidate could be Theta* or JPS. The latter was used in [57] to achieve safe and fast collision-free flights. Another potential solution could be smoothing paths by interpolating them using splines along with gradient-based obstacle avoidance. Such a solution was used in [58] to achieve collision-free flights in unknown environments.

Currently, path searching is performed using the A* algorithm. Replanning is performed by repeating the path search from the current position. This is very inefficient. In future research, we will investigate path planning and navigation efficiency improvements using algorithms that do not have to recalculate the whole path after obstacle encounters, e.g., [36,37]. Also, for better planning efficiency in high-complexity environments, we will investigate sampling-based algorithms, as it was performed in [38–40].

Image generation also came with its challenges as some outliers ended up in the dataset. For example, interiors of the buildings and objects with transparent walls captured from inside or underground. These outliers were included due to limitations of the tools that AirSim provides to verify if the coordinates of the teleported drone are not inside an object or outside the map. The simulator only indicates collisions if the drone becomes stuck inside thinner objects or walls. A more sophisticated data generation script could be developed to detect and exclude mentioned outliers. This could be performed by either filtering the outliers from the dataset or preventing them from being generated in the first place by obtaining the correct sizes and coordinates of objects inside the environment and excluding generated coordinates in those areas. Another option would be generating the dataset by manually flying the drone and capturing the images in specific time intervals. Furthermore, all the images were generated under similar weather conditions which presents limitations if different weather conditions were introduced into the simulation environment. To improve the robustness of depth estimation and train models that generalize well, images with varying illumination (weather conditions) could be introduced for the datasets. Also, in further research, the performance of a single model trained on a larger combined dataset needs to be investigated.

The monocular single-shot depth estimation model Depth Anything V2, despite being among the most accurate ones, failed to keep consistent depth estimations across multiple frames. This could be improved by integrating more accurately known distances to certain key points obtained by SLAM (or other feature-based approaches) into the depth estimation model or utilizing another depth estimation approach better suited for UAV navigation (multi-frame estimation, taking video sequences as input). However, this would require more computational resources that could not be spared in an autonomous standalone system. Also, the monocular single-shot depth estimation model introduced problems when facing or getting too close to a wall. Since depth information could not be extracted from such images due to a lack of details, the model would highly underestimate depth and tend to crash.

The frame processing rate observed in the simulation was low, indicating slow response times and an increased risk that the drone will not be able to replan its path in time before colliding with objects that are yet to be identified. However, the Unreal Engine environment itself used a lot of resources during these simulations, and the FPS is lower than it would be in a real-world environment when the processing power is used only on gathering RGB images, depth estimation, and path planning. To better assess the frame processing rate using the current equipment or to the determine system requirements that would satisfy the desired FPS, future tests should be designed to reduce resource usage from external sources. Rendering of the Unreal Engine environment should be minimized only to the camera's field of view to reduce the usage of the view that is used to navigate the drone. Modifications to the testing environment could include the utilization of two GPUs: one dedicated to rendering the environment and another responsible for image retrieval, depth estimation, and path planning. This solution would reduce the computational load of the rendering process, making performance closer to real-world conditions. Image processing FPS will increase in real-world applications when the resources are not occupied by the simulation environment, which in conducted tests used about 25–30% of the GPU resources. Furthermore, image extraction from the AirSim environment had some latency due to the API, where using an onboard processing unit on a drone with a monocular camera would be a more direct, faster approach to extract images. All in all, the main problem influencing FPS in the conducted research is the environment itself and its framework, which would not influence the FPS in real-world applications and the FPS could improve by two or more times.

Furthermore, the current implementation of the system is synchronized between image generation and navigation. Concurrency and better performance could be achieved if mapping, navigation, and depth processing calculations were parallelized.

Additionally, three simulation environments were used in the simulations with varying complexities. One environment was relatively simple, while the other two were highly complex. Further research will focus on improving navigation in complex environments. With more optimal navigation solutions, additional challenges such as moving objects like cars and animals, and more challenging weather effects like clouds, fog, or other weather conditions, will be explored. These additional environmental conditions will increase the complexity of the environment and assess how well the proposed solution adapts to dynamically moving objects, brightness, visibility, etc.

Lastly, all tests were performed using simulations on a desktop computer. For real-life applications, all calculations must be performed on a miniature onboard computer with CUDA capabilities. NVIDIA Jetson TX2 with 256 CUDA cores, 1.3 GHz dual-core CPU, and 8GB of memory should be up for this task. However, this assumption must be evaluated practically.

6. Conclusions

The proposed UAV navigation framework was developed for autonomous flight simulation. It utilizes the Depth Anything V2 deep learning model to estimate depth images from single RGB images captured by a monocular front-facing camera. These images are used to dynamically build a probability-based occupancy grid map. The A* algorithm is then used to plan optimal paths in the built map for the drone to navigate towards the coordinates of a target.

The navigation solution was tested across three different open-source environments in an AirSim simulator with proposed scenarios. Results showed successful autonomous drone navigation in low-complexity environments, while highly complex settings presented challenges such as inconsistent depth estimations and drone clipping into objects with trickier geometries.

Finetuning depth estimation models for specific environments improved the accuracy and reliability of the system, reducing the likelihood of navigation failures caused by mismatched depth predictions. The results demonstrate a substantial decrease in MAPE across different environments after the finetuning process. The MAPE was reduced from 120.45% to 33.41% in AirSimNH, from 70.09% to 8.04% in Blocks, and from 121.94% to 32.86% in MSBuild2018.

Navigation performance improvements were also observed when utilizing fine-tuned models instead of pretrained. The number of goals reached in all scenarios increased by 3.33% in AirSimNH and 72.22% in Blocks. Furthermore, the distance to goal positions in unsuccessful flights decreased by 36 m in AirSimNH, 56 m in Blocks, and 81 m in MSBuild2018. These findings emphasize the importance of domain-specific model adaptation, demonstrating that finetuning with synthetic data that is tailored to a given environment allows for more accurate depth estimation and better overall system performance in autonomous UAV navigation.

However, the frame processing rate (ranging from 2.7 to 4.4 FPS) indicates slow response times, which could lead to collisions with objects along previously planned and outdated paths. The simulation conditions suggest that these results, which appear worse than they would be in real-life conditions, could be misleading. Despite this limitation, modifications of the simulation environment setup to provide more system resources to the algorithm would improve the performance and response time of the proposed solution.

Further refinement is required for depth estimation methods, navigation algorithms, and overall system architecture to achieve more reliable navigation in broader and higher-complexity environments. Further research is also needed on the navigation solution's robustness against different illumination conditions and environment feature levels. Despite this, the authors hope that the proposed framework and simulation scenarios will be useful for researchers to compare and improve methodologies used in monocular vision-based autonomous navigation.

Author Contributions: Conceptualization, J.G., H.G., T.K., L.P. and A.K.; methodology, J.G., H.G., T.K., L.P. and A.K.; software, J.G., T.K. and L.P.; validation, J.G., H.G., T.K., L.P. and A.K.; formal analysis, J.G., H.G., T.K., L.P. and A.K.; investigation, J.G., H.G., T.K., L.P. and A.K.; resources, J.G., T.K. and L.P.; data curation, H.G., J.G., T.K. and L.P.; writing—original draft preparation, H.G.; writing—review and editing, J.G., H.G., T.K., L.P. and A.K.; visualization, H.G., J.G., T.K. and L.P.; supervision, A.K.; project administration, H.G., J.G., T.K., L.P. and A.K. All authors have read and agreed to the published version of the manuscript.

Funding: This research received no external funding.

Data Availability Statement: The original data presented in the study are openly available at https://github.com/jonasctrl/monocular-slam-drone (accessed on 22 March 2025).

Conflicts of Interest: The authors declare no conflicts of interest.

Appendix A

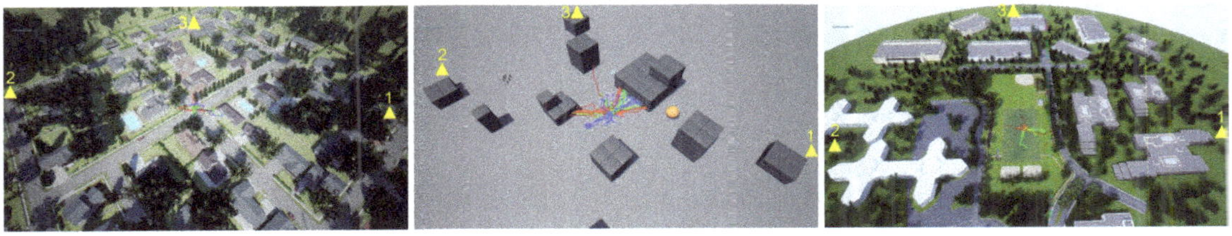

(a) (b) (c)

Figure A1. Paths of flight during experimentation using pre-trained Depth Anything V2 model for depth image estimation. Environments: (**a**) AirSimNH; (**b**) Blocks; and (**c**) MSBuild2018. Different path colors mean different voxel sizes: red lines 0.5, green lines 1.0, blue lines 2.0. Yellow triangles show end points for different scenarios. Scenarios are numbered accordingly.

Table A1. Image processing rates (FPS) for AirsimNH, Blocks, and MSBuild2018 environments across voxel sizes of 0.5, 1, and 2, as well as an additional overall FPS measure. Comparison of image processing rate with different depth retrieval methods: true depth image from AirSim, fine-tuned, and pretrained depth estimation models.

Environment	Voxel Size	AirSim Depth Images	Fine-Tuned Model	Pretrained Model
AirsimNH	0.5	2.72	2.77	3.10
	1	2.99	3.03	3.15
	2	3.32	3.15	3.39
	Overall	3.01	2.98	3.21
Blocks	0.5	3.88	3.59	4.00
	1	4.32	3.86	4.30
	2	4.35	4.00	4.38
	Overall	4.19	3.82	4.22
MSBuild2018	0.5	3.03	3.57	3.67
	1	3.64	3.79	3.91
	2	4.08	3.91	4.17
	Overall	3.58	3.76	3.92

Appendix B

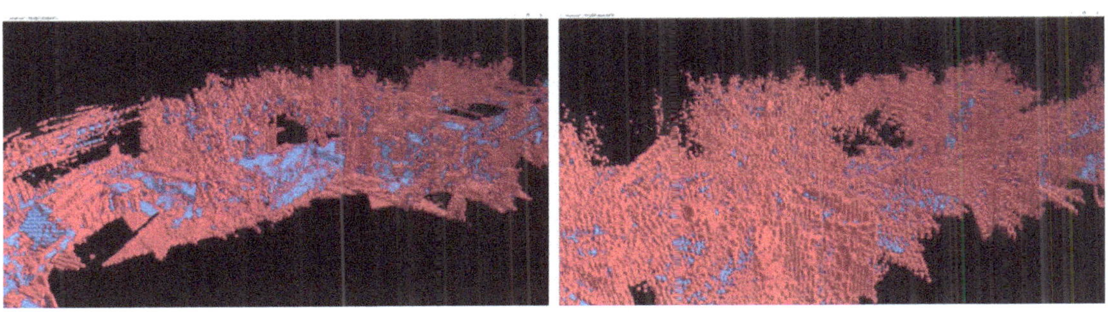

(a) (b)

Figure A2. *Cont.*

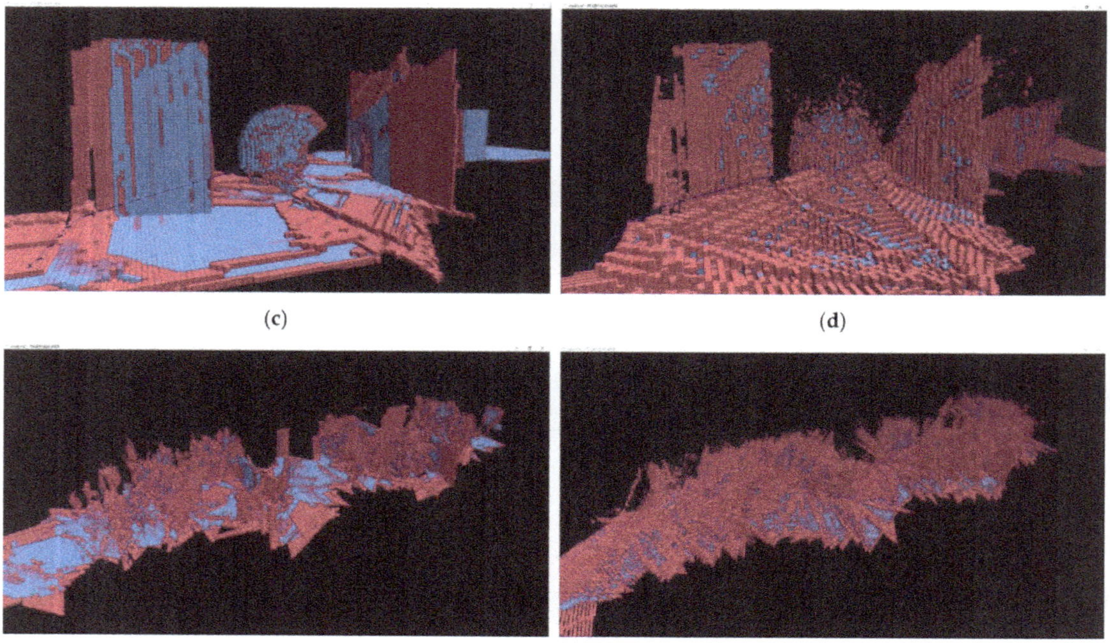

(c) (d) (e) (f)

Figure A2. Snapshots of occupancy grid maps built during autonomous navigation simulations and visualized in rviz. Depth images acquired from AirSim and estimated with finetuned models were used for different environments and scenarios, respectively: AirSimNH scenario no.2 (**a**,**b**); Blocks scenario no.1 (**c**,**d**); and MSBuild2018 scenario no.1 (**e**,**f**). Voxels of different colors depict different certainty of known occupied space: blue means higher certainty of occupancy; red means lower certainty. Map size is 800 × 800 (except for **e** and **f** its 1200 × 1200) and voxel size is 0.5.

References

1. Shakhatreh, H.; Sawalmeh, A.H.; Al-Fuqaha, A.; Dou, Z.; Almaita, E.; Khalil, I.; Othman, N.S.; Khreishah, A.; Guizani, M. Unmanned Aerial Vehicles (UAVs): A Survey on Civil Applications and Key Research Challenges. *IEEE Access* **2019**, *7*, 48572–48634. [CrossRef]
2. Gyagenda, N.; Hatilima, J.V.; Roth, H.; Zhmud, V. A Review of GNSS-Independent UAV Navigation Techniques. *Rob. Auton. Syst.* **2022**, *152*, 104069. [CrossRef]
3. Yang, N.; Wang, R.; Gao, X.; Cremers, D. Challenges in Monocular Visual Odometry: Photometric Calibration, Motion Bias and Rolling Shutter Effect. *arXiv* **2018**, arXiv:1705.04300. [CrossRef]
4. Arampatzakis, V.; Pavlidis, G.; Mitianoudis, N.; Papamarkos, N. Monocular Depth Estimation: A Thorough Review. *IEEE Trans. Pattern Anal. Mach. Intell.* **2024**, *46*, 2396–2414 [CrossRef]
5. Wang, C.; Ma, H.; Chen, W.; Liu, L.; Meng, M.Q.-H. Efficient Autonomous Exploration with Incrementally Built Topological Map in 3-D Environments. *IEEE Trans. Instrum. Meas.* **2020**, *69*, 9853–9865. [CrossRef]
6. Faria, M.; Ferreira, A.S.; Pérez-Leon, H.; Maza, I.; Viguria, A. Autonomous 3D Exploration of Large Structures Using an UAV Equipped with a 2D LIDAR. *Sensors* **2019**, *19*, 4849. [CrossRef] [PubMed]
7. Vanegas, F.; Gaston, K.J.; Roberts, J.; Gonzalez, F. A Framework for UAV Navigation and Exploration in GPS-Denied Environments. In Proceedings of the 2019 IEEE Aerospace Conference, Big Sky, MT, USA, 2–9 March 2019; pp. 1–6.
8. Chang, Y.; Cheng, Y.; Manzoor, U.; Murray, J. A Review of UAV Autonomous Navigation in GPS-Denied Environments. *Rob. Auton. Syst.* **2023**, *170*, 104533. [CrossRef]
9. Shah, S.; Dey, D.; Lovett, C.; Kapoor, A. AirSim: High-Fidelity Visual and Physical Simulation for Autonomous Vehicles. In Proceedings of the Field and Service Robotics, Zürich, Switzerland, 13–15 September 2017.
10. Lu, Y.; Xue, Z.; Xia, G.-S.; Zhang, L. A Survey on Vision-Based UAV Navigation. *Geo-Spat. Inf. Sci.* **2018**, *21*, 21–32. [CrossRef]
11. Arafat, M.Y.; Alam, M.M.; Moh, S. Vision-Based Navigation Techniques for Unmanned Aerial Vehicles: Review and Challenges. *Drones* **2023**, *7*, 89. [CrossRef]

12. Horn, B.K.P.; Schunck, B.G. Determining Optical Flow. *Artif. Intell.* **1981**, *17*, 185–203. [CrossRef]
13. Lucas, B.D.; Kanade, T. An Iterative Image Registration Technique with an Application to Stereo Vision. In Proceedings of the IJCAI'81: 7th International Joint Conference on Artificial Intelligence, Vancouver, BC, Canada, 24–28 August 1981; Volume 2, pp. 674–679.
14. Santos-Victor, J.; Sandini, G.; Curotto, F.; Garibaldi, S. Divergent Stereo for Robot Navigation: Learning from Bees. In Proceedings of the Proceedings of IEEE Conference on Computer Vision and Pattern Recognition, New York, NY, USA, 15–17 June 1993; pp. 434–439.
15. Agrawal, P.; Ratnoo, A.; Ghose, D. Inverse Optical Flow Based Guidance for UAV Navigation through Urban Canyons. *Aerosp. Sci. Technol.* **2017**, *68*, 163–178. [CrossRef]
16. Couturier, A.; Akhloufi, M.A. A Review on Absolute Visual Localization for UAV. *Rob. Auton. Syst.* **2021**, *135*, 103666. [CrossRef]
17. Harris, C.; Stephens, M. A Combined Corner and Edge Detector. In Proceedings of the Proc. AVC, Manchester, UK, 15–17 September 1988; pp. 23.1–23.6.
18. Edward, R.; Drummond, T. Machine Learning for High-Speed Corner Detection. In Proceedings of the Computer Vision—ECCV, Graz, Austria, 7–13 May 2006; Leonardis, A., Bischof, H., Pinz, A., Eds.; Springer: Berlin/Heidelberg, Germany, 2006; pp. 430–443.
19. Lowe, D.G. Distinctive Image Features from Scale-Invariant Keypoints. *Int. J. Comput. Vis.* **2004**, *60*, 91–110. [CrossRef]
20. Calonder, M.; Lepetit, V.; Strecha, C.; Fua, P. BRIEF: Binary Robust Independent Elementary Features. In Proceedings of the Computer Vision—ECCV, Heraklion, Crete, 5–11 September 2010; Daniilidis, K., Maragos, P., Paragios, N., Eds.; Springer: Berlin/Heidelberg, Germany, 2010; pp. 778–792.
21. Bay, H.; Tuytelaars, T.; Van Gool, L. SURF: Speeded Up Robust Features. In Proceedings of the Computer Vision—ECCV, Graz, Austria, 7–13 May 2006; Leonardis, A., Bischof, H., Pinz, A., Eds.; Springer: Berlin/Heidelberg, Germany, 2006; pp. 404–417.
22. Rublee, E.; Rabaud, V.; Konolige, K.; Bradski, G. ORB: An Efficient Alternative to SIFT or SURF. In Proceedings of the 2011 International Conference on Computer Vision, Barcelona, Spain, 6–13 November 2011; pp. 2564–2571.
23. Hornung, A.; Wurm, K.M.; Bennewitz, M.; Stachniss, C.; Burgard, W. OctoMap: An Efficient Probabilistic 3D Mapping Framework Based on Octrees. *Auton. Robot.* **2013**, *34*, 189–206. [CrossRef]
24. Dryanovski, I.; Morris, W.; Xiao, J. Multi-Volume Occupancy Grids: An Efficient Probabilistic 3D Mapping Model for Micro Aerial Vehicles. In Proceedings of the 2010 IEEE/RSJ International Conference on Intelligent Robots and Systems, Taipei, Taiwan, 18–22 October 2010; pp. 1553–1559.
25. Herrera-Granda, E.P.; Torres-Cantero, J.C.; Peluffo-Ordóñez, D.H. Monocular Visual SLAM, Visual Odometry, and Structure from Motion Methods Applied to 3D Reconstruction: A Comprehensive Survey. *Heliyon* **2024**, *10*, e37356. [CrossRef]
26. Campos, C.; Elvira, R.; Rodríguez, J.J.G.; Montiel, J.M.; Tardós, J.D. ORB-SLAM3: An Accurate Open-Source Library for Visual, Visual–Inertial, and Multimap SLAM. *IEEE Trans. Robot.* **2021**, *37*, 1874–1890. [CrossRef]
27. Davison, A.J.; Reid, I.D.; Molton, N.D.; Stasse, O. MonoSLAM: Real-Time Single Camera SLAM. *IEEE Trans. Pattern Anal. Mach Intell.* **2007**, *29*, 1052–1067. [CrossRef]
28. Engel, J.; Schöps, T.; Cremers, D. LSD-SLAM: Large-Scale Direct Monocular SLAM. In Proceedings of the Computer Vision—ECCV, Zurich, Switzerland, 6–12 September 2014; Fleet, D., Pajdla, T., Schiele, B., Tuytelaars, T., Eds.; Springer International Publishing: Cham, Switzerland, 2014; pp. 834–849.
29. Yang, L.; Kang, B.; Huang, Z.; Zhao, Z.; Xu, X.; Feng, J.; Zhao, H. Depth Anything V2. *arXiv* **2024**, arXiv:2406.09414.
30. Godard, C.; Mac Aodha, O.; Firman, M.; Brostow, G.J. Digging into Self-Supervised Monocular Depth Prediction. In Proceedings of the International Conference on Computer Vision (ICCV), Seoul, Republic of Korea, 27 October–2 November 2019.
31. Guizilini, V.; Ambrus, R.; Pillai, S.; Raventos, A.; Gaidon, A. 3D Packing for Self-Supervised Monocular Depth Estimation. *arXiv* **2020**, arXiv:1905.02693.
32. Dhulkefl, E.; Durdu, A.; Terzioğlu, H. Dijkstra algorithm using uav path planning. *Konya J. Eng. Sci.* **2020**, *8*, 92–105. [CrossRef]
33. Zhibo, Z. A Review of Unmanned Aerial Vehicle Path Planning Techniques. *Appl. Comput. Eng.* **2024**, *33*, 234–241. [CrossRef]
34. Mandloi, D.; Arya, R.; Verma, A.K. Unmanned Aerial Vehicle Path Planning Based on A* Algorithm and Its Variants in 3d Environment. *Int. J. Syst. Assur. Eng. Manag.* **2021**, *12*, 990–1000. [CrossRef]
35. Luo, Y.; Lu, J.; Zhang, Y.; Qin, Q.; Liu, Y. 3D JPS Path Optimization Algorithm and Dynamic-Obstacle Avoidance Design Based on Near-Ground Search Drone. *Appl. Sci.* **2022**, *12*, 7333. [CrossRef]
36. Suanpang, P.; Jamjuntr, P. Optimizing Autonomous UAV Navigation with D* Algorithm for Sustainable Development. *Sustainability* **2024**, *16*, 7867. [CrossRef]
37. Zhu, X.; Yan, B.; Yue, Y. Path Planning and Collision Avoidance in Unknown Environments for USVs Based on an Improved D* Lite. *Appl. Sci.* **2021**, *11*, 7863. [CrossRef]
38. Yang, K.; Gan, S.K.; Sukkarieh, S. A Gaussian Process-Based RRT Planner for the Exploration of an Unknown and Cluttered Environment with a UAV. *Adv. Robot.* **2013**, *27*, 431–443. [CrossRef]
39. Yan, F.; Zhuang, Y.; Xiao, J. 3D PRM Based Real-Time Path Planning for UAV in Complex Environment. In Proceedings of the 2012 IEEE International Conference on Robotics and Biomimetics (ROBIO), Guangzhou, China, 11–14 December 2012; pp. 1135–1140.

40. Candeloro, M.; Lekkas, A.M.; Hegde, J.; Sørensen, A.J. A 3D Dynamic Voronoi Diagram-Based Path-Planning System for UUVs. In Proceedings of the OCEANS 2016 MTS/IEEE Monterey, Monterey, CA, USA, 19–23 September 2016; pp. 1–8.
41. Bi, Y.; Lan, M.; Li, J.; Zhang, K.; Qin, H.; Lai, S.; Chen, B.M. Robust Autonomous Flight and Mission Management for MAVs in GPS-Denied Environments. In Proceedings of the 2017 11th Asian Control Conference (ASCC), Gold Coast, Australia, 17–20 December 2017; pp. 67–72.
42. Sampedro, C.; Rodriguez-Ramos, A.; Bavle, H.; Carrio, A.; de la Puente, P.; Campoy, P. A Fully-Autonomous Aerial Robot for Search and Rescue Applications in Indoor Environments Using Learning-Based Techniques. *J. Intell. Robot. Syst.* **2019**, *95*, 601–627. [CrossRef]
43. Mohta, K.; Watterson, M.; Mulgaonkar, Y.; Liu, S.; Qu, C.; Makineni, A.; Saulnier, K.; Sun, K.; Zhu, A.; Delmerico, J.; et al. Fast, Autonomous Flight in GPS-Denied and Cluttered Environments. *J. Field Robot.* **2018**, *35*, 101–120. [CrossRef]
44. Bachrach, A.; Prentice, S.; He, R.; Roy, N. RANGE–Robust Autonomous Navigation in GPS-Denied Environments. *J. Field Robot.* **2011**, *28*, 644–666. [CrossRef]
45. Leishman, R.C.; McLain, T.W.; Beard, R.W. Relative Navigation Approach for Vision-Based Aerial GPS-Denied Navigation. *J. Intell. Robot. Syst.* **2014**, *74*, 97–111. [CrossRef]
46. Li, Q.; Li, D.-C.; Wu, Q.; Tang, L.; Huo, Y.; Zhang, Y.; Cheng, N. Autonomous Navigation and Environment Modeling for MAVs in 3-D Enclosed Industrial Environments. *Comput. Ind.* **2013**, *64*, 1161–1177. [CrossRef]
47. Perez-Grau, F.J.; Ragel, R.; Caballero, F.; Viguria. A.; Ollero, A. An Architecture for Robust UAV Navigation in GPS-Denied Areas. *J. Field Robot.* **2018**, *35*, 121–145. [CrossRef]
48. Valenti, F.; Giaquinto, D.; Musto, L.; Zinelli, A.; Bertozzi, M.; Broggi, A. Enabling Computer Vision-Based Autonomous Navigation for Unmanned Aerial Vehicles in Cluttered GPS-Denied Environments. In Proceedings of the 2018 21st International Conference on Intelligent Transportation Systems (ITSC), Maui, HI, USA, 4–7 November 2018; pp. 3886–3891.
49. Schmid, K.; Lutz, P.; Tomić, T.; Mair, E.; Hirschmüller, H. Autonomous Vision-Based Micro Air Vehicle for Indoor and Outdoor Navigation. *J. Field Robot.* **2014**, *31*, 537–570. [CrossRef]
50. Lutz, P.; Müller, M.G.; Maier, M.; Stoneman, S.; Tomić, T.; von Bargen, I.; Schuster, M.J.; Steidle, F.; Wedler, A.; Stürzl, W.; et al. ARDEA—An MAV with Skills for Future Planetary Missions. *J. Field Robot.* **2020**, *37*, 515–551. [CrossRef]
51. Lin, Y.; Gao, F.; Qin, T.; Gao, W.; Liu, T.; Wu, W.; Yang, Z.; Shen, S. Autonomous Aerial Navigation Using Monocular Visual-Inertial Fusion. *J. Field Robot.* **2018**, *35*, 23–51. [CrossRef]
52. Esrafilian, O.; Taghirad, H.D. Autonomous Flight and Obstacle Avoidance of a Quadrotor by Monocular SLAM. In Proceedings of the 2016 4th International Conference on Robotics and Mechatronics (ICROM), Tehran, Iran, 26–28 October 2016; pp. 240–245.
53. von Stumberg, L.; Usenko, V.; Engel, J.; Stückler, J.; Cremers, D. From Monocular SLAM to Autonomous Drone Exploration. In Proceedings of the 2017 European Conference on Mobile Robots (ECMR), Paris, France, 6–8 September 2017; pp. 1–8.
54. Laina, I.; Rupprecht, C.; Belagiannis, V.; Tombari, F.; Navab, N. Deeper Depth Prediction with Fully Convolutional Residual Networks. In Proceedings of the 2016 Fourth International Conference on 3D Vision (3DV), Stanford, CA, USA, 25–28 October 2016; pp. 239–248.
55. Bresenham, J.E. Algorithm for Computer Control of a Digital Plotter. *IBM Syst. J.* **1965**, *4*, 25–30. [CrossRef]
56. Hart, P.E.; Nilsson, N.J.; Raphael, B. A Formal Basis for the Heuristic Determination of Minimum Cost Paths. *IEEE Trans. Syst. Sci. Cybern.* **1968**, *4*, 100–107. [CrossRef]
57. Tordesillas, J.; Lopez, B.T.; Everett, M.; How, J.P. FASTER: Fast and Safe Trajectory Planner for Navigation in Unknown Environments. *IEEE Trans. Robot.* **2022**, *38*, 922–938. [CrossRef]
58. Sun, W.; Sun, P.; Ding, W.; Zhao, J.; Li, Y. Gradient-Based Autonomous Obstacle Avoidance Trajectory Planning for B-Spline UAVs. *Sci. Rep.* **2024**, *14*, 14458. [CrossRef]

drones

MDPI

Article

An AI-Based Deep Learning with K-Mean Approach for Enhancing Altitude Estimation Accuracy in Unmanned Aerial Vehicles

Prot Piyakawanich [1] and Pattarapong Phasukkit [1,2,*]

[1] School of Engineering, King Mongkut's Institute of Technology Ladkrabang, Bangkok 10520, Thailand; 65016154@kmitl.ac.th
[2] King Mongkut Chaokhun Thahan Hospital, King Mongkut's Institute of Technology Ladkrabang, Bangkok 10520, Thailand
* Correspondence: pattarapong.ph@kmitl.ac.th; Tel.: +66-86045-5445

Abstract: In the rapidly evolving domain of Unmanned Aerial Vehicles (UAVs), precise altitude estimation remains a significant challenge, particularly for lightweight UAVs. This research presents an innovative approach to enhance altitude estimation accuracy for UAVs weighing under 2 kg without cameras, utilizing advanced AI Deep Learning algorithms. The primary novelty of this study lies in its unique integration of unsupervised and supervised learning techniques. By synergistically combining K-Means Clustering with a multiple-input deep learning regression-based model (DL-KMA), we have achieved substantial improvements in altitude estimation accuracy. This methodology represents a significant advancement over conventional approaches in UAV technology. Our experimental design involved comprehensive field data collection across two distinct altitude environments, employing a high-precision Digital Laser Distance Meter as the reference standard (Class II). This rigorous approach facilitated a thorough evaluation of our model's performance across varied terrains, ensuring robust and reliable results. The outcomes of our study are particularly noteworthy, with the model demonstrating remarkably low Mean Squared Error (MSE) values across all data clusters, ranging from 0.011 to 0.072. These results not only indicate significant improvements over traditional methods, but also establish a new benchmark in UAVs altitude estimation accuracy. A key innovation in our approach is the elimination of costly additional hardware such as Light Detection and Ranging (LiDAR), offering a cost-effective, software-based solution. This advancement has broad implications, enhancing the accessibility of advanced UAVs technology and expanding its potential applications across diverse sectors including precision agriculture, urban planning, and emergency response. This research represents a significant contribution to the integration of AI and UAVs technology, potentially unlocking new possibilities in UAVs applications. By enhancing the capabilities of lightweight UAVs, we are not merely improving a technical aspect, but revolutionizing the potential applications of UAVs across industries. Our work sets the stage for safer, more reliable, and precise UAVs operations, marking a pivotal moment in the evolution of aerial technology in an increasingly UAV-dependent world.

Keywords: Unmanned Aerial Vehicles (UAVs); drones; altitude estimation; deep learning; sensor fusion; K-means clustering; lightweight UAVs; lightweight drones

1. Introduction

Unmanned Aerial Vehicles (UAVs), commonly known as drones, have revolutionized various sectors, from precision agriculture to urban planning and emergency response. As these aerial platforms become increasingly ubiquitous, the demand for more precise and reliable operational capabilities has intensified. Among the critical challenges facing lightweight UAVs, particularly those under 2 kg, is the need for accurate altitude estimation.

This capability is fundamental to ensuring safe flight operations, optimizing data collection, and expanding the potential applications of these versatile machines.

Despite significant advancements in UAV technology, the challenge of accurate altitude estimation in lightweight drones remains a significant hurdle. Traditional high-precision altitude estimation methods for lightweight UAVs (under 2 kg) heavily rely on hardware solutions such as LiDAR, which present significant disadvantages including prohibitive costs, added weight burden, reduced payload capacity, and increased power consumption from the battery system. These limitations substantially impact the operational effectiveness and practical applications of lightweight drones.

This paper proposes a novel software-based altitude estimation method using deep learning with K-means clustering (DL-KMA) that processes data from existing onboard sensors while maintaining high accuracy. Our approach leverages the synergy between unsupervised and supervised learning techniques to achieve significant improvements in altitude estimation without requiring additional expensive hardware.

The selection of K-means clustering as our primary clustering algorithm is justified through comprehensive comparative analysis with alternative methods. The K-means clustering demonstrates superior performance compared to hierarchical clustering methods in handling multidimensional UAV altitude data, particularly in terms of computational complexity (O(n) versus $O(n^2 \log n)$). Furthermore, when compared to DBSCAN, K-means offers enhanced scalability and more distinct cluster boundaries, which is crucial for precise altitude range segmentation. Unlike Gaussian Mixture Models (GMMs), K-means provide more interpretable results with well-defined cluster boundaries, making it particularly suitable for altitude-based decision making in UAV applications.

The effectiveness of our proposed method is validated by experimental results, achieving MSE values ranging from 0.011 to 0.072 across all data clusters. These results establish DL-KMA as a practical alternative for drone height estimation, combining high accuracy with cost-effectiveness. This approach not only addresses current challenges in UAV altitude estimation, but also opens new avenues for research in lightweight drone development.

The field of Unmanned Aerial Vehicle (UAV) technology has seen significant advancements in recent years, particularly in altitude estimation for lightweight drones. This review examines the current state of research and highlights key developments in this area.

Recent advancements, however, have explored alternative approaches. Ref. [1] investigated the application of deep learning techniques to enhance object detection in UAV imagery, potentially improving altitude estimation accuracy without the need for additional hardware. Similarly, Liu et al. [2] demonstrated the integration of AI-based approaches for optimizing UAV altitude while considering power constraints in Internet of Vehicles applications, highlighting the growing importance of efficient altitude management in complex delivery systems. This trend is further developed in the research of Wang et al. [3] on multi-UAV path planning, which showcases how AI applications can coordinate multiple UAVs while maintaining optimal altitude for urban delivery scenarios. Nevertheless, as Feiyu et al. [4] revealed in their study of UAV Position Optimization for Servicing Ground Users Based on Deep Reinforcement Learning., which can prove challenging for lightweight drones weighing less than 2 kg due to computational constraints.

This limitation highlights the need for more cost-effective and weight-efficient solutions in altitude estimation for small UAVs. Recent research has shifted towards software-based solutions leveraging artificial intelligence. Chen et al. [5] and Sanjar and Sulaym [6] proposed adaptive systems for object detection and image classification, highlighting the potential of AI in enhancing UAV capabilities without additional hardware. This trend is further supported by the work of Odabas Yildirim et al. [7] and Panthakkan et al. [8] in automated vehicle detection using deep learning models.

Deep learning techniques have shown promise in improving UAV performance. Zeng et al. [9] and Chen et al. [10] developed various approaches for UAV navigation and mapping using deep learning. While effective, some of these methods still rely on LiDAR, limiting their applicability to lightweight drones.

Addressing the specific needs of lightweight UAVs, several studies have focused on optimizing algorithms for resource-constrained platforms [8,11]. The integration of unsupervised and supervised learning techniques has emerged as a promising direction, as demonstrated by Makrigiorgis et al. [12,13] and Cheng et al. [14].

Altitude estimation without additional hardware has been explored by Taame et al. [15] using Kalman filters, Yang et al. [16] with altitude-guided detection, and Aslani and Saberinia [17] in the context of energy-efficient UAV networks.

Furthermore, despite these advancements, there remains a gap in research specifically addressing altitude estimation for UAVs under 2 kg without relying on additional hardware. The work of Zhenyu et al. [18] on global localization of UAVs and the multi-feature approach of Weiguang et al. [19] suggest potential directions for improvement.

Our study contributes to the field of UAV technology in several significant ways. Firstly, it introduces an innovative AI-based methodology for enhancing altitude estimation accuracy in lightweight UAVs. Secondly, we conduct a comprehensive evaluation of the proposed method across a diverse range of environmental conditions and altitude ranges, ensuring its applicability in various operational scenarios. Finally, our research demonstrates a marked improvement in altitude estimation accuracy without necessitating the incorporation of additional expensive hardware.

These advancements represent a significant leap forward in the development of altitude estimation systems for lightweight UAVs, underscoring the potential of AI-driven solutions in enhancing the performance and capabilities of these aircraft. The findings of this research have important implications for the broader field of autonomous aerial systems and contribute to ongoing efforts to improve the efficiency and reliability of UAVs operations.

2. Materials and Research Methodology

The attached diagram (Figure 1) illustrates a method for measuring the altitude of a drone using a digital laser distance meter. The setup includes a drone in flight, a digital laser distance meter positioned on a tripod, and a remote pilot operating the drone. The digital laser distance meter is aimed vertically at the drone, measuring the direct distance from the device to the drone's undercarriage. This direct measurement is referred to as the "Measured Distance". The diagram also indicates the "Offset Distance", which is the vertical distance from the ground to the laser meter's sensor. The combination of these distances allows the calculation of the drone's absolute altitude above ground level. This method is essential for applications requiring precise altitude measurements, such as surveying, mapping, or aerial photography. The clear, labeled diagram assists in understanding the procedure, emphasizing the roles of each component in obtaining accurate altitude data.

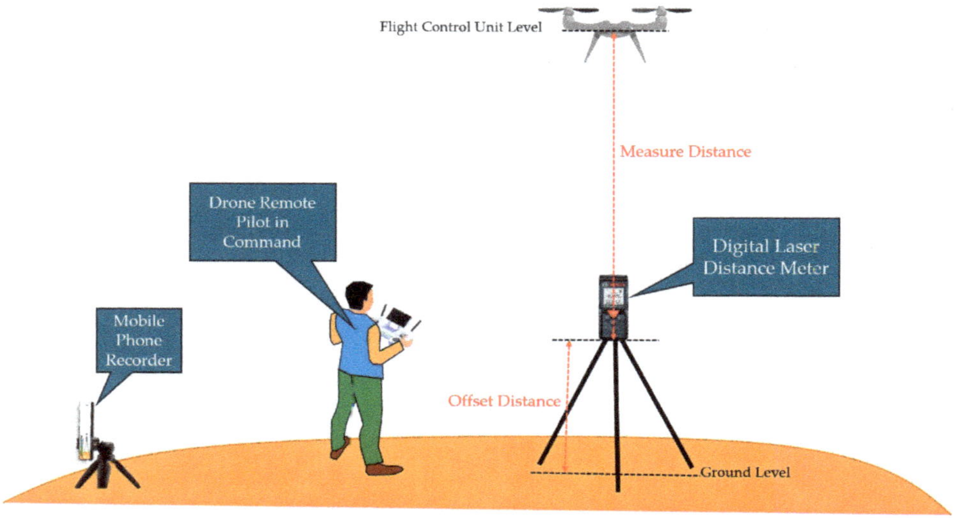

Figure 1. A method for measuring the altitude of a drone using a digital laser distance meter.

2.1. Materials and Methods

The study utilized an Unmanned Aerial Vehicle (UAV) equipped with the following components (Figure 2):

- Flight Control Unit (FCU): 32 bit ARM Cortex M4 core with Floating Point Unit (FPU) (168 MHz/256 KB, RAM 2 MB Flash) (Pixhawk, an internationally developed open-hardware project.) Sensors: MPU6000 (primary accelerometer and gyroscope), ST Micro 16-bit gyroscope, ST Micro 14-bit accelerometer/compass, MEAS barometer (InvenSense [now part of TDK Corporation], a U.S.-based company, San Jose, CA, USA)
- Interfaces: 5× UART serial ports, Spektrum DSM/DSM2/DSM-X Satellite input, Futaba S.BUS input, PPM sum signal, RSSI input, I2C, SPI, 2× CAN, USB
- Dimensions: 38 g weight, 50 mm width, 15.5 mm height, 81.5 mm length

GPS Module (Neo 3): (u-blox, A Swiss-based company, Zurich, Switzerland)

- Processor: STM32F412
- GNSS Receiver: Ublox M9N
- Supported GNSS Bands: GPS/QZSS L1 C/A, GLONASS L10F, BeiDou B1I, Galileo E1B/C, SBAS L1 C/A
- Navigation Update Rate: Up to 25 Hz (RTK)
- Position Accuracy: Up to 1.5 m
- Dimensions: 60 × 60 × 16 mm, 33 g weight

Propulsion System: (Xrotor, Shenzhen, China)

- Motors: 4 x Xrotor pro 50A 380 KV
- Propellers: 15″ Carbon Fiber, 12 mm hole size

Reference Measurement Device: (CZDANG, Shenzhen, China)

- Digital Laser Distance Meter (Figure 3)
- Measurement Range: 120 m
- Accuracy: ±3 mm
- Laser Class: Class II

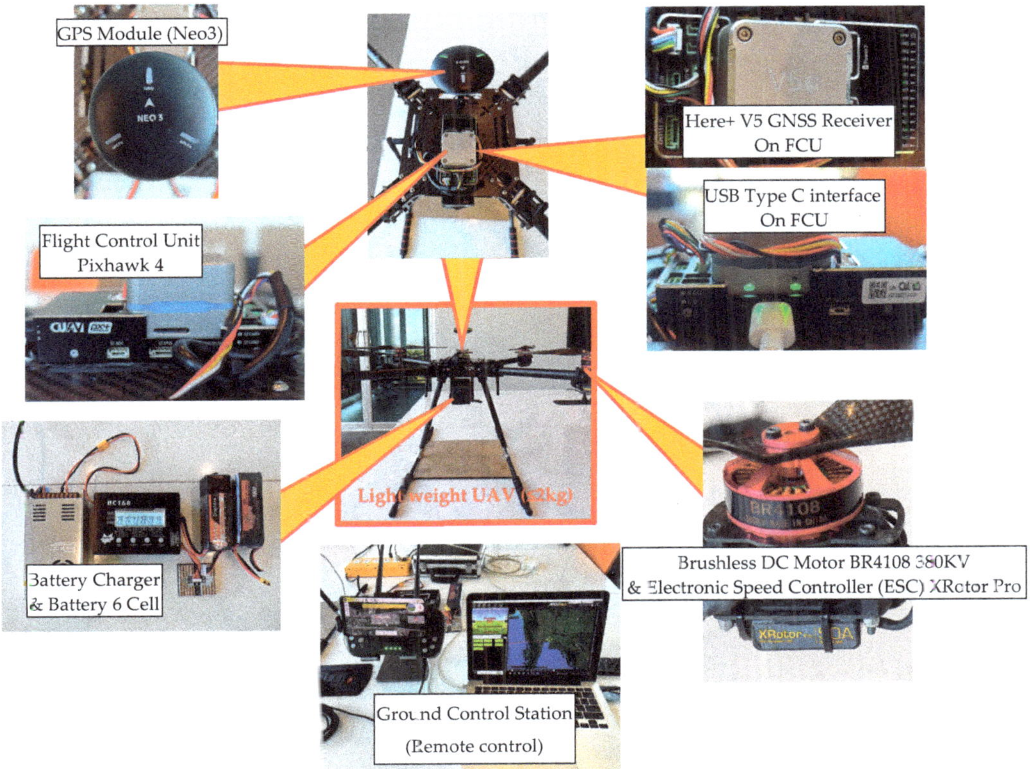

Figure 2. Hardware Components and System Architecture of the Experimental UAV Platform.

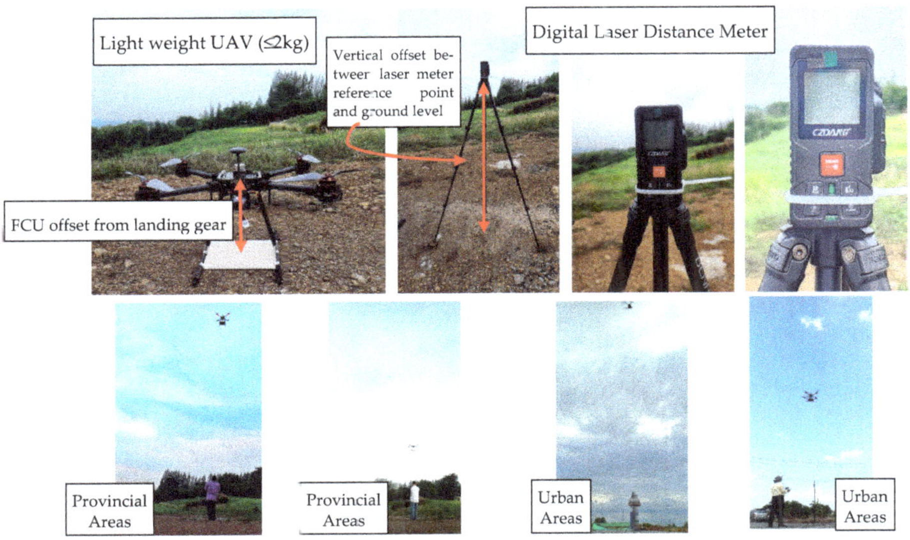

Figure 3. Experimental Setup for UAV Altitude Data Collection and Validation Using Digital Laser Measurement.

2.1.1. Preparation of Experimental Altitude Ranges for Comprehensive UAV Performance Analysis

In the rapidly evolving field of unmanned aerial vehicle (UAV) research, precise altitude measurements across various operational ranges are crucial for developing accurate and reliable systems. The DL-KMA method aims to provide comprehensive coverage while facilitating detailed analysis of altitude-dependent factors affecting UAV performance.

The initial design of altitude ranges for this research was developed to measure UAVs operating within a height range of 1 to 30 m, segmented into the following four distinct altitude intervals: 1–10 m, 10–20 m, 20–25 m, and 25–30 m. These proposed altitude ranges are strategically crafted to ensure continuous data collection across the entire span, aligning with previous work on multi-altitude aerial vehicle datasets, which emphasizes the importance of defined altitude intervals for consistent data analysis and model training [13]. This study builds upon such datasets through data extraction and validation methodology, as shown in Figure 4, utilizing sensor fusion algorithms for UAV altitude estimation to identify critical transition points where estimation accuracy significantly changes, enhancing adaptive control system development.

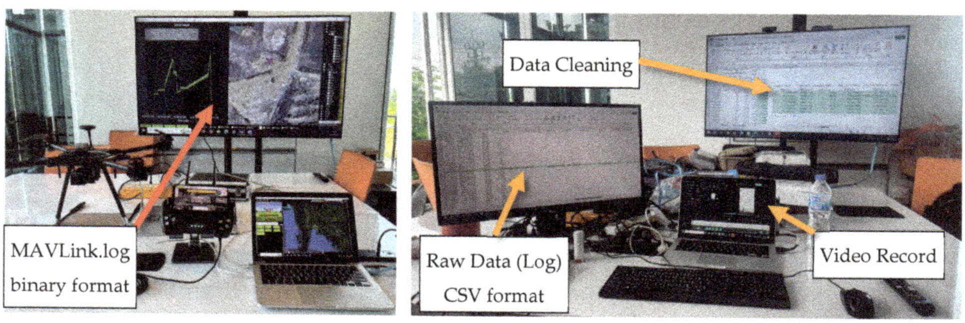

Figure 4. UAV Flight Data Extraction and Validation Process Using Video-synchronized Log Analysis.

The K-Means Clustering, a widely used unsupervised machine learning technique, offers significant advantages in identifying natural groupings within multidimensional data. Recent advancements, such as the K-Means Clustering approach based on Chebyshev Polynomial Graph Filtering, have further enhanced the effectiveness of this technique in handling complex datasets, as demonstrated [20]. By applying K-Means to altitude-related features extracted from UAV flight logs, we can determine data-driven altitude ranges that reflect the inherent structure of the collected measurements. This approach allows for a more nuanced understanding of how environmental factors impact altitude estimation accuracy at different heights, as compared to predetermined fixed ranges.

Recent studies have demonstrated the effectiveness of clustering techniques in UAV-related applications. For instance, Ren et al. (2022) [21] successfully employed K-Means Clustering to optimize UAV deployment for wireless communications, showing improved coverage and energy efficiency.

The application of K-Means Clustering to UAV altitude range determination offers data-driven range selection, forming clusters based on the natural distribution of altitude measurements. This approach potentially reveals critical transition points where estimation accuracy undergoes significant changes, thus enhancing adaptive control system development for UAVs [20]. Secondly, an adaptive approach for UAV altitude estimation automatically adjusts to environmental influences such as wind patterns and air density variations in complex indoor settings, enhancing accuracy under varying conditions (Pritzl et al., 2022) [22].

Furthermore, Kolarik et al. (2023) [23] demonstrate that clustering techniques optimize sensor performance evaluation by identifying altitude-specific trends or anomalies in estimation accuracy, facilitating more precise calibration of algorithms and enhancing overall measurement reliability in complex environments. Additionally, comparative analysis between adjacent altitude brackets provides valuable insights into the continuity and consistency of UAS photogrammetry-based altitude estimation methods across diverse terrain sites, enhancing the understanding of method performance at different heights (Liu et al., 2024) [24].

To implement this approach, comprehensive flight data are collected, including altitude measurements and related sensor readings. The K-Means algorithm is then applied to this multidimensional dataset, with the optimal number of clusters determined through methods such as the elbow method. Previous studies have demonstrated the effectiveness of integrating the elbow method with K-Means clustering for identifying optimal cluster configurations, which can be particularly useful in our context for determining the most appropriate number of altitude ranges [25].

The fundamental equations for K-Means Clustering are as follows:

Cluster Assignment Equation:

$$c_i = argmin_j||x_i - \mu_j||^2 \tag{1}$$

where c_i denotes the cluster assigned to data point x_i and μ_j represents the centroid of cluster j.

Centroid Update Equation:

$$\mu_j = \frac{1}{|C_j|} \sum_{x_i \in C_j} x_i \tag{2}$$

where C_j is the set of all data points in cluster j, and $|C_j|$ is the cardinality of this set.

Objective Function:

$$J = \sum_{i=1}^{n} \sum_{j=1}^{k} w_{ij}||x_i - \mu_j||^2 \tag{3}$$

where J is the objective function to be minimized, n is the total number of data points, k is the number of clusters, and w_{ij} equals 1 if data point i belongs to cluster j, and 0 otherwise.

$$WCSS = \sum_{j=1}^{k} \sum_{X_i \in C_j} ||x_i - \mu_j||^2 \tag{4}$$

where $WCSS$ represents the Within-Cluster Sum of Squares.

In the context of integrating K-Means with C_j Learning for UAV applications, the following equations are pertinent:

Cluster-Specific Model Training: For each cluster C_j, train the following separate model f_j:

$$f_j(X) = W_j \cdot X + B_j \tag{5}$$

where W_j and B_j are the model parameters for cluster j.

Prediction in Clustered Context: For a new data point, determine its nearest cluster centroid μ_j, then apply the following model for that cluster f_j:

$$\hat{Y}_j = f_j(X) = W_j \cdot X + B_j \tag{6}$$

where $\hat{Y}_j$ is the prediction output of the model specific to cluster j.

This data-driven clustering approach to altitude range determination represents a significant advancement in UAV research methodology. By leveraging machine learning techniques, researchers can gain deeper insights into the multifaceted factors influencing altitude estimation accuracy. This strategy not only enhances the robustness of UAV research, but also provides critical insights for improving drone performance and safety across diverse applications and environments.

As the field of UAV technology continues to evolve, such data-driven approaches to experimental design and analysis will play a pivotal role in advancing the capabilities and reliability of these aerial systems. The integration of machine learning techniques like K-Means Clustering into UAV altitude analysis promises to unlock new possibilities in precision flight control, environmental monitoring, and autonomous navigation.

2.1.2. Preparation for Testing the Digital Laser Distance Meter

A methodical approach was implemented to assess the digital laser distance meter, in accordance with the stratified altitude measurement protocol for unmanned aerial vehicle (UAV) research. The experimental design encompassed the following four distinct altitude ranges: 1–10, 10–20, 20–25, and 25–30 m, providing a comprehensive evaluation across various operational heights.

The calibration of digital laser distance meters measuring ground-to-UAV altitude for UAV altimetry requires meticulous consideration of environmental variables to ensure precision in established controlled testing environments for UAV operations. Previous research has evaluated laser distance sensors in geomatics, highlighting the importance of environmental calibration to achieve accurate measurements [26]. Reference points and critical altitudes were determined utilizing high-precision surveying instrumentation, a methodology analogous to that employed by Taddia et al. (2020) [27] in their UAV topographic surveys, thus ensuring consistency and reliability in data collection.

The digital laser distance meter was calibrated in accordance with manufacturer specifications, with a particular emphasis on performance within the 1–30 m range (Figure 5). This instrument was subsequently utilized for establishing reference points and critical altitudes, serving as a crucial component in the data acquisition process for this research study.

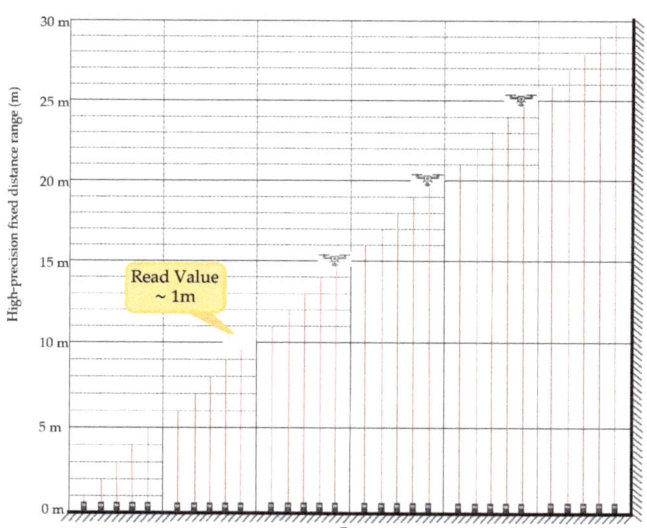

Figure 5. The image presents a schematic representation of a high-precision fixed distance range measurement experiment, likely designed for unmanned aerial vehicle (UAV) altitude calibration.

The vertical axis denotes the distance range from 0 to 30 m, while the horizontal axis represents time progression. The graph is characterized by a grid system with precise intervals, facilitating accurate measurements.

Red vertical lines at regular intervals indicate measurement points, with a note specifying "Read Value $\sim$ 1 m". This suggests a systematic data collection process at predefined increments. The UAV icons positioned at 10 m, 15 m, 20 m, and 25 m heights illustrate the experimental setup at various altitudes.

The experiment appears to involve incremental altitude increases for the UAV, allowing for comprehensive data collection across the specified range. This methodical approach emphasizes the importance of precision and repeatability in UAV altitude measurements. The graph's design implies a rigorous, controlled experimental environment, crucial for validating or calibrating distance measurement equipment in UAV applications. Such a setup is essential for ensuring high accuracy across various altitudes within the operational range of UAVs.

By structuring the testing process to mirror the proposed altitude strata, this preparation phase aims to provide comprehensive data that can contribute to the development of robust altitude estimation algorithms for UAVs. The careful consideration of overlapping ranges and environmental factors in the testing setup reflects the complex nature of UAV-based altitude measurement and aligns with the research objectives of improving drone performance across diverse operational conditions.

2.2. Deep Learning Regression-Based Models

Figure 6 illustrates a DL-KMA in each cluster (4 clusters) designed to enhance altitude estimation accuracy in Unmanned Aerial Vehicles (UAVs). The model's input features are derived from an array of operational data logs that exhibit strong correlations with UAVs altitude such as barometric readings, atmospheric pressure measurements, and other relevant parameters. The model's output target is optimized to produce altitude estimations that closely align with the ground truth measurements obtained via a digital laser distance meter, thus maximizing accuracy (±3 mm).

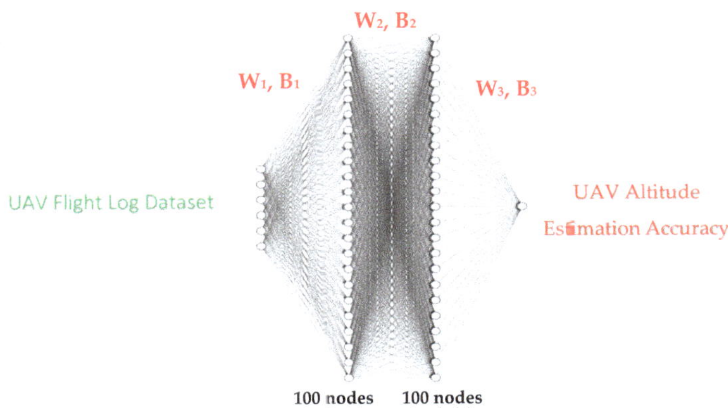

Figure 6. The DL-KMA predicts altitude estimation accuracy in each cluster of Unmanned Aerial Vehicles (UAVs).

The algorithmic architecture composed of two hidden layers, each containing 100 nodes. This doubling of node quantity in both hidden layers, compared to the single-input deep learning model, serves to augment the multi-input model's predictive capabilities. The weight matrices (W) and bias vectors (B) interconnecting the layers (W_1, B_1, W_2, B_2, W_3, and B_3) undergo optimization through a gradient descent iterative algorithm. For the activation functions, the hidden layers employ the hyperbolic tangent function (tanh(z)), while the output layer utilizes the Rectified Linear Unit (ReLU) activation function.

This sophisticated model structure is designed from scratch to effectively process and integrate multiple input parameters, thereby enhancing the robustness and precision of UAV altitude estimations across diverse operational scenarios.

In this study, altitude measurements were categorized into four distinct altitude ranges to ensure comprehensive data coverage and to assess the various impacts observed following UAV flights, as previously discussed in Section 2.1.2. A digital laser distance meter, meticulously calibrated for each range, was employed as the reference device for altitude measurement. Given the extensive data recorded from UAV flights, the initial dataset for training and testing the multiple-input algorithm model comprised 48,000 datasets, derived from 6 altitude-related features measured 8000 times.

The dataset with altitude-related features (48,000 datasets) was divided into four groups based on the specified altitude ranges, each containing 12,000 data points. Normalization was applied to the full dataset initially to maintain uniform scaling, preventing any discrepancies in normalization scales between training and testing data. Following normalization, within each group, 80% (9600 datasets) were allocated for training and 20% (2400 datasets) for testing.

The training dataset comprised X_{train} (normalized training input data) and Y_{train} (normalized training output data), while the test dataset included X_{test} (normalized test input data) and Y_{test} (normalized test output data). The variables W_1, B_1, W_2, B_2, W_3, and B_3 were randomly initialized. These variables were optimized using a gradient descent algorithm with a learning rate (α) of 0.01 over 3000 epochs. Furthermore, L1-norm regularization was used to avoid overfitting. The iteration is terminated once divergence occurs between the MSE of the training and testing datasets.

Prior to training and testing the DL-KMA, standardized altitude data from the UAV, as measured by the digital laser distance meter, were required as a reference signal (target) to predict UAV altitude estimation accuracy. The training and testing input and output datasets (X_{train}, Y_{train}, X_{test}, and Y_{test}) under four distinct altitude ranges, are normalized using min–max normalization in Equation (7).

$$\text{Data}_{normalization} = \frac{\text{Dataset} - \text{Training\_Dataset}_{min}}{\text{Training\_Dataset}_{max} - \text{Training\_Dataset}_{min}} \tag{7}$$

where Dataset represents any input or output data to be (X_{train}, Y_{train}, X_{test}, and Y_{test}); $\text{Training\_Dataset}_{min}$ is the minimum value from the training dataset only; and $\text{Training\_Dataset}_{max}$ is the maximum value from the training dataset only. The value of the normalized datasets ($\text{Data}_{normalization}$) ranges between 0 and 1 [0, 1].

In the feedforward of the deep learning models, the activation function between hidden layers is a hyperbolic tangent function ($\tanh(Z)$), as shown in Equation (8), where $\tanh(Z) = [-1, 1]$. The activation function ReLU(Z) is used in the output layer, as shown in Equation (9), where Z is the linear combination. The predicted output $\hat{Y}_j$ of the DL-KMA is the **UAV altitude estimation accuracy.**

$$\tanh(Z) = \frac{e^{Z} - e^{-Z}}{e^{Z} + e^{-Z}} \tag{8}$$

$$\text{Relu}(Z) = \left\{ \begin{matrix} 0, & Z < 0 \\ Z, & Z \geq 0 \end{matrix} \right\} \tag{9}$$

1. Layer Specific Equations:
    - First Hidden Layer Equations

$$Z^{(1)} = W_1 \cdot X + B_1 \tag{10}$$

where W_1 and B_1 represent the weights and biases of the first hidden layer

- Second Hidden Layer.

$$Z^{(2)} = W_2 \cdot \tanh\left(Z^{(1)}\right) + B_2 \tag{11}$$

where W_2 and B_2 are the weights and biases of the second hidden layer, and tanh is applied as the activation function for the output of the first layer.

2. Output Layer Equations:

$$\hat{Y} = W_3 \cdot \tanh\left(Z^{(2)}\right) + B_3 \tag{12}$$

where W_3 and B_3 are the weights and biases for the output layer, with a chosen activation function like ReLU if applicable for the final prediction.

3. Activation Functions:

Clearly state that the first and second hidden layers use the tanh activation function to handle non-linear transformations, while the output layer uses a ReLU or other appropriate activation depending on the target prediction (e.g., linear if regression).

In the back propagation of deep learning models, the mean squared error (MSE) between the normalized training output dataset (Y_{train};$\mathrm{Y_i}$) and predicted normalized output $\hat{\mathrm{Y}}_\mathrm{i}$ is first calculated using Equation (13), and the gradient descent iterative optimization algorithm is subsequently applied to fine-tune W and B by using the chain rule derivative.

Mean Squared Error (MSE):

$$MSE = \frac{1}{n}\sum_{i=1}^{n}(Y_i - \hat{Y}_i)^2 \tag{13}$$

The prediction performance (i.e., predictive ability) of DL-KMA is assessed by mean squared error (MSE; Equation (13)), mean absolute error (MAE; Equation (14)), and coefficient of determination (R^2; Equation (15))

$$MAE = \frac{1}{n}\sum_{i=1}^{n}|Y_i - \hat{Y}_i| \tag{14}$$

where Y_i is the normalized testing output dataset (Y_{test}), $\hat{Y}_i$ is predicted normalized output ($Y_{predict}$), and n is the number of datasets.

The coefficient of determination (R^2) is a goodness-of-fit measure for regression models and the dependent variables on a scale of 0–1

$$R^2 = \frac{(\mathrm{Var}(Y_{true}) - \mathrm{MSE})}{\mathrm{Var}(Y_{true})} \tag{15}$$

where $\mathrm{Var}(Y_{true})$ is the variance of actual values, MSE is the mean squared error between predicted and actual values.

Figure 7 illustrates an advanced deep learning pipeline for enhancing UAV altitude precision. The process begins with UAV flight log datasets, which serve as input features. These data undergo preprocessing through an Elbow method, followed by K-Means clustering to identify key patterns. The clustered data are then normalized to ensure consistent scaling across features.

The core of the system is a deep neural network, represented by the complex interconnected structure. This network processes the normalized data, learning intricate relationships between input features and altitude estimates. The output undergoes denormalization to convert it back to real-world units.

The final step produces a "Precision Altitude" estimate, which is compared to the "True Altitude" measured from ground level. This comparison allows for continuous refinement of the model's accuracy. The system aims to significantly improve UAV altitude estimation by leveraging deep learning techniques on rich flight log data, potentially enhancing the precision and reliability of drone operations in various applications.

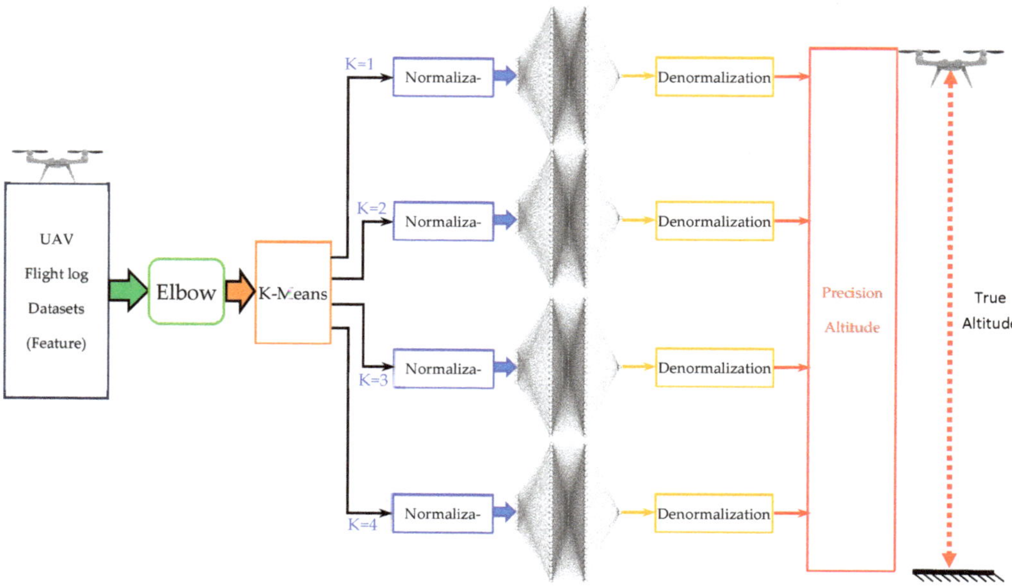

Figure 7. An advanced deep learning pipeline for enhancing UAV alti-tude precision.

In the proposed DL-KMA, the features after Elbow and K-Means method for clustering already are used. Each clustering will be normalized prior to feedforwarding for the normalized target, given the optimized W_1, B_1, W_2, B_2, W_3, and B_3. The normalized target is subsequently denormalized to obtain the altitude estimation.

3. Experimental Setup and Data Analysis

Figure 8 depicts a system for precise drone altitude control using DL-KMA. The workflow begins with a drone equipped with telemetry and MAVLink logging capabilities. Data are transmitted via MAVLink to both a remote-control device and a computer running an AI DL-KMA.

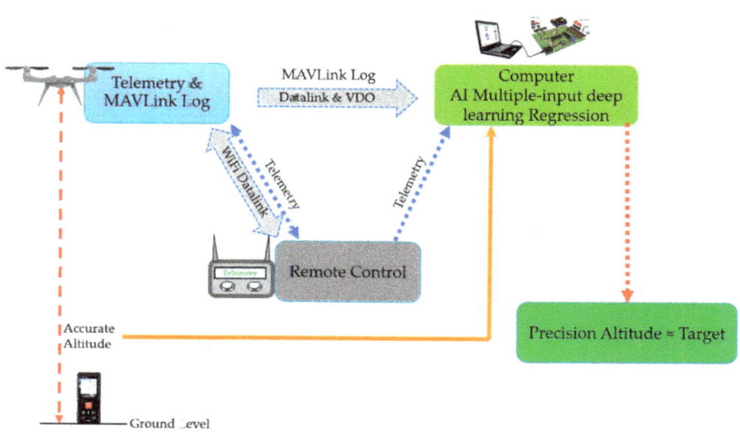

Figure 8. A system for precise drone altitude control using DL-KMA.

The remote control receives telemetry data through WiFi/Datalink, while the computer processes MAVLink log data including datalink and video (VDO) information. Additionally, accurate altitude measurements are obtained from ground level using a separate device, likely a laser rangefinder.

The AI model on the computer integrates the following inputs: telemetry from the remote control, MAVLink log data, and the precise ground-level altitude measurement. Using deep learning regression techniques, it processes this information to estimate a "Precision Height ≈ Target" value.

This system aims to achieve highly accurate altitude control by combining various data sources and leveraging AI capabilities. The integration of ground truth measurements with onboard sensors and control inputs allows for more precise hovering and altitude maintenance than traditional methods relying solely on onboard systems.

3.1. Experimental Design, Instrumentation Configuration

3.1.1. Area Setup

The study was conducted across two distinct geographical locations to ensure diverse environmental conditions:

Area A, Terrain (Provincial): Muak Lek, Saraburi Province Thailand (14.782831° N, 101.253410° E, elevation: approximately 430 m).

Area B, Terrain (Metropolis): MiniRC Airfield, Bangkok Thailand (13.895999° N, 100.779172° E, elevation: approximately 3 m).

3.1.2. UAVs Setup

The experimental setup for Unmanned Aerial Vehicles (UAVs) was meticulously designed to ensure seamless operation and data collection. The primary components comprising a customized lightweight drone (>2 kg) without an integrated camera, lithium-polymer batteries, a remote-control unit, a smartphone for real-time Bluetooth connectivity, and "QGroundControl" software for advanced flight control.

The selected UAV underwent comprehensive pre-flight checks, including propeller inspection, frame integrity verification, and sensor calibration. Battery management protocols evolved throughout the study. In Area A, a single battery limited flight time and necessitated extensive charging periods (75–90 min). This experience informed the utilization of two batteries in Area B, significantly extending data collection duration.

Battery safety protocols were rigorously implemented, as demonstrated in Figure 9, including full balance-cell charging, adherence to specified current limitations, and the incorporation of cooling periods between charging cycles to ensure safety and longevity. The remote-control unit employed a mobile screen for telemetry display, paired with a smartphone via Bluetooth for real-time command transmission.

The "QGroundControl" software, installed on a Google Pixel 6 Android smartphone, facilitated advanced flight planning and control. To mitigate electromagnetic interference, all electronic devices were strategically positioned.

Prior to commencing the research, it is essential to prepare the equipment meticulously. The distance laser distance meter should be mounted on a tripod with three legs to ensure stability, and a bubble level should be employed to achieve symmetry, ensuring that the laser beam is projected in a straight line. Additionally, the device's automatic distance measurement function must be tested before every flight to verify its accuracy.

Before UAV deployment, weather conditions should be assessed using forecasting applications such as UAV Forecast or Windy. This is crucial to avoid the adverse effects of gusty winds or excessive wind speeds, which could compromise the accuracy of altitude measurements.

Furthermore, it is important to prepare a mobile device to capture images and a wireless microphone to record both the readings from the digital laser distance meter and the altitude data displayed by the UAV. To ensure a smooth and complete data collection process, it is advisable to conduct a preliminary flight test lasting 1 to 2 min. This test will confirm the accuracy and reliability of the recorded measurements, providing confidence in the data acquisition for each session.

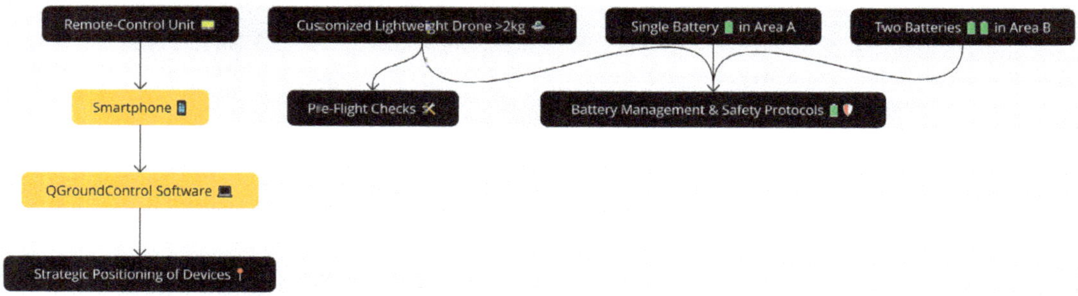

Figure 9. Diagram Illustrating Area Configuration and UAV Deployment Setup.

3.2. Data Analysis

3.2.1. Data Acquisition

The data collection process from UAVs aims to obtain high-quality data for analysis. In the preliminary research phase, the investigator collects data from MAVLink logs in binary format and converts these binary files to CSV (Comma-Separated Values) format (Figure 10).

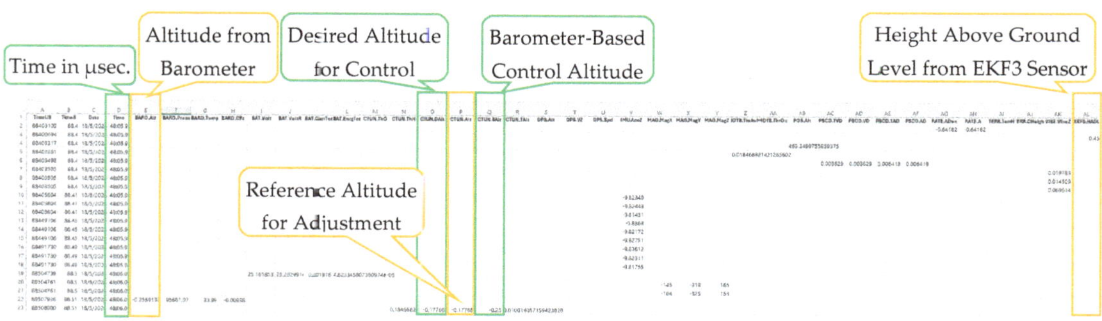

Figure 10. Example Raw Data from UAV MAVLink logs binary format and converts to CSV files.

Each flight session generates hundreds of thousands of data rows due to the simultaneous recording of multiple sensor inputs at millisecond intervals. Detailed information on these data logs can be found at "https://ardupilot.org/copter/docs/logmessages.html" (accessed on 24 November 2024).

3.2.2. Data Cleaning and Selecting Feature

The study employed an innovative approach to enhance altitude estimation accuracy for lightweight UAVs using deep learning regression-based models with altitude compensation. The methodology combined unsupervised and supervised learning techniques to process and analyze data from multiple onboard sensors.

Following the initial data collection, a comprehensive feature selection process was implemented to identify the most relevant parameters for the altitude prediction model (detailed procedures and sample data are shown in Figure 11). This process is crucial for enhancing model performance, reducing computational complexity, and mitigating potential overfitting issues.

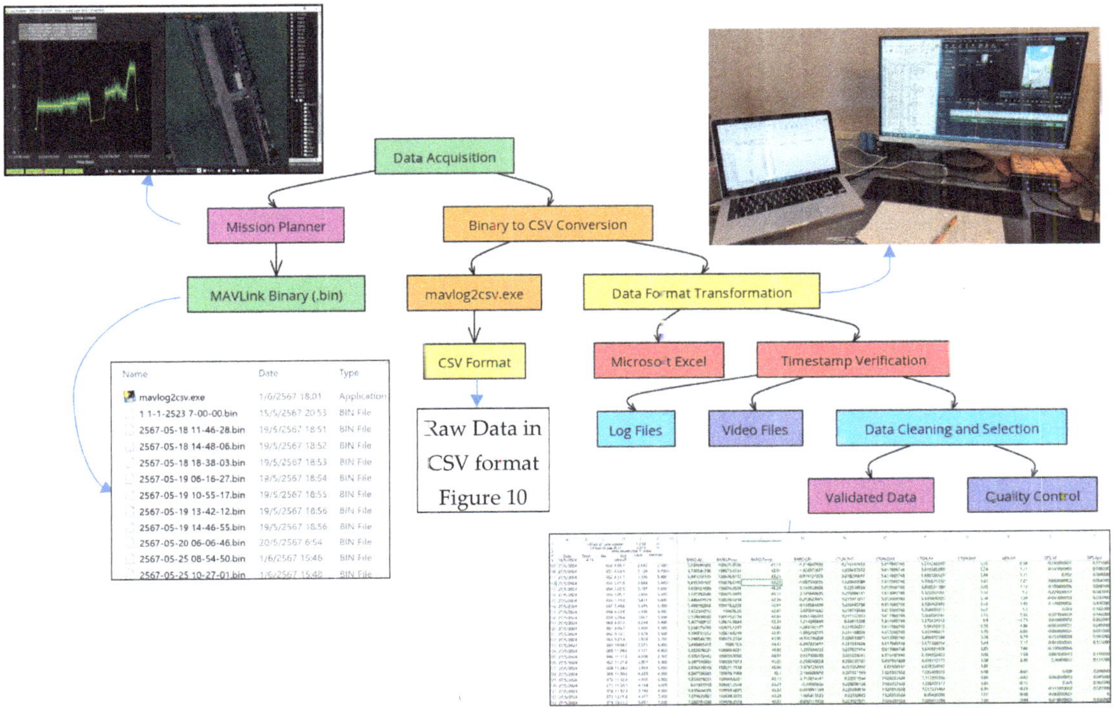

Figure 11. Data Collection and Preprocessing Workflow for UAV Flight Log Analysis with VDO Integration.

The feature selection methodology incorporated both statistical analysis and domain expertise. Correlation analysis, K-Means Clustering were employed to assess the significance and independence of each feature.

Through this rigorous analytical process, 6 predictor variables and 1 target variable were identified. These selected features demonstrated significant variability and strong correlations with the target variable, making them ideal candidates for the regression model. The refined feature set is composed of the following:

1. Remote Control Altitude
    - **Description**: Altitude displayed on the UAV's remote control, calculated by the onboard Flight Control Unit (FCU) using data from barometric sensors, IMU, and GPS.
    - **Relevance**: Provides the operator with a real-time, integrated altitude estimate, essential for manual adjustments, especially in low-visibility or GPS-limited scenarios.
    - **Example**: During flight, the remote control combines data from the barometer, IMU, and GPS, enabling altitude monitoring and adjustments to maintain the UAV's height above ground. For instance, if the UAV ascends beyond the desired altitude, the operator can adjust controls to descend back to the set level.
2. BARO.Alt (Barometric Altitude)
    - **Description**: Altitude based on barometric pressure readings, which decrease predictably with increasing altitude.
    - **Relevance**: Crucial for altitude stability, especially in variable weather conditions, by providing a consistent atmospheric pressure reference.

- **Example**: As the UAV ascends, barometric readings decrease, indicating an increase in altitude. Conversely, if the UAV descends, barometric readings increase, helping to confirm the reduction in altitude. This trend stabilizes altitude estimation even in windy conditions.

3. CTUN.DAlt (Desired Altitude for Control)
   - **Description**: The target altitude set by the operator or flight controller.
   - **Relevance**: Serves as a benchmark for altitude correction, enabling the control system to adjust the UAV's position if deviations from the target occur.
   - **Example**: If programmed to maintain at 10 m, the UAV adjusts to return to this set altitude upon detecting any deviation. If the UAV descends below 10 m due to a gust of wind, it will adjust to ascend back to the desired altitude.
4. CTUN.Alt (Reference Altitude for Adjustment)
   - **Description**: A dynamically updated altitude base level for comparison with the desired altitude.
   - **Relevance**: Acts as a feedback loop, helping correct altitude changes in response to environmental factors.
   - **Example**: During ascent, this reference helps stabilize the climb by correcting for altitude shifts caused by wind. Likewise, during descent, it provides a stable reference point to ensure controlled lowering, avoiding sudden drops or altitude fluctuations.
5. CTUN.BAlt (Barometer-Based Control Altitude)
   - **Description**: Altitude calculated purely from barometric readings, which the control system uses to ensure stability.
   - **Relevance**: Ensures accurate altitude holding, especially at high altitudes or in conditions where GPS signals are unreliable.
   - **Example**: During high-altitude operations, the UAV relies on this reading to maintain stability when GPS accuracy is compromised. If the UAV descends unexpectedly, barometric control altitude helps stabilize the descent until the target altitude is regained.
6. XKF5.HAGL (Height Above Ground Level from EKF3 Sensor)
   - **Description**: Height above ground level, calculated from the Extended Kalman Filter (EKF3) sensor, considering immediate terrain conditions.
   - **Relevance**: Vital for terrain navigation, especially in areas with varied topography, by maintaining a safe distance above the ground.
   - **Example**: When flying over hilly terrain, this measurement ensures consistent altitude above ground, preventing possible collisions with obstacles. For instance, if the UAV ascends over rising ground, HAGL helps it maintain safe clearance, while during descent over descending terrain, it ensures the UAV keeps the necessary distance to avoid collisions.

The target variable is defined as follows:

1. Target: Reference altitude from digital laser distance meter

This dimensionality reduction not only streamlines the model's input, but also enhances its interpretability and generalization capabilities. The selected features represent a balanced mix of sensor data and control parameters, providing a comprehensive yet concise representation of the UAV's altimetric state.

The feature selection process aligns with best practices in data analytics, ensuring that the model focuses on the most informative aspects of the UAV's operation while minimizing redundancy and noise in the input data. This refined dataset forms the foundation for developing a robust and efficient altitude prediction model.

3.2.3. Elbow and K-Mean Method

Figure 12 illustrates the application of the Elbow Method, a technique used in cluster analysis to determine the optimal number of clusters (K) in a dataset. The graph plots the Within–Cluster Sum of Squares (WCSS) against the number of clusters (K).

	Telemetry from drone	BARO.Alt	CTUN.DAlt	CTUN.Alt	CTUN.BAlt	XKF5.HAGL
0	0.986126	2.017928	1.938615	6.670662	2.381847	2.430090
1	4.158758	1.222261	1.563349	4.153438	2.039141	3.059479
2	1.250687	1.499990	1.175041	1.445505	1.603899	2.741992
3	2.352219	2.946180	3.313249	2.323771	2.040426	3.133715
4	3.037643	0.597440	2.144905	1.860795	1.618477	2.387396

	Target
0	1.112000
1	1.115610
2	1.119221
3	1.122831
4	1.126442

Python-based analysis of the elbow method yielded K=4 as the optimal number of clusters for the UAV flight log dataset.

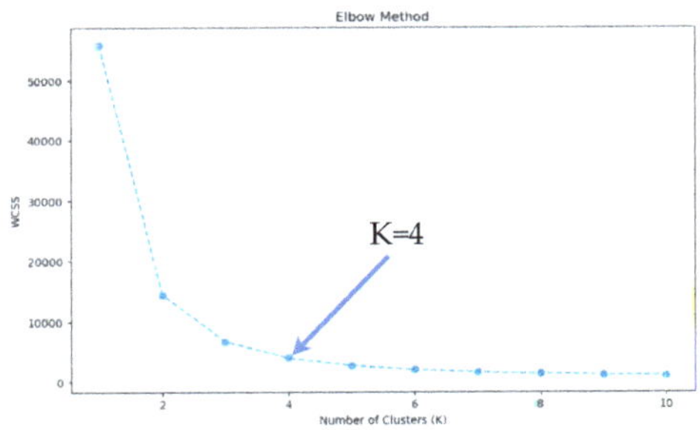

Figure 12. Elbow method analysis of UAV telemetry data demonstrating optimal K=4 clusters based on WCSS (Within-Cluster Sum of Squares) metric and corresponding flight parameters.

The x-axis represents the number of clusters, ranging from 1 to 10, while the y-axis denotes the WCSS, which is a measure of the compactness of the clusters. As the number of clusters increases, the WCSS generally decreases.

The plot exhibits a characteristic "elbow" shape, where the rate of decrease in WCSS slows significantly after a certain point. This inflection point is typically considered the optimal number of clusters, as adding more clusters beyond this point yields diminishing returns in terms of explaining the variance in the data.

In this specific graph, the elbow appears to occur at K = 4, as indicated by the arrow and label. This suggests that four clusters may be the optimal choice for this dataset, balancing between model complexity and explanatory power.

The use of a dashed blue line connecting the data points aids in visualizing the trend, while the solid blue arrow emphasizes the location of the elbow. This graphical representation serves as a valuable tool for researchers in determining the most appropriate number of clusters for their analysis.

3.3. Preparation of Training and Testing Datasets

The model training was developed using a comprehensive dataset comprising 8000 instances, each characterized by six distinct features. This dataset served as the foundation for our approach, which integrated both supervised and unsupervised learning methodologies. High-precision measurements obtained from a Digital Laser Distance Meter were utilized as ground truth for target values, ensuring the model's accuracy and reliability.

To optimize model performance and ensure robust generalization, we implemented a rigorous training strategy encompassing the following key elements:

Data Preprocessing and Partitioning: The raw data underwent meticulous cleaning and normalization procedures to mitigate potential biases and ensure consistency across features. Subsequently, the dataset was partitioned into training (80%) and testing (20%) subsets, with careful attention paid to maintaining the distribution of target values across both sets.

Loss Function and Regularization: Mean Squared Error (MSE) was selected as the primary loss function due to its demonstrated effectiveness in regression tasks. To mitigate overfitting and enhance model generalization, L1 regularization was applied. This technique promotes sparsity in model parameters, effectively reducing model complexity while preserving predictive power.

Early Stopping Mechanism: An early stopping protocol was implemented to further prevent overfitting and optimize computational resources. This approach involves monitoring the model's performance on a validation set during training and halting the process when no significant improvement is observed over a predefined number of iterations.

Model Evaluation and Testing: The model's performance was rigorously assessed using the held-out test set, employing metrics such as Mean Absolute Error (MAE), Mean Squared Error (MSE), and R-squared (R^2) to provide a comprehensive evaluation of predictive accuracy and generalization capability.

This methodical training approach was designed to develop a robust model capable of accurate altitude predictions across diverse operational scenarios. The resulting model aims to significantly enhance UAV navigation and control systems, contributing to the advancement of autonomous aerial vehicle technology.

3.4. DL-KMA

The DL-KMA was developed from scratch to enhance altitude estimation accuracy for lightweight UAVs (under 2 kg) without cameras. This innovative approach integrates both unsupervised and supervised learning techniques, combining K-Means Clustering with a deep learning regression model.

Data Preprocessing and Normalization:

The model was implemented entirely from scratch using Python, allowing for customized optimization tailored to UAV altitude estimation requirements.

Altitude Range Clustering:

To address variations in altitude estimation accuracy across different flight conditions, the dataset was divided into four distinct altitude ranges using K-Means Clustering.

This clustering approach allows for targeted altitude estimation across different flight scenarios, enhancing the model's versatility and applicability.

Each dataset, comprising about 2000 instances with six distinct features, underwent thorough cleaning and normalization. The normalization process, crucial for ensuring consistency across features, was performed using the min-max normalization technique as described in Section 2.2 Equation (7).

In this equation, the dataset represents the training and testing input and output datasets, and Dataset_{min} and Dataset_{max} are the minimum and maximum values, respectively.

Model Architecture:

The model architecture consists of:

1. Input Layer: Accepting six selected features derived from UAV flight logs.
2. Hidden Layers: Two hidden layers, each containing 100 nodes, with hyperbolic tangent (tanh) activation functions.
3. Output Layer: A single neuron with ReLU activation for altitude prediction.

Training Process:

The dataset was split into 80% for training and 20% for testing, maintaining the distribution of target values. The training process involved the following:

1. Loss Function: Mean Squared Error (MSE) was used as the primary loss function.
2. Regularization: L1 regularization was applied to prevent overfitting.
3. Optimization: A gradient descent iterative algorithm was used to fine-tune the weight matrices (W) and bias vectors (B).
4. Early Stopping: Implemented to prevent overfitting and optimize training time.

The training was conducted over 3000 epochs with a learning rate (α) of 0.01.

Altitude Compensation Mechanism:

An altitude compensation mechanism was incorporated to adjust the model's predictions based on the identified cluster from the K-Means algorithm. This enhances accuracy within particular altitude ranges or flight conditions.

Experimental Validation:

The model was validated using real-world data collected across two distinct geographical locations in Thailand, ensuring diverse environmental conditions. Data collection spanned from 08:30 to 18:00, capturing a full range of daily weather variations characteristic of tropical climates.

The response time of the DL-KMA model was evaluated using a MacBook Pro (13-inch, Mid 2012) equipped with a 2.5 GHz Intel Core i5 processor. The computational performance, implemented in Python programming language, produced the approximate response times as shown in Table 1 below. These timing metrics serve as baseline indicators for future real-world implementations and provide insights into the model's practical deployment capabilities in UAV applications.

Table 1 Show Response time per input of model DL-KMA in each cluster specifications for the MacBook Pro 2012 models:

Table 1. Performance comparison of K-means clustering (K = 1–4) showing dataset sizes, response times, and processing efficiency for UAV telemetry data analysis.

Cluster	Datasets (Before Normalize)	Response Time (msec)	Response Time/Input
K = 1	1998 rows	84,000	42.04 ms
K = 2	1978 rows	84,000	42.76 ms
K = 3	1994 rows	85,000	42.62 ms
K = 4	2030 rows	89,000	43.84 ms

- Processor: 2.5 GHz Intel Core i5 (Ivy Bridge)
- Memory: DDR3 RAM (upgradeable)
- Storage: Traditional hard drive with option for SSD
- Display: 13.3-inch standard display
- Ports: USB 3.0, Thunderbolt
- Optical Drive: DVD SuperDrive

This comprehensive approach, combining advanced AI techniques with rigorous real-world testing, aims to significantly enhance UAV navigation and control systems, contributing to the advancement of autonomous aerial vehicle technology.

4. Results and Discussion

The implementation of our DL-KMA with altitude compensation demonstrated significant improvements in altitude estimation accuracy for lightweight UAVs. The results are presented in terms of key performance metrics and comparative analysis with traditional methods.

4.1. Cluster-Specific Performance

The K-means clustering approach facilitated targeted altitude estimation across diverse flight scenarios by partitioning data into homogeneous subgroups. This method enabled the identification of latent patterns within flight data, allowing for the optimization of altitude estimation algorithms tailored to specific conditions. Cluster-specific performance metrics elucidated the strengths and limitations of estimation methodologies across different scenarios, informing the selection of appropriate models or parameters for each cluster to achieve optimal results. This approach enhanced the understanding of altitude estimation performance in varied flight contexts.

4.2. Model Performance Metrics

Figure 13 illustrates the Model Performance Metrics for DL-KMA in altitude estimation. We evaluated the algorithm across K values (1–4) using Mean Squared Error (MSE),

Mean Absolute Error (MAE), and coefficient of determination (R^2), optimizing clustering configurations for accurate drone altitude determination.

This rigorous assessment framework enabled a comprehensive analysis of the algorithm's efficacy in altitude estimation across various clustering configurations, providing insights into the optimal parameterization for drone altitude determination.

This systematic evaluation across the four pre-clustered datasets reveals valuable insights into the model's performance characteristics. The progressive improvement in performance metrics from the first to the fourth cluster demonstrates the varying degrees of pattern complexity and predictability within each data segment. These findings highlight the importance of understanding cluster-specific performance in drone altitude estimation systems, as different data segments may exhibit distinct characteristics that influence model accuracy.

K = 1: MSE = 0.072, MAE = 0.013, R^2 = 0.999

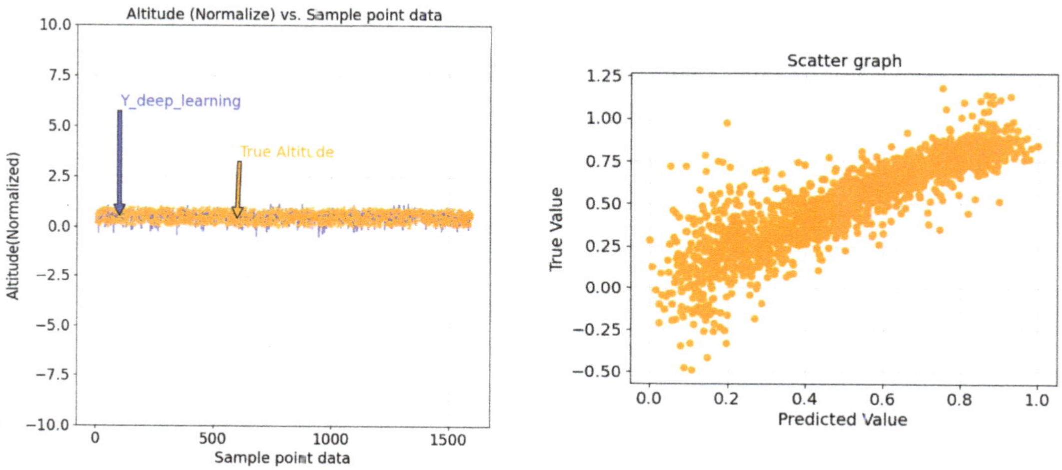

K = 2: MSE= 0.034, MAE = 0.116, R^2 = 0.999

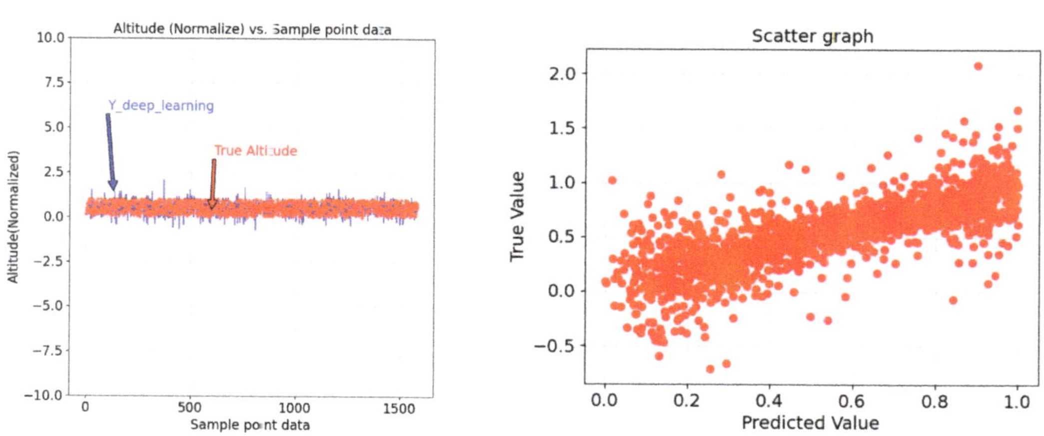

Figure 13. *Cont.*

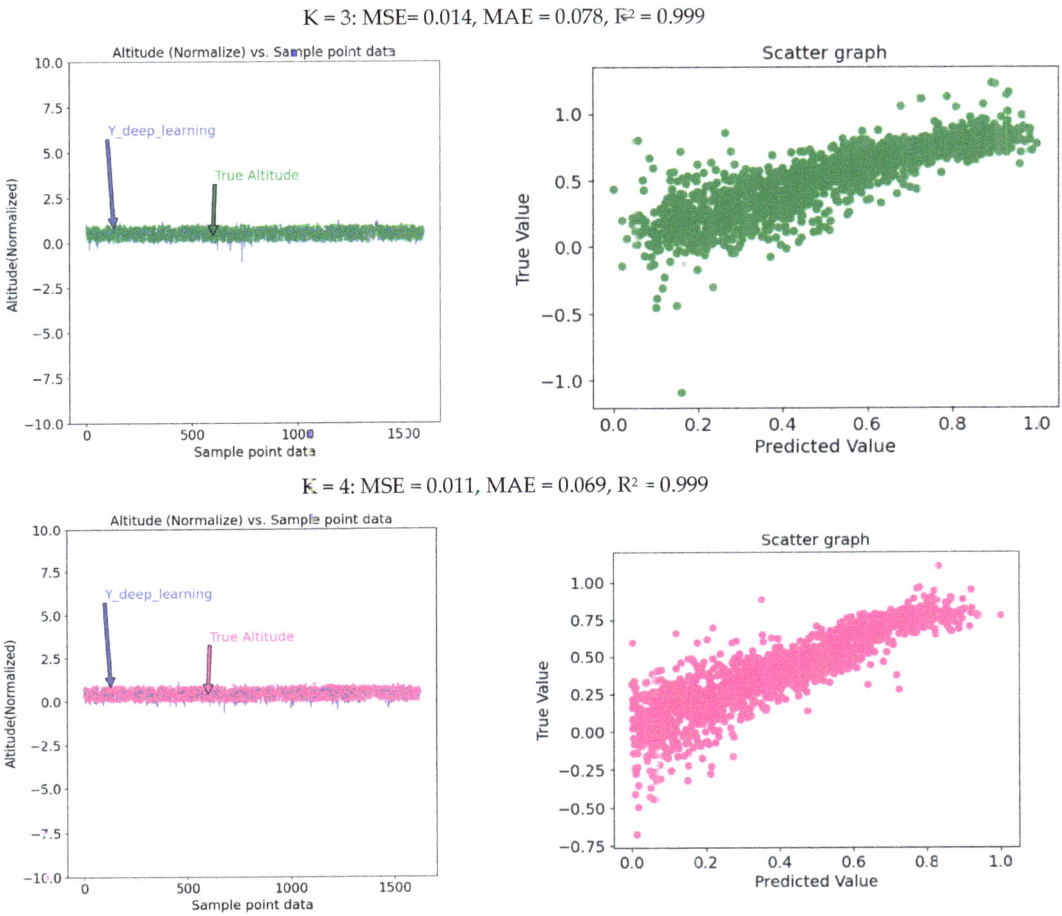

Figure 13. Model Performance Metrics for DL-KMA in altitude estimation.

The comprehensive analysis across all clusters provides crucial insights into the DL-KMA model's robustness in handling diverse altitude estimation scenarios, with each cluster potentially representing different flight conditions or altitude ranges within the drone's operational envelope.

For the first cluster (K = 1), the model demonstrated baseline performance with an MSE of 0.072 and MAE of 0.013. This evaluation using the first clustered dataset established a fundamental benchmark for altitude estimation patterns, with the relatively higher MSE indicating potential variations within this specific data partition.

When evaluating the second cluster (K = 2), we observed substantial improvement in model accuracy, with MSE decreasing to 0.034 and MAE to 0.116. This analysis of the second data cluster revealed enhanced prediction capabilities, suggesting that the characteristics of altitude-related patterns within this group were more conducive to precise height estimations.

The third cluster analysis (K = 3) yielded further improvements with an MSE of 0.014 and MAE of 0.078. This evaluation demonstrated the model's enhanced capability in capturing altitude-specific patterns within this data segment, effectively optimizing the balance between model complexity and performance for this cluster's characteristics.

Most significantly, the fourth cluster evaluation (K = 4) achieved optimal performance metrics, recording the lowest MSE of 0.011 and an improved MAE of 0.069. This final cluster's analysis proved most effective in processing altitude-related data patterns, suggesting that the inherent characteristics of this data segment were particularly well-suited for precise altitude estimation. Our results highlight the potential of integrating K-means clustering with Deep Learning regression models in the drone industry, particularly for precise height estimation tasks. This approach could significantly enhance the accuracy of drone-based measurements in various applications, including terrain mapping, building inspection, and environmental monitoring.

Furthermore, this study contributes to the broader discourse on data preprocessing techniques in machine learning, emphasizing the tangible benefits of thoughtful dataset partitioning. The marked improvement in model performance achieved through optimal clustering underscores the value of this methodology in refining predictive algorithms for drone-related applications.

In conclusion, our findings not only demonstrate the efficacy of the proposed model, but also provide valuable insights into the optimization of Deep Learning algorithms for height estimation in drone technology. The analysis of different K values reveals the importance of finding the right balance in data segmentation, with K = 4 emerging as the optimal choice for this dataset and application. This research paves the way for further exploration of advanced data segmentation techniques to enhance the precision and reliability of drone-based measurements across diverse operational contexts.

4.3. Comparative Analysis

Our proposed DL-KMA for UAVs height estimation represents a significant advancement over current methods. While traditional techniques like barometric sensors, GPS, LiDAR, and computer vision each have their merits, they also face limitations such as weather sensitivity, signal obstruction, high costs, or varying accuracy.

Based on recent literature, the following is a detailed comparison that could be added to the manuscript for a more in-depth analysis of altitude estimation accuracy between the developed DL-KMA model and traditional sensors like barometers, LiDAR, GPS, and flow sensors:

Barometric sensors, while cost-effective and easily integrated, offer altitude accuracy between ±0.5 and ±2 m, with controlled methods reaching up to ±0.05 m. In comparison, the DL-KMA model achieves significantly higher precision, with a Mean Squared Error (MSE) of 0.011, Mean Absolute Error (MAE) of 0.013, and R^2 value of 0.999. This level of accuracy, less sensitive to atmospheric fluctuations, demonstrates the DL-KMA model's robustness and reliability, surpassing standard barometric sensors in providing stable altitude estimates under varying conditions.

LiDAR Sensors: LiDAR offers centimeter-level accuracy but is generally unsuitable for lightweight UAVs due to its power and weight requirements. The LiDAR provides an accuracy of about ±0.1 m, making it highly precise but impractical for low-cost or weight-sensitive applications. The DL-KMA model, achieving an MSE of 0.011, offers competitive accuracy without the added hardware demands of LiDAR.

GPS-Based Altitude Estimation: GPS is effective for general outdoor altitude estimation but can have deviations of 1–3 m in dense or obstructed areas due to satellite signal limitations. The DL-KMA model, which integrates multi-sensor data, achieves a stable altitude estimation with an R^2 value of 0.999, making it more reliable in GPS-limited environments.

Optical Flow Sensors: Suitable for indoor or low-altitude flight, optical flow sensors can detect altitude changes without GPS. However, their accuracy is reduced on low-texture surfaces or in environments with rapid motion. The DL-KMA model, which does not rely solely on optical flow, addresses this limitation by integrating additional sensor inputs, making it more adaptable and robust in varied conditions.

Table 2 Comparative Analysis of DL-KMA System versus Traditional Hardware-Based Approaches for UAV Height Estimation.

Table 2. Performance Comparison between DL-KMA and Traditional LiDAR Systems for UAV Applications.

Comparison Aspects	Proposed System (DL-KMA)	Traditional Hardware System (LiDAR)
Cost	USD (No additional hardware installation required, uses existing sensors in FCU, cost-effective in terms of equipment)	Based on our analysis, 3 LiDAR models were identified that meet or closely align with the following research specifications: 1. 399 USD: Benewake TF03-100 LiDAR 2. 7700 USD: DJI Zenmuse L1 3. 1599 USD: Livox AVIA
Computational Cost	Software-based processing Minimal computational overhead	- Requires LiDAR data processing, high power consumption during operation
Response Time	Approximately 42.815 ms	*LiDAR Sensors:* • Response time: 0.02–0.1 ms • Common drone LiDAR sensors like TFmini Plus have update rates of 1000 Hz (1 ms response time) *GPS Receivers:* • Update rate: 1–10 Hz (typical for drone applications) • Response time: 100–1000 milliseconds • Note: Modern GPS modules optimized for UAVs can achieve higher rates up to 20 Hz (50 ms) *Barometric Pressure Sensors:* • Response time: 5–10 milliseconds • Common drone barometers (like BMP388) have response times around 5.5 ms • Update rates typically 50–200 Hz
Error (Accuracy)	- MSE = 0.011 ≈ ±0.105 m -MAE = 0.069 ≈ ±0.069 m (Accuracy comparable to Digital Laser distance meter [±3 mm])	- LiDAR: Accuracy ±0.1 m - Barometric: ±0.5 to ±2 m - GPS: Error margin 1–3 m
Limitations	- Requires model training - Depends on training data quality	- Increased equipment weight, higher power consumption, not suitable for lightweight UAVs

By adding these comparative results to the manuscript, it emphasizes the advantages of the DL-KMA model, especially in terms of flexibility, hardware efficiency, and accuracy across different environments, which is particularly beneficial for lightweight UAVs. This detailed comparison helps readers see the practical relevance of the DL-KMA approach across various UAV applications. By integrating multiple data inputs and leveraging K-means clustering, the model achieves unparalleled adaptability and robustness. It mitigates the weaknesses of single-sensor systems, potentially reducing costs compared to high-end LiDAR while maintaining accuracy. This innovation promises to enhance UAV operations across various applications, from precision agriculture to urban planning, by providing more reliable and precise height estimations in real-time.

4.4. Data Collection Limitations and Challenges

The process of gathering data in real-world environments presented the following significant challenges:

Battery Constraints: Each flight was limited to 15 min, with a 75–90 min charging interval between flights. This restriction allowed for only six data collection sessions per day (08:30–18:00), highlighting the challenges in accumulating comprehensive and sufficient data.

Altitude Measurement Precision: The digital laser distance meter's narrow laser beam (7 mm radius) posed difficulties in targeting UAVs at altitudes exceeding 10 m, particularly when confronted with visibility limitations and unpredictable wind conditions at higher altitudes.

Varying Visibility Conditions: The midday period (11:40–14:50) presented the most challenging visibility due to intense sunlight, affecting the ability to maintain precise UAV positioning.

This real-world data collection approach significantly enhances the reliability and applicability of our model. It ensures that our altitude estimation techniques are robust across a diverse range of actual operational scenarios, including daily weather fluctuations, seasonal weather patterns characteristic of tropical climates, and various altitude levels and flight patterns.

The integration of multi-sensor data with this comprehensive real-world dataset enables more accurate and reliable altitude predictions, particularly in challenging environments. The model's ability to adapt to different flight situations through clustering enhances its robustness and applicability across diverse operational contexts.

4.5. *Environmental Impact Analysis on Model Accuracy*

Our research deliberately incorporated data from diverse environmental conditions to ensure model robustness. Data collection was conducted as follows across two distinct locations with significant environmental variations:

Geographical Diversity:

- Area A (Muak Lek, Saraburi): elevation approximately 430 m
- Area B (Bangkok): elevation approximately 3 m

Environmental Challenges:

- Varying wind conditions at different altitudes
- Atmospheric pressure differences due to elevation disparities
- Diverse lighting conditions throughout the day (08:30–18:00)
- Critical midday period (11:40–14:50) with intense sunlight

A key strength of our approach lies in the intentional combination of datasets from these varying conditions. The DL-KMA model processes these environmental variations implicitly through the input features, without explicit knowledge of the environmental conditions during data collection. This design choice enhances the model's robustness and generalization capabilities.

The optimized model demonstrates remarkable consistency in accuracy (MSE: 0.011, R^2: 0.999) across the following:

- Different elevations and atmospheric pressures
- Various times of day
- Changing wind conditions
- Diverse lighting conditions

This performance validates our hypothesis that a well-designed deep learning model can effectively handle environmental variability without requiring explicit environmental condition inputs, making it more practical for real-world applications.

5. Conclusions

This study presents significant advancements in UAVs altitude estimation through the development of a custom Deep Learning architecture implemented entirely from scratch. The research introduces an innovative approach by integrating K-means Clustering for data segmentation, enabling targeted altitude estimation across diverse flight conditions. This integration enhances the model's adaptability to various environmental factors and flight patterns, significantly improving its versatility and accuracy.

The model underwent comprehensive real-world validation across a spectrum of geographical and environmental conditions, ensuring its robustness and applicability in practical scenarios. Notably, this solution eliminates the necessity for additional costly hardware such as LiDAR, offering a cost-effective, software-based alternative for precise altitude estimation in lightweight UAVs.

The proposed model exhibits exceptional performance metrics. In its optimal configuration, it achieved a Mean Squared Error (MSE) of 0.011, a Mean Absolute Error (MAE) of 0.013, and a coefficient of determination (R^2) of 0.999, underscoring the model's high accuracy and reliability. These findings collectively represent a significant advancement in UAV altitude estimation techniques, offering both theoretical insights and practical applications for the drone industry, particularly in scenarios where lightweight and cost-effective solutions are crucial.

The incorporation of K-means Clustering represents a significant advancement in UAVs altitude estimation technology. By segmenting the altitude data into distinct clusters, we enable the model to recognize and adapt to specific patterns within each altitude range. This approach not only improves overall accuracy but also enhances the model's ability to handle varying atmospheric conditions and flight dynamics unique to different altitude brackets.

Our research demonstrates the efficacy of combining advanced machine learning techniques with traditional sensor data to overcome the limitations of conventional altitude estimation methods. The model's performance metrics underscore its remarkable accuracy and consistency across diverse operational scenarios. These results significantly surpass traditional methods such as barometric sensors, GPS, and computer vision techniques in terms of accuracy, consistency, and adaptability to varying environmental conditions.

Nonetheless, this study also highlights areas for future research. The challenges encountered in data collection, particularly related to battery constraints and visibility issues at higher altitudes, underscore the need for further technological advancements in UAVs hardware. Future studies could explore the development of more energy-efficient UAVs systems or alternative power sources to extend flight times and data collection capabilities. Additionally, the integration of our DL-KMA model with other onboard systems, such as obstacle avoidance or navigation algorithms, could lead to more comprehensive and robust UAVs control systems.

Furthermore, while our model demonstrates excellent performance in the tested scenarios, future research opportunities exist to enhance the model's capabilities under extreme conditions. Regarding battery limitations, subsequent studies could investigate energy-efficient systems, particularly autonomous charging systems for extended operations, as proposed by Barrile et al. (2024) [28]. The DL-KMA model shows significant potential for specialized applications, such as precision agriculture monitoring, as demonstrated by Xu et al. (2024) [29] and Raj et al. (2024) [30] in their crop height estimation research. Additionally, Han (2024) [31] further validated the effectiveness of UAV remote sensing in agricultural applications. Moreover, the model could be adapted for urban infrastructure inspection applications.

In conclusion, this research successfully achieved its primary objective of enhancing UAV flight control through the development of the DL-KMA model, demonstrating significant improvements in altitude estimation accuracy. The experimental results validated our initial hypothesis that artificial intelligence-driven approaches could optimize critical flight parameters beyond traditional methods. As targeted in our research goals, the model proved its reliability across diverse operational conditions, aligning with practical applications such as disaster response scenarios, as explored by Chandran and Vipin (2024) [32] in their multi-UAV network architectures. This study fulfills its intended purpose of establishing a robust foundation for future autonomous aerial systems, particularly advancing our original aim of enhancing lightweight drone applications in our increasingly UAV-dependent society.

Author Contributions: Conceptualization, P.P. (Pattarapong Phasukkit); methodology, P.P. (Pattarapong Phasukkit); validation, P.P. (Pattarapong Phasukkit); formal analysis, P.P. (Prot Piyakawanich), P.P.; investigation, P.P. (Prot Piyakawanich) and P.P. (Pattarapong Phasukkit); writing—original draft P.P. (Prot Piyakawanich); funding acquisition, P.P. (Pattarapong Phasukkit). All authors have read and agreed to the published version of the manuscript.

Funding: King Mongkut's Institute Technology Ladkrabang (KMITL).

Institutional Review Board Statement: Not applicable.

Informed Consent Statement: Not applicable.

Data Availability Statement: Data available in a publicly accessible repository.

Acknowledgments: The authors extend their profound gratitude to the BURNLAB within the Electronics Department, School of Engineering at King Mongkut's Institute Technology Ladkrabang (KMITL).

Conflicts of Interest: The authors declare no conflicts of interest.

References

1. Poudel, G. Improved Object Detection in UAV Images using Deep Learning. *Cogniz. J. Multidiscip. Stud.* **2024**, *4*, 83–109. [CrossRef]
2. Qin, P.; Wu, X.; Cai, Z.; Zhao, X.; Fu, Y.; Wang, M. Joint Optimization of UAV's Flying Altitude and Power Allocation for UAV-Enabled Internet of Vehicles. *IEEE Trans. Intell. Transp. Syst.* **2023**, *10*, 3421–3434.
3. Pesci, A.; Teza, G.; Fabris, M. Editorial of Special Issue Unconventional Drone-Based Surveying. *Drones* **2023**, *7*, 175. [CrossRef]
4. Gao, F.; Wang, Z.; Liu, X.; Liu, S. UAV Position Optimization for Servicing Ground Users Based on Deep Reinforcement Learning. *J. Phys. Conf. Ser.* **2024**, *2861*, 012011. [CrossRef]
5. Wang, W.; Chen, H.; Zhang, X.; Zhou, W.; Shi, W. Aprus: An Airborne Altitude-Adaptive Purpose-Related UAV System for Object Detection. In Proceedings of the 2022 IEEE 24th International Conference on High Performance Computing & Communications; 8th International Conference on Data Science & Systems; 20th International Conference on Smart City; 8th International Conference on Dependability in Sensor, Cloud & Big Data Systems & Application (HPCC/DSS/SmartCity/DependSys), Hainan, China, 18–20 December 2022. [CrossRef]
6. Sanjar, S.; Sulaym, S. Optimal Deep Learning-Based Image Classification for IoT-Enabled UAVs in Remote Sensing Applications. *Int. J. Adv. Appl. Comput. Intell.* **2024**, *6*, 1–12. [CrossRef]
7. Yildirim, E.O.; Sefercik, U.G.; Kavzoglu, T. Automated identification of vehicles in very high-resolution UAV orthomosaics using YOLOv7 deep learning model. *Turk. J. Electr. Eng. Comput. Sci.* **2024**, *32*, 144–165. [CrossRef]
8. Panthakkan, A.; Mansoor, W.; Al Ahmad, H. Accurate UAV-Based Vehicle Detection: The Cutting-Edge YOLOv7 Approach. In Proceedings of the 2023 International Symposium on Image and Signal Processing and Analysis (ISPA), Rome, Italy, 18–19 September 2023. [CrossRef]
9. Zeng, H.; Li, J.; Qu, L. Lightweight Low-Altitude UAV Object Detection Based on Improved YOLOv5s. *Int. J. Adv. Netw. Monit. Control* **2024**, *9*, 87–99. [CrossRef]
10. Chen, Z.; Li, J.; Li, Q.; Dong, Z.; Yang, B. DeepAAT: Deep Automated Aerial Triangulation for Fast UAV-based mapping. *arXiv* **2024**, arXiv:2402.01134. [CrossRef]
11. Wang, C.; Li, Z.; Gao, Q.; Cui, T.; Sun, D.; Jiang, W. Lightweight and Efficient Air-to-Air Unmanned Aerial Vehicle Detection Neural Networks. In Proceedings of the 2023 IEEE International Conference on Unmanned Systems (ICUS), Hefei, China, 13–15 October 2023. [CrossRef]
12. Makrigiorgis, R.; Kyrkou, C.; Kolios, P. How High can you Detect? Improved accuracy and efficiency at varying altitudes for Aerial Vehicle Detection. In Proceedings of the 2023 International Conference on Unmanned Aircraft Systems (ICUAS), Warsaw, Poland, 6–9 June 2023. [CrossRef]
13. Makrigiorgis, R.; Kyrkou, C.; Kolios, P. *Multi-Altitude Aerial Vehicles Dataset (Version 1.0)*; Zenodo: Geneva, Switzerland, 2023. [CrossRef]
14. Cheng, Q.; Wang, Y.; He, W.; Bai, Y. Lightweight air-to-air unmanned aerial vehicle target detection model. *Sci. Rep.* **2024**, *14*, 2609. [CrossRef] [PubMed]
15. Taame, A.; Lachkar, I.; Abouloifa, A.; Mouchrif, I. UAV Altitude Estimation Using Kalman Filter and Extended Kalman Filter. In *Automatic Control and Emerging Technologies, Proceedings of ACET 2023, Kenitra, Morocco, 11–13 July 2023*; El Fadil, H., Zhang, W., Eds.; Lecture Notes in Electrical Engineering; Springer: Singapore, 2024; p. 1141. [CrossRef]
16. Yang, Z.; Xie, F.; Zhou, J.; Yao, Y.; Hu, C.; Zhou, B. AIGDet: Altitude-Information-Guided Vehicle Target Detection in UAV-Based Images. *IEEE Sens. J.* **2024**, *24*, 22672–22684. [CrossRef]
17. Aslani, R.; Saberinia, E. Joint Power Control and Altitude Planning for Energy-Efficient UAV-Assisted Vehicular Networks. In *Advances in Systems Engineering*; Selvaraj, H., Chmaj, G., Zydek, D., Eds.; Springer: Cham, Switezerland, 2023; p. 761. [CrossRef]

18. Li, Z.; Jiang, X.; Ma, S.; Ma, X.; Lv, Z.; Ding, H.; Ji, H.; Sun, Z. Expediting the Convergence of Global Localization of UAVs through Forward-Facing Camera Observation. *Drones* **2024**, *8*, 335. [CrossRef]
19. Zhai, W.; Li, C.; Cheng, Q.; Mao, B.; Li, Z.; Li, Y.; Ding, F.; Qin, S. Fei, S.; Chen, Z. Enhancing Wheat Above-Ground Biomass Estimation Using UAV RGB Images and Machine Learning: Multi-Feature Combinations, Flight Height, and Algorithm Implications. *Remote Sens.* **2023**, *15*, 3653. [CrossRef]
20. Du, L.; Liang, Y.; Mian, I.A.; Zhou, P. K-Means Clustering Based on Chebyshev Polynomial Graph Filtering. In Proceedings of the ICASSP 2024 IEEE International Conference on Acoustics, Speech and Signal Processing (ICASSP), Seoul, Republic of Korea, 14–19 April 2024. Available online: https://ieeexplore.ieee.org/document/10446384 (accessed on 24 November 2024).
21. Ren, H.; Pan, C.; Wang, K.; Deng, Y.; Elkashlan, M.; Nallanathan, A. Achievable Data Rate for URLLC-Enabled UAV Systems With 3-D Channel Model. *IEEE Wirel. Commun. Lett.* **2019**, *8*, 1587–1590. [CrossRef]
22. Pritzl, V.; Vrba, M.; Tortorici, C.; Ashour, R.; Saska, M. Adaptive estimation of UAV altitude in complex indoor environments using degraded and time-delayed measurements with time-varying uncertainties. *Robot. Auton. Syst.* **2022**, *156*, 104315. [CrossRef]
23. Kolarik, J.; Lyng, N.L.; Bossi, R.; Li, R.; Witterseh, T.; Smith, K.M.; Wargocki, P. Application of Cluster Analysis to Examine the Performance of Low-Cost Volatile Organic Compound Sensors. *Buildings* **2023**, *13*, 2070. [CrossRef]
24. Liu, Y.; Han, K.; Rasdorf, W. A Comparison of Accuracy of UAS Photogrammetry in Different Terrain Sites. In *Construction Research Congress 2024*; American Society of Civil Engineers: Reston, VA, USA, 2024; pp. 367–376. [CrossRef]
25. Humaira, H.; Rasyidah, R. Determining the Appropiate Cluster Number Using Elbow Method for K-Means Algorithm. In Proceedings of the 2nd Workshop on Multidisciplinary and Applications (WMA) 2018, Padang, Indonesia, 24–25 January 2018; EAI: Gent, Belgium, 2020. [CrossRef]
26. Khalaf, A.Z.; Alyasery, B.H. Laser Distance Sensors Evaluation for Geomatics Researches. *Iraqi J. Sci.* **2020**, *61*, 1831–1841. [CrossRef]
27. Taddia, Y.; Stecchi, F.; Pellegrinelli, A. Coastal Mapping Using DJI Phantom 4 RTK in Post-Processing Kinematic Mode. *Drones* **2020**, *4*, 9. [CrossRef]
28. Barrile, V.; La Foresta, F.; Genovese, E. Optimizing Unmanned Aerial Vehicle Electronics: Advanced Charging Systems and Data Transmission Solutions. *Electronics* **2024**, *13*, 3208. [CrossRef]
29. Xu, T.; Damron, E.; Silvestri, S. High-Precision Crop Monitoring Through UAV-Aided Sensor Data Collection. In Proceedings of the ICC 2024—IEEE International Conference on Communications, Denver, CO, USA, 20 August 2024. [CrossRef]
30. Raj, A.S.; Karthekeyan, S.G.; Kaviyarasu, A.; Henrietta, H.M. Revolutionizing Precision Agriculture with Drone-Based Imaging and Fuzzy Intelligent Algorithms. *Qeios* 2024, *preprint*. [CrossRef]
31. Han, Y. Application of Unmanned Aerial Vehicle Remote Sensing for Agricultural Monitoring. E3S Web of Conferences. In Proceedings of the 2024 International Conference on Ecological Protection and Environmental Chemistry Budapest, Hungary, 21–23 June 2024. [CrossRef]
32. Chandran, I.; Vipin, K. Comparative Analysis of Stand-alone and Hybrid Multi-UAV Network Architectures for Disaster Response Missions. In Proceedings of the 2024 International Conference on Advancements in Power, Communication and Intelligent Systems (APCI), Kannur, India, 21–22 June 2024. [CrossRef]

drones

MDPI

Article

Usability Comparison between 2D and 3D Control Methods for the Operation of Hovering Objects

Daeseong Lee [1], Hajun Kim [2], Heesoo Yoon [3] and Wonsup Lee [4,*]

[1] Swen, Inc., Seoul 06178, Republic of Korea; dave@swencorp.com
[2] AI Research Center, Neowiz, Inc., Seongnam 13487, Republic of Korea; hajune28@neowiz.com
[3] Meissa, Inc., Seoul 06127, Republic of Korea
[4] School of Global Entrepreneurship and Information Communication Technology, Handong Global University, Pohang 37554, Republic of Korea
* Correspondence: w.lee@handong.edu; Tel.: +82-54-260-1511

Abstract: This paper experimentally analyzed the cognitive load of users based on different methods of operating hovering objects, such as drones. The traditional gamepad-type control method (2D) was compared with a control method that mapped the movement directions of the drone to the natural manipulation gestures of the user using a Leap Motion device (3D). Twenty participants operated the drone on an obstacle course using the two control methods. The drone's trajectory was measured using motion-capture equipment with a reflective marker. The distance traveled by the drone, operation time, and trajectory smoothness were calculated and compared between the two control methods. The results showed that when the drone's movements were mapped to the user's natural directional gestures, the drone's 3D movements were perceived as more natural and smoother. A more intuitive drone control method can reduce cognitive load and minimize operational errors, making it more user friendly and efficient. However, due to the users' lack of familiarity with Leap Motion, it resulted in longer distance and time and lower subjective satisfaction; therefore, a more improved 3D control method over Leap Motion is needed to address the limitations.

Keywords: hovering object; 3D movement; depth camera; hand–gesture interface; eye–hand coordination

1. Introduction

Vertical take-off and landing (VTOL) aircraft such as helicopters or drones move by using propulsion devices, such as propellers, while hovering in the air. Unlike fixed-wing aircraft, which gain propulsion in one direction, VTOL aircrafts can move in six directions (up, down, left, right, forward, and backward) or rotate around three axes (longitudinal, lateral, and vertical) while hovering in one place. The rotations of VTOL aircrafts along the axes are referred to as roll (rotation along the longitudinal axis), pitch (rotation along the lateral axis), and yaw (rotation along the vertical axis). VTOL aircrafts are typically operated through four controls [1–3]: (1) moving vertically by controlling the rotation speed of the propellers, (2) moving forward and backward by tilting the body through pitch rotation, (3) moving left and right by tilting the body through roll rotation, and (4) rotating around the vertical axis through yaw rotation. Since VTOL aircrafts can perform precise hovering in the air using propellers, they are effectively used for military operations, observation and surveillance activities, firefighting, and so forth [4]. In particular, as VTOL aircrafts with multiple propellers can achieve vertical take-off and landing by precisely controlling the propellers, which enables stable and smooth flight, they can quickly respond to emergency situations through a sudden stop or sharp turn; they are expected to be used as a means of transportation in urban air mobility (UAM), the next generation of urban transportation.

The existing method of controlling VTOL aircrafts, which moves in four ways (up–down movement, front–back movement, left–right movement, and yaw rotation) in 3D space, has nonintuitive eye–hand coordination because the mapping between the movement of the body and the interface is unnatural [5–8]. In the case of a helicopter, collective control (which can be controlled up and down with the left hand) is used to control the body up and down. Cyclic control (which can be controlled front–back and left–right with the right hand) is used to control the body front–back and left–right. Two rudder pedals are used to control yaw rotation on both feet. The movement directions of the helicopter are mapped to the arms and legs of the pilot, and the pilot needs to have good eye–hand coordination to control the helicopter's movements precisely while moving quickly in the air [9]. However, the problem of eye–hand coordination is more complex in the case of a drone because one of the main causes is the pilot's point of view. In the case of a helicopter, as the pilot is inside the aircraft, the direction of control by the pilot is the same as the direction of movement of the helicopter. However, drones are controlled by the operator from an external (third-person) perspective. Therefore, if the drone rotates its yaw to change direction, the control direction and the drone's movement direction become mismatched (e.g., when the drone rotates 90 degrees clockwise if the operator tries to move the drone forward by manipulating the joystick upwards, the drone moves to the right not forward) [10]. Usability issues related to eye–hand coordination based on the unnatural mapping between the drone's movement and the drone–control interface are reported, even excluding problems related to perspective. In general, a drone is controlled using a device having two joysticks by manipulating the left joystick with the left hand to move the drone in the forward, backward, leftward, and rightward directions; and manipulating the right joystick with the right hand to control the drone's altitude (by controlling the joystick upward and downward) and yaw rotation (by controlling the joystick leftward and rightward). Both joysticks are manipulated in the same manner (say, two-dimensional (2D) control), but the mapping of the upward, downward, leftward, and rightward directions of the joystick to the drone's movement is different between the two joysticks. For example, manipulating the left joystick to the left direction causes the drone to fly leftward while manipulating the right joystick to the left direction causes the drone to rotate in a counterclockwise direction. Therefore, due to the lack of intuitive mapping between the direction of the drone's movement and the way the controller is manipulated, cognitive errors can occur in controlling the drone and it takes a lot of practice time to develop eye–hand coordination [2,10–12].

The previous studies mainly focused on the interface for intuitive drone control but there is a lack of experimental studies that analyze the performance of drone control in terms of usability of drone–control interface (mapping between the drone's movements and the drone–control interface). To solve the problem that two-handed drone control is difficult, more intuitive one-handed-control controllers were studied from the aspect of eye–hand coordination [2,12,13] and related products have also been introduced in the market (e.g., Shift, this is engineering Inc., Seongnam, Republic of Korea; FT Aviator, Fluidity Technologies Inc., Houston, TX, USA). Previous studies [14–19] have proposed methods utilizing hand gestures to verify the assumption that one-handed controllers or hand-gesture-based control approaches can provide intuitive control over the up–down, front–back, and left–right movements of drones in three-dimensional (3D) space. However, there are few studies that quantitatively compare and analyze the usability difference of these intuitive drone-control methods compared to two-handed controllers.

This study compares the performance of drone control using two different methods by experimentally analyzing the effect of mapping between the manipulation and movement of a drone. This study implemented an interface that controls the drone to be intuitively controlled by hand gestures using Leap Motion (Ultraleap Inc., Mountain View, CA, USA), a hand-gesture recognition device. The usability difference between the Leap Motion (3D controller) and a generally used drone-control device (a gamepad-type device consisting of two joysticks; 2D controller) was compared from the perspective of the drone's movement trajectory with 20 participants. The drone's trajectory was measured using a motion analysis system and the distance traveled, the time of movement, and the smoothness of the trajectory were calculated using the trajectory data. Since the mapping between the direction of interface operation and the direction in which the drone moves becomes unnatural in the drone's yaw rotation [7,8,13], this study only analyzed the remaining three movements (up–down, front–back, and left–right) among the four movements of the drone.

2. Methods

2.1. Participants

The participants in the experiment were 20 university students (age range: 20 to 27), with 10 males and 10 females. To exclude the effect of previous drone piloting experience and proficiency on the experiment, participants who had no prior drone piloting experience were recruited.

2.2. Apparatus

This study used a program developed to control the drone using two controllers. The drone and controllers used for the experiment were a Tello (DJI Technology Co. Ltd., Shenzhen, Guangdong, China; Figure 1a), a DualShock 4 gamepad (Sony Corp., Tokyo, Japan; Figure 1b), and a Leap Motion hand-gesture recognition sensor (Figure 1c), respectively. The drone utilized in this study can hover in place precisely under windless and uncontrolled conditions using the Vision Positioning System (VPS) developed by the manufacturer. The VPS can operate effectively as long as the drone flies at speeds below 5 m/s and at altitudes above 3.5 ft (1 m), avoiding inappropriate surfaces like monochrome, highly reflective, transparent, or surfaces with identical repeating textures (e.g., tiles), and ensuring suitable ambient brightness levels between 10 lux and 100,000 lux, without drastic light changes. The drone-control program was developed using Gobot (https://github.com/hybridgroup/gobot/, accessed on 3 August 2023), which is a library based on the Go language and provides various functions for sensors and controllers. Using an in-house program, the drone was controlled with different input methods while issuing the same commands using the same algorithm. The gamepad (2D controller) was used to control the drone in the forward and backward directions and left and right directions while pressing the up, down, left, and right keys on the left joystick; and in the up and down directions (altitude) while pressing the up and down keys on the right joystick. Note that the drone's yaw rotation was not implemented. When the hand is released from the direction keys, the drone's speed decreases and eventually stops its movement (hovering). The Leap Motion device (3D controller) is used to control the drone by tracking the absolute position (x, y, and z coordinates) and gesture of the hand within a specific cuboid space (left–right: $-200 < x < 200$ mm, front–rear: $-200 < z < 200$ mm, and top–bottom: $150 < y < 350$ mm) at the top of the device. If the hand moves forward on the device, the z coordinate of the palm increases, making the drone move forward. The drone can be switched to hovering mode with the gamepad by releasing the hand from the joysticks, or with the Leap Motion device by moving the hand to the center area in the 3D space. Therefore, a fist gesture was used to immediately switch the drone to hovering mode anywhere on the device. Other actions such as takeoff and landing can be performed via keyboard commands. The maximum flight speed of the drone was set to a particular level (0.5 m/s = 6.25% of its maximum speed, 8 m/s) suitable for beginners to navigate obstacles.

The experiment was conducted by setting up three types of obstacle courses and collecting the drone's movement trajectory information using the OptiTrack motion-capture system (NaturalPoint Inc., Corvallis, OR, USA; Figure 1d), which is composed of 12 cameras at a sampling rate of 60 Hz. The obstacle course consisting of three sections was arranged in a rectangular formation (Figure 2) and the drone was flown clockwise once to pass through those three sections. The course consisted of the A section, which had the drone pass through rings in a straight line toward the 12 o'clock direction from the participant; the B section, where the drone flew toward the 3 o'clock direction from the end of the A section and passed through a high obstacle; the C section, where the drone flew toward the 6 o'clock direction from the end of the B section and avoided obstacles in a zigzag manner;

and, lastly, the section, which had no obstacles and brought the drone back to the starting position toward the 9 o'clock direction from the end of the C section. This study excluded yaw rotation, so the drone was flown to the right during the B section and backward during the C section. The 3D position of the drone was tracked using a motion-capture system with a reflective marker attached to the drone. The trajectory refers to the sequence of 3D positions (x, y, and z) of the drone that changes over time.

Figure 1. Apparatus of this study: (**a**) Drone (DJI's Tello), (**b**) Gamepad (Sony DualShock 4), (**c**) 3D controller (Leap Motion), and (**d**) setup of the experiment place with a 12-camera motion-capture system (OptiTrack).

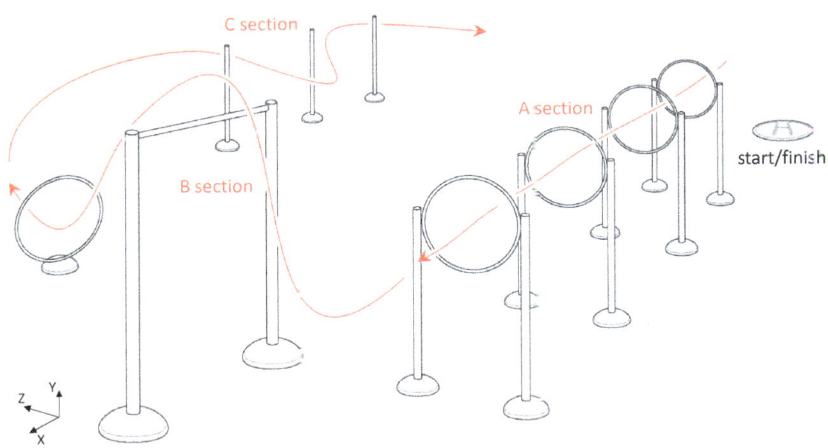

Figure 2. The obstacle course consists of three sections.

2.3. Experiment Design

The experiment was conducted by a four-step procedure (S1: introduction, S2: practice, S3: main experiment, and S4: debriefing). First, the purpose and process of the experiment were introduced to a participant and an informed consent was obtained. Second, the participant got familiar with the two drone-control methods for about 30 min, without flying the drone in the course during the practice. Third, the main experiment was conducted after the participants responded that they were familiar with the operation of the two controllers. The evaluation order of the two controllers was counterbalanced. Lastly, the participant responded to the survey on the subjective usability scores of each controller. Then, their participation was compensated.

2.4. Analysis Methods

This study divided the drone's trajectory data measured by the motion-capture system into three sections (A, B, C) and analyzed each section based on three performance aspects (movement distance, movement time, and trajectory smoothness) and subjective usability scores based on six aspects (controllability, intuitiveness, accuracy, mapping, learnability, and overall satisfaction) (Table 1) using a 7-point Likert scale (1: very dissatisfied, 4:

neutral, 7: very satisfied). The smoothness of the movement trajectory was defined as the cumulative angle change in each section divided by the movement distance. Among the participants, there were instances where the drone was temporarily hovering between sections and the data that was not related to drone operation was excluded from the analysis. The statistical differences between the two controllers for the three quantitative performance measures and the six subjective usability scores were analyzed using a paired *t*-test ($\alpha = 0.05$).

Table 1. Performance and usability measures for drone control.

Measure		Description
objective measures	movement distance	the accumulation of the Euclidean distances between the positions of the drone in consecutive frames
	movement time	the cumulative duration of the drone's movement, as calculated by the number of frames in which the drone moves
	trajectory smoothness	The level of smoothness of the drone's movement, as determined by the accumulated angle change divided by the distance traveled in each axis
subjective measures	controllability	the degree to which the drone can be controlled and directed by the operator in a precise manner.
	intuitiveness	the degree to which the operation of the drone is understandable to the operator while it is controlled
	accuracy	the degree to which the drone's movement matches the intended movement, as specified by the operator
	mapping	the degree to which the relationship between the operator's inputs and the drone's resulting movements is established and understood
	learnability	the degree to which the operator can acquire the skills necessary to control the drone effectively through experience or training
	overall satisfaction	the degree to which the operator is overall satisfied with the drone operation

3. Results

Figure 3 illustrates the trajectories of the drone measured by the motion-capture system when controlled by 20 participants using the gamepad (represented by blue lines) and the Leap Motion (represented by red lines). The top view is visualized for the A and C sections, while the front view is shown for the B section. In the A section, the Leap Motion controlled the drone to follow a straighter trajectory compared to the gamepad.

The average movement distances produced by the gamepad and Leap Motion differed slightly by section (Figure 4a and Table 2). In the A section, the average movement distance produced by the Leap Motion was significantly shorter (mean difference = 0.6 m) compared to that produced by the gamepad (*p*-value = 0.04). In contrast, the B section showed a significantly longer average movement distance (mean difference = 1.2 m) produced by the Leap Motion (*p*-value = 0.03). The C section also showed a longer average movement distance (mean difference = 0.3 m) produced by the Leap Motion, though the difference was not significant (*p*-value = 0.49). As for the average movement distance in a total of three sections, there was no significant difference observed (mean difference = 0.8 m, *p*-value = 0.35).

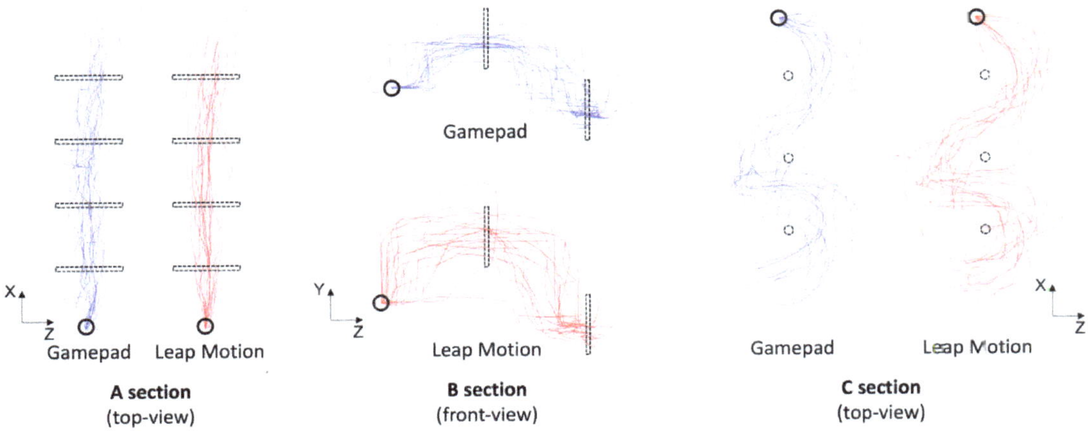

Figure 3. Movement trajectory of the drone controlled by the gamepad (blue) and the Leap Motion device (red) shown with obstacles (dotted line). In the B section, the Leap Motion exhibited smoother trajectory compared to the gamepad. Lastly, in the C section, there was no visually distinct difference between the Leap Motion and the gamepad.

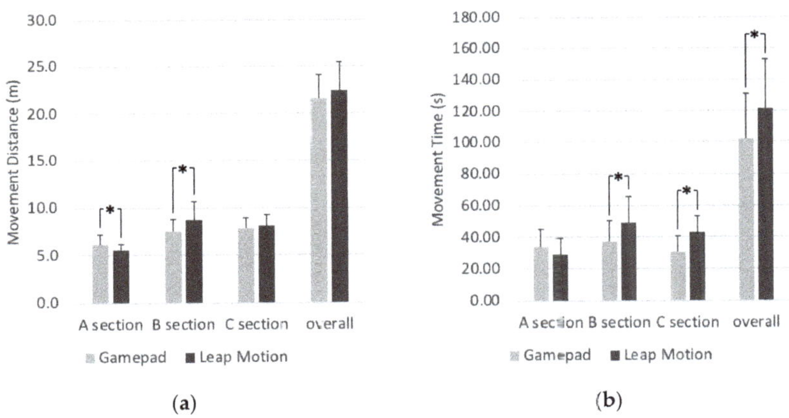

Figure 4. Performance analysis results in (**a**) movement distance and (**b**) movement time (* $p < 0.05$).

The results indicate that the Leap Motion had a longer average movement time compared to the gamepad (Figure 4b). In the A section, the average movement time produced by the Leap Motion was shorter (mean difference = 4.9 s) on average, though the difference was not statistically significant (p-value = 0.16). In contrast, the B and C sections showed a significantly longer movement time on average for the Leap Motion, with mean differences of 12.0 s and 12.4 s, respectively (p-value = 0.02 and p-value < 0.001). The overall average movement time across all sections also showed a longer duration for the Leap Motion compared to the gamepad (mean difference = 19.5 s; p-value = 0.048).

The results regarding the smoothness of the trajectory, as defined by the accumulated angle change divided by the movement distance, showed that the Leap Motion made different results by course (Figure 5 and Table 2). The trajectory smoothness shows a high value when the drone makes sharp turns. In the A and B sections, the Leap Motion produced a smoother trajectory on average with a smaller accumulated angle change. Significant differences were observed in the X- and Z-axes in the A section and the Z-axis in the B section. In contrast, in the C section, the Leap Motion showed a less smooth trajectory on average with a significant difference in the *Y*- and *Z*-axes compared to the gamepad. While there are no significant differences between the two methods, the overall average smoothness across all sections showed a smoother trajectory for the Leap Motion compared to the gamepad, except for the *Y*-axis. This was because two out of twenty participants using Leap Motion in the C section made unnecessary movements in the *Y*-axis. They were not treated as outliers, but their data were included in the results.

Table 2. Performance analysis results by measure.

	Measure		A Section		B Section		C Section	
			Mean Difference	*p*-Value	Mean Difference	*p*-Value	Mean Difference	*p*-Value
objective measures	movement distance		0.6	0.04	−1.2	0.03	−0.3	0.49
	movement time		4.9	0.16	−12.0	0.02	−12.4	<0.001
	trajectory smoothness	*X*-axis	194.8	0.02	97.7	0.14	−87.5	0.03
		Y-axis	178.8	0.13	83.4	0.13	−264.6	0.08
		Z-axis	167.5	0.04	81.4	0.03	−76.3	0.01
	Measure		**mean difference**				***p*-value**	
subjective measures	controllability		1.4				<0.001	
	intuitiveness		0.3				0.28	
	accuracy		1.0				0.03	
	mapping		1.4				<0.001	
	learnability		0.9				0.02	
	overall satisfaction		1.0				0.01	

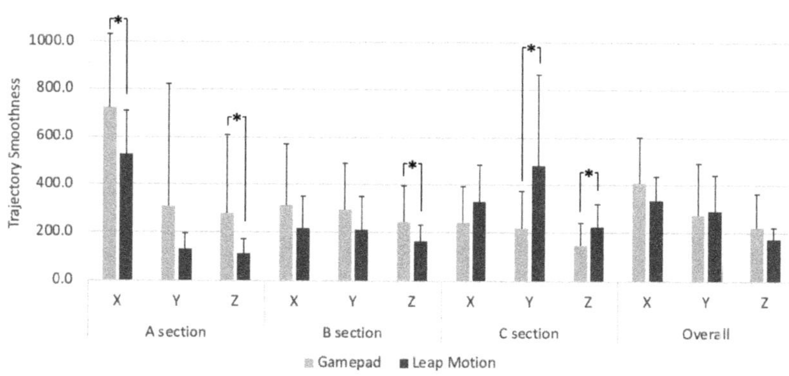

Figure 5. Performance analysis results in trajectory smoothness (* $p < 0.05$).

Lastly, regarding the subjective evaluation, the overall scores for the gamepad interface were 5% to 29% higher on average compared to the Leap Motion (Figure 6 and Table 2). The differences in scores were statistically significant at $\alpha = 0.05$ in most of the usability measures (controllability, accuracy, mapping, and learnability), except for intuitiveness.

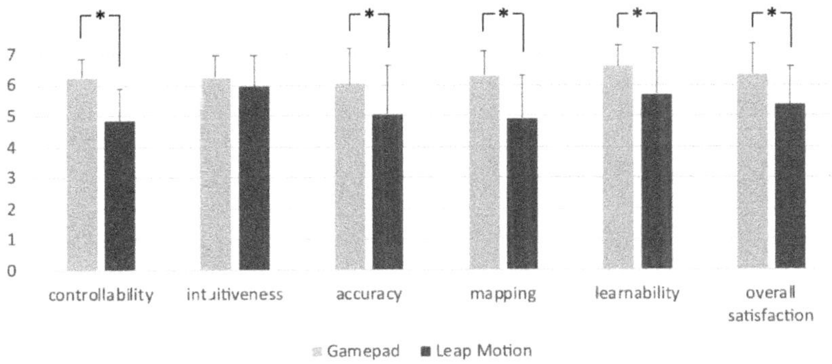

Figure 6. Subjective usability analysis results (* $p < 0.05$).

4. Discussion

This study aimed to compare the performance of two interfaces, the 2D (gamepad) and 3D (Leap Motion), in controlling a VTOL aircraft. The results showed that the 3D control method, which is an intuitive interface for controlling an object in three-dimensional space, was superior in several critical aspects. First, it can reduce the operator's cognitive fatigue by providing a more natural and ergonomic control experience, resulting in decreased cognitive errors. It may demonstrate a quicker response compared to the two-joystick interface. This highlights the ability of an intuitive interface to respond rapidly to changing circumstances, which is particularly important for ensuring the safe operation of a VTOL aircraft. Second, the 3D method can provide improved performance in terms of the drone's movement smoothness, as the drone is operated by following natural human motion. Lastly, the 3D interface also can provide a more immersive experience for the operator, making a better visualization of the VTOL aircraft's movements in 3D space. Based on these findings, it can be concluded that an intuitive 3D interface (e.g., Leap Motion) with a clear control-manipulation mapping would be not only a more effective and efficient solution but also a more intuitive and safer solution for controlling VTOL aircrafts compared to 2D interfaces [6,13].

The results of this study provide insights into the advantages of using a 3D controller, such as the Leap Motion, for controlling VTOL aircrafts in 3D space. The Leap Motion interface demonstrated several benefits in terms of mapping and control. By mapping the drone's movements to the user's natural directional gestures, the Leap Motion interface allows for more intuitive and precise control. The natural mapping of the Leap Motion interface resulted in smoother control and lower angle changes during both straight (A section) and up–down (B section) movements. In previous studies, it has been also mentioned that when the direction of the drone is mapped to the control inputs, it allows for intuitive and precise control of the drone. Kim and Chang (2021) invented a handheld controller where the user holds and moves it in three dimensions to control the drone and found that the developed controller provided more natural control compared to the gamepad, resulting in fewer failures in drone-control tasks, such as collisions [13]. And Di Vincenzo et al. (2022) argued that using hand gestures enables faster and more accurate control of the drone in terms of eye–hand coordination [20]. A 3D controller that effectively maps the direction of manipulation to the movement of the drone enhances the overall user experience and facilitates more accurate controlling of the drone, which is particularly important in scenarios that require precise control and smooth movements. The advantages of the 3D controller highlight its potential for enhancing the control capabilities of VTOL aircrafts and opening up new possibilities for applications in various industries and domains.

The use of the Leap Motion for controlling a drone presents certain limitations compared to using the gamepad. One limitation is the slower speed of control with the Leap Motion, which can be attributed to the users' unfamiliarity and difficulty in manipulating the drone's movements using hand gestures within a specific range in the air. This resulted in a lower subjective usability score, indicating challenges in user adaptation and proficiency with the Leap Motion interface. Particularly, there was a discrepancy between the users' subjective opinions of the control method and the actual drone-control results in intuitiveness. The Leap Motion interface was shown to be more intuitive for controlling the drone by showing better trajectory smoothness, while the Leap Motion control method was perceived as less intuitive in terms of subjective satisfaction ratings. Moreover, the performance of the Leap Motion was hindered in zigzag movements (C section) due to users' lack of proficiency, leading to suboptimal control characterized by sharp turns and unnecessary up–down movements. Participants also reported that the Leap Motion was not easy to learn and use, especially for novice users, as it lacks physical feedback mechanisms [6,13,20]. These limitations highlight the necessity for further exploration and the development of effective and user-friendly interfaces to enhance the control of VTOL aircraft s. It is crucial to focus on optimizing the speed, accuracy, and user familiarity of gesture-based drone control, thereby improving overall performance and usability. To address these limitations, future research can explore the use of physical devices that provide intuitive control of the drone in 3D space by incorporating handheld or manipulative actions in six directions. Moreover, consideration for different types of hand gestures is necessary. In this study, the drone was controlled using the Leap Motion by extending the palm and performing movements in six directions. However, the limited space for arm movement above the Leap Motion sometimes caused the hand position to go beyond its sensing range, leading to control errors. Previous studies have explored alternative methods for drone control, such as utilizing wrist rotations (e.g., abduction, adduction, flexion, extension, and rotation) or employing specific hand shapes (e.g., clenched fist, thumb-up, pointing, and V gesture), without the need for extensive arm movement [3,17].

This study is limited to analyzing the influence of eye–hand coordination (specifically mapping) on the control of hovering objects by novice users. This study does not compare various controllers and does not analyze other hovering objects besides drones; thus, the results of this study cannot be hastily generalized. Furthermore, advanced control laws used in real-world drone operations were not considered. By focusing on simply moving the drone in the vertical, horizontal, and lateral directions, this study aimed to analyze the impact of eye–hand coordination on the smoothness and precision of hovering object control. Therefore, other factors than controlling the drone's directions were fixed in the experiment in order to eliminate unexpected influences on the results. Also, this study aimed to analyze the effects of control methods on users without prior drone-operating experience. Lastly, while this study focused on participants who had no bias from previous drone-control experience, it is noted that participants with familiarity with drone control using various methods may show different results.

Future research should address the limitations of this study and further explore the best interfaces for controlling VTOL aircrafts. One potential area of investigation is to examine the issue of yaw rotation, which was excluded from this study for a specific reason. A possible solution could be to include yaw rotation in the mapping of the interface to assess how it impacts the performance and experience of the operator. This could be achieved by combining hand gestures with body movements, such as rotation of the head or body toward the direction of the drone's head-up orientation. Or, a VR device can be utilized to offer users a first-person perspective and the motion of wrist abduction/adduction can be naturally mapped to yaw rotation for controlling the drone [3,6,21]. Additionally, the comparison between different interfaces could be extended to include more devices or methods, such as touchscreen [22], hand gestures [3,16,17,23,24], body gestures [5], electromyography [21], gaze [20,25,26], brain signal [1], and speech [16,24] interfaces. This would provide a more comprehensive understanding of the strengths and weaknesses of different interfaces for controlling VTOL aircrafts. Moreover, as previous studies emphasize the importance of training for safe drone operation [27,28], future research needs to investigate not only eye–hand coordination but also the learnability aspect.

The findings of this study have important implications for the design and development of intuitive interfaces for controlling various hovering objects. The superiority of the Leap Motion in terms of reducing cognitive fatigue, providing better performance, quicker response in emergency situations, and a more immersive experience, suggests that an intuitive interface could be useful in various domains. These domains could include the control of VTOL aircrafts such as drones and helicopters, underwater vehicles, construction robots, and robot arms in a manufacturing setting, or dangerous tasks such as bomb disposal, as well as spacecraft. The precise mapping of the interface to the object's movement is critical for the successful handling of such tasks and requires further exploration.

5. Conclusions

This study compared the use of the 2D and 3D methods for controlling a drone, a type of vertical take-off and landing (VTOL) aircraft. The results showed that the 3D controller provided smoother and more intuitive control compared to the 2D controller. The intuitive hand gestures used with the 3D controller allowed for natural movements of the VTOL aircrafts in 3D space; therefore, the 3D controller has the potential to enhance the user experience and improve the overall control of VTOL aircrafts. It is important to note that further research on 3D control methods is needed to overcome the limitations of the Leap Motion device and provide a more user-friendly experience. Further research will optimize the ease of control and feedback mechanisms to enhance user proficiency for better control for VTOL aircrafts. Overall, the findings of this study highlight the importance of exploring and developing innovative interfaces for controlling VTOL aircrafts to achieve smoother, more precise, and more satisfying flight performance.

Author Contributions: Conceptualization, H.K.; methodology, all coauthors; software, H.K.; formal analysis, D.L. and H.Y.; investigation, H.K. and H.Y.; data curation, D.L.; writing—original draft preparation, D.L., H.K. and H.Y.; writing—review and editing, W.L.; visualization, D.L.; supervision, W.L.; project administration, W.L.; funding acquisition, W.L. All authors have read and agreed to the published version of the manuscript.

Funding: This research was jointly supported by Handong Global University Research Grants (No. 202300900001) and the National Research Foundation of Korea (NRF) grants funded by the Korean Government (2020R1F1A1050076).

Data Availability Statement: The dataset containing the performance (e.g., movement distance, movement time, trajectory smoothness) and subjective measures of 20 participants for two drone control interfaces can be found at https://doi.org/10.6084/m9.figshare.23822760 (accessed on 6 June 2023).

Conflicts of Interest: The authors declare no conflict of interest.

References

1. Kim, S.; Lee, S.; Kang, H.; Kim, S.; Ahn, M. P300 Brain–Computer Interface-Based Drone Control in Virtual and Augmented Reality. *Sensors* **2021**, *21*, 5765. [CrossRef] [PubMed]
2. Seok, J.; Han, J.; Baek, J.; Chang, W.; Kim, H. Development of an intuitive motion-basd dron controller. *J. Korea Soc. Die Mold Eng.* **2017**, *11*, 41–45.
3. Konstantoudakis, K.; Christaki, K.; Tsiakmakis, D.; Sainidis, D.; Albanis, G.; Dimou, A.; Daras, P. Drone Control in AR: An Intuitive System for Single-Handed Gesture Control, Drone Tracking, and Contextualized Camera Feed Visualization in Augmented Reality. *Drones* **2022**, *6*, 43. [CrossRef]
4. Park, N.; Ahn, Y.; Hwang, Y. A study on the development of a remote control drone for disaster response. *J. Soc. Disaster Inf.* **2019**, *15*, 578–589.
5. Sanna, A.; Lamberti, F.; Paravati, G.; Manuri, F. A Kinect-based natural interface for quadrotor control. *Entertain. Comput.* **2013**, *4*, 179–185. [CrossRef]
6. Tezza, D.; Andujar, M. The State-of-the-Art of Human–Drone Interaction: A Survey. *IEEE Access* **2019**, *7*, 167438–167454. [CrossRef]
7. Aretz, A.J.; Wickens, C.D. The Mental Rotation of Map Displays. *Hum. Perform.* **2009**, *5*, 303–328. [CrossRef]
8. Gugerty, L.; Brooks, J. Reference-Frame Misalignment and Cardinal Direction Judgments: Group Differences and Strategies. *J. Exp. Psychol. Appl.* **2004**, *10*, 75–88. [CrossRef]

9. Park, C.; Kim, S.; Tak, H.; Shin, S.; Choi, Y. The correlation between flight training factors in helicopter pilot training course and learning achievement. *J. Korean Soc. Aviat. Aeronaut.* **2019**, *27*, 45–53. [CrossRef]
10. Williams, K.W. *Human Factors Implications of Unmanned Aircraft Accidents: Flight-Control Problems*; DOT/FAA/AM-06/8; Federal Aviation Administration: Washington, DC, USA, 2006.
11. Kim, H.; Lim, S.; Lee, W. Comparative analysis of controlling drone. In Proceedings of the 10th International Conference on Applied Human Factors and Ergonomics (AHFE 2019), Washington, DC, USA, 24–28 July 2019.
12. Jeon, J.; Yu, S.; Cho, K. The embodiment of the drone controller: Single-handed drone controller to fly a drone safely and freely. In Proceedings of the Human-Computer Interaction (HCI) Korea 2017, Jeongseon, Republic of Korea, 8–10 February 2017.
13. Kim, H.; Chang, W. Intuitive Drone Control using Motion Matching between a Controller and a Drone. *Arch. Des. Res.* **2022**, *35*, 93–113. [CrossRef]
14. Zhao, Z.; Luo, H.; Song, G.-H.; Chen, Z.; Lu, Z.-M.; Wu, X. Web-based interactive drone control using hand gesture. *Rev. Sci. Instrum.* **2018**, *89*, 014707. [CrossRef]
15. Smeragliuolo, A.H.; Hill, N.J.; Disla, L.; Putrino, D. Validation of the Leap Motion Controller using markered motion capture technology. *J. Biomech.* **2016**, *49*, 1742–1750. [CrossRef]
16. Fernández, R.A.S.; Sanchez-Lopez, J.L.; Sampedro, C.; Bavle, H.; Molina, M.; Campoy, P. Natural user interfaces for human-drone multi-modal interaction. In Proceedings of the 2016 International Conference on Unmanned Aircraft Systems (ICUAS), Arlington, VA, USA, 7–10 June 2016.
17. Yoo, M.; Na, Y.; Song, H.; Kim, G.; Yun, J.; Kim, S.; Moon, C.; Jo, K. Motion Estimation and Hand Gesture Recognition-Based Human–UAV Interaction Approach in Real Time. *Sensors* **2022**, *22*, 2513. [CrossRef]
18. Budiyanto, A.; Ramadhan, M.I.; Burhanudin, I.; Triharminto, H.H.; Santoso, B. Navigation control of drone using hand gesture based on complementary filter algorithm. In Proceedings of the 5th International Conference on Advanced Material for Better Future (ICAMBF 2020), Surakarta, Indonesia, 13–14 October 2020.
19. Mutalib, M.K.Z.B.A.; Mohd, M.N.H.; Tomari, M.R.B.M.; Sari, S.B.; Ambar, R.B. Flying Drone Controller by Hand Gesture Using Leap Motion. *Int. J. Adv. Trends Comput. Sci. Eng.* **2020**, *9*, 111–116. [CrossRef]
20. Di Vincenzo, M.; Palini, F.; De Marsico, M.; Borghi, A.M.; Baldassarre, G. A Natural Human-Drone Embodied Interface: Empirical Comparison With a Traditional Interface. *Front. Neurorobotics* **2022**, *16*, 898859. [CrossRef]
21. Miehlbradt, J.; Cherpillod, A.; Mintchev, S.; Coscia, M.; Artoni, F.; Floreano, D.; Micera, S. Data-driven body–machine interface for the accurate control of drones. *Proc. Natl. Acad. Sci. USA* **2018**, *115*, 7913–7918. [CrossRef]
22. De Marsico, M.; Spagnoli, A. Using hands as an easy UAV joystick for entertainment applications. In Proceedings of the 13th Biannual Conference of the Italian SIGCHI Chapter: Designing the next interaction, Padova, Italy, 23–25 September 2019; pp. 1–9.
23. Bachmann, D.; Weichert, F.; Rinkenauer, G. Review of Three-Dimensional Human-Computer Interaction with Focus on the Leap Motion Controller. *Sensors* **2018**, *18*, 2194. [CrossRef]
24. Xiang, X.; Tan, Q.; Zhou, H.; Tang, D.; Lai, J. Multimodal Fusion of Voice and Gesture Data for UAV Control. *Drones* **2022**, *6*, 201. [CrossRef]
25. Yuan, L.; Reardon, C.; Warnell, G.; Loianno, G. Human Gaze-Driven Spatial Tasking of an Autonomous MAV. *IEEE Robot. Autom. Lett.* **2019**, *4*, 1343–1350. [CrossRef]
26. Zhou, T.; Li, B.; Liu, Y.; Chen, S.; Wang, Y. Gaze-assisted remote control for quad-rotor UAV. In Proceedings of the 2021 International Conference on High Performance Computing and Communications (HPCCE 2021), Guangzhou, China, 3–5 December 2021.
27. Covaciu, F.; Iordan, A.-E. Control of a Drone in Virtual Reality Using MEMS Sensor Technology and Machine Learning. *Micromachines* **2022**, *13*, 521. [CrossRef]
28. Koç, D.; Seçkin, A.Ç.; Satı, Z.E. Evaluation of Participant Success in Gamified Drone Training Simulator Using Brain Signals and Key Logs. *Brain Sci.* **2021**, *11*, 1024. [CrossRef] [PubMed]

drones

MDPI

Article

Finite-Time Neuro-Sliding-Mode Controller Design for Quadrotor UAVs Carrying Suspended Payload

Özhan Bingöl [1,*] and Hacı Mehmet Güzey [2,†]

1 Department of Electrical and Electronics Engineering, Gumushane University, Baglarbası Str., 29100 Gümüşhane, Turkey
2 Department of Electrical and Electronics Engineering, Sivas University of Science and Technology, Kardesler Str. No: 7/1, 58100 Sivas, Turkey
* Correspondence: ozhan.bingol@gumushane.edu.tr
† These authors contributed equally to this work.

Abstract: Due to the quadrotor's underactuated nature, suspended payload dynamics, parametric uncertainties, and external disturbances, designing a controller for tracking the desired trajectories for a quadrotor that carries a suspended payload is a challenging task. Furthermore, one of the most significant disadvantages of designing a controller for nonlinear systems is the infinite-time convergence to the desired trajectory. In this paper, a finite-time neuro-sliding mode controller (FTNSMC) for a quadrotor with a suspended payload that is subject to parametric uncertainties and external disturbances is designed. By constructing a finite-time sliding mode controller, the quadrotor can follow the reference trajectories in finite time. Furthermore, despite time-varying nonlinear dynamics, parametric uncertainties, and external disturbances, a neural network structure is added to the controller to effectively reduce chattering phenomena caused by high switching gains, and significantly reduce the size of the control signals. Following the completion of the controller design, the system's stability is demonstrated using the Lyapunov stability criterion. Extensive numerical simulations with various scenarios are run to demonstrate the effectiveness of the proposed controller.

Keywords: finite-time stability; neural network; quadrotor UAV; sliding mode control

1. Introduction

Unmanned system research is expanding quickly as a result of recent technology developments that have led to lower costs and smaller equipment [1]. The most popular of these technologies is unmanned aerial vehicles (UAVs), which have a larger range of civil and military applications [2,3]. The quadrotor is a type of UAV that shows great potential and is recognized as an ideal UAV by most research studies because of features such as simple construction, excellent maneuverability, and low cost [4]. Studies on UAVs and load-carrying applications are quickly expanding as a result of their widespread deployment. Due to its great maneuverability, ability to take off and land vertically, and capacity to carry payloads almost as heavy as its own body weight, the quadrotor is an excellent choice for autonomous transportation [5]. There are numerous ways to attach loads to quadrotor UAVs, including using a robotic arm to pick up the weight [6], mounting it to the quadrotor [7], and attaching it to the quadrotor using a cable [8,9]. The final technique, a quadrotor with a payload, offers a number of benefits over the first two: the construction is simpler, there are fewer restrictions on the size and form of the cargo, and it does not require landing for loading or unloading, which can save time and energy throughout the transport process. Moreover, using a robotic arm or attaching the payload directly to the quadrotor can severely limit the quadrotor's ability to land vertically.

Due to the aforementioned benefits, the relevance of carrying a suspended payload is increased by its usage in civil applications, including the delivery of first-aid supplies to disaster zones, cargo delivery, and agricultural spraying [5].

Despite its advantages, the suspended load-carrying quadrotor system is nonlinear, strongly coupled, and underactuated [10]. Additionally, unknown external disturbances and parametric uncertainties degrade the system's control performance. Considering these difficulties, developing a powerful controller for the the suspended load-carrying quadrotor system is difficult.

1.1. Related Work

Different control strategies were proposed in the literature to control a quadrotor payload system. In [11], a PID controller was proposed, feedback linearization-based controller was established in [9], and a backstepping-based nonlinear controller was presented in [12]. To actively regulate the position of the load, [8] suggested a nonlinear geometric control technique. In [13], the construction of a PID-based geometric controller was described that allowed for the payload to asymptotically follow a predetermined trajectory for both the payload position and attitude. In order to resolve the challenges brought on by the intense interaction of the system states and underactuated features of the dynamic system, in [14], the controller design was split into two parts, the UAV's attitude control design, and the UAV's position control and payload's swing motion control design. The second part's control rule, which takes into account the suppression of the payload's swing motion, was developed using the partial feedback linearization process. With an underactuated quadrotor payload system, wind disturbances were also considered in [15]. A path-following controller based on an uncertainty and disturbance estimator was suggested. In addition to wind disturbances, load variations were also taken into account in [16]. To deal with transient disturbance and estimate the system characteristics, an adaptive robust controller for dynamic subsystems was designed. For the kinematic subsystem, a global sliding mode controller is used to produce the necessary attitude angles for following the intended 3D trajectory.

The above control strategies are founded on theories of exponential or asymptotic stability. In reality, obtaining rapid tracking may not be possible with such approaches. To overcome this, a finite-time control strategy was devised for several systems in the literature [17]. For a class of nonstrict feedback nonlinear systems that contain input saturation, unidentified smooth functions, and error limitations, the challenge of finite-time control using neural networks is studied in [18]. The stochastically finite-time control problem for uncertain multiple-input, multiple-output (MIMO) stochastic nonlinear systems is addressed in [19] using a nontriangular solution. In [20], a finite-time control system using the nonsingular fast terminal sliding mode (NFTSM) and finite-time disturbance observer (FDO) approaches is developed to tackle the precise trajectory tracking issue of a surface vehicle affected by complicated marine environments. The adaptive finite-time decentralized control issue for nonlinear large-scale systems with time-varying output constraints and input saturation is discussed in [21]. According to the aforementioned research, finite-time control is a practical and efficient approach that, in theory, ensures the controlled system's rapid transient response and resilience and exactly satisfies the demand for the transient and robust performance of nonlinear systems.

1.2. Contributions of This Work

The sliding mode controller (SMC) is a robust control method that is regularly applied to nonlinear systems due to its benefits, such as its insensitivity to parameter variations, the avoidance of external disturbances, and quick dynamic responses [22]. Despite these advantages, it is extremely difficult to design an efficient controller due to drawbacks such as the chattering effect in the control signal, the requirement to accept uncertainties within a certain range, and the controller's adaptability to significant parameter changes and external disturbances [23]. Many different methods were developed in the literature to successfully eliminate this phenomenon [24–26]. With the recent developments in the area of artificial intelligence, artificial-intelligence-based controllers are being developed for different nonlinear systems in order to eliminate these disadvantages of SMC [27–31]. Any function may be approximated using neural networks [32,33]. It can both estimate the nonlinearity of the systems and lower the amplitude of the SMC's chattering.

In a previous work [34], we designed a novel neural-network-based SMC to control the quadrotor payload system and eliminate the disadvantages of SMC. In order to improve our previous work by using the above-mentioned advantages regarding finite-time stability, a finite-time neuro-sliding mode controller structure for quadrotor UAV carrying a suspended payload is proposed. The main contributions of this study are listed as follows:

1. Compared to many studies in the literature [10,35,36], a comprehensive nonlinear mathematical model of the quadrotor payload system was established by taking into account external disturbances and parameter uncertainties.
2. A novel finite-time neuro-sliding mode controller design is proposed for a quadrotor transporting a suspended payload. The proposed controller also takes into account suspended payload dynamics, unlike the system discussed in [37]. It also has a neural-network component when compared to the study in [38]. Thus, a more comprehensive control structure was obtained compared to systems using only SMC as in [39]. While the proposed controller uses the robust structure of the SMC, it also successfully learns the unknown dynamics with the help of the neural-network component, overcomes the disadvantages of the SMC, unlike in studies such as [40], and ensures that the system states reach the desired trajectory, or their errors reach zero or a close neighbourhood of zero in finite time. Without finite-time analysis, stability is shown mathematically when time goes to infinity; we show that neuro-sliding mode control converges in finite time to be able to use it in practice.
3. Comprehensive stability analysis with Lyapunov stability theory for the proposed controller is derived to demonstrate the payload-carrying quadrotor's finite-time stability.

The paper is structured as follows. Section 2 provides a basic overview of quadrotor dynamics, payload dynamics, neural networks, and finite-time stability. In Section 3, the suggested neural-network-based finite-time sliding-mode controller for quadrotor UAV carrying suspended payload is established, and the efficacy of the technique is demonstrated using numerical simulations in Section 4. Section 5 concludes with some remarks.

Next, the background and preliminaries are given.

2. Theoretical Background and Preliminaries

This section provides a theoretical overview of quadrotor dynamics, payload dynamics, finite-time stability, and neural networks. Moreover, some useful definitions and lemmas are given to construct the proposed controller properly.

2.1. Preliminaries

In this section, the following definitions and lemmas are provided to properly model the proposed control structure in a later section.

Definition 1 ([41]). *Nonlinear system $\dot{x} = f(x,u)$ is semiglobal practical finite-time stable (SGPFS) in equilibrium point $x = 0$ if there is a constant $\mu > 0$ with $x(t_0) = x_0$ and settling time $T(\mu, x_0) < \infty$ ensuring $\|x(t)\| < \mu$ for all $t > t_0 + T$.*

Lemma 1 ([41]). *Nonlinear system $\dot{x} = f(x,u)$ is SGPFS if there is a positive definitive function $V(x)$ and constants $c > 0$, $0 < \lambda < 1$, and $\delta > 0$ that ensure:*

$$\dot{V}(x) \leq -cV^{\lambda}(x) + \delta, t \geq 0 \tag{1}$$

Reaching time is given for $\forall 0 < \gamma \leq 1$ as follows:

$$T_r = \frac{1}{(1-\lambda)\gamma c}\left(V^{1-\lambda}(x(0)) - \left(\frac{\delta}{(1-\gamma)c}\right)^{\frac{1-\lambda}{\lambda}}\right) \tag{2}$$

Lemma 2 **([42]).** *For $\zeta_i \in \Re$, $i = 1, \ldots, n$, $0 < p < 1$, the following inequality holds:*

$$\left(\sum_{i=1}^{n} |\zeta_i|\right)^p \leq \sum_{i=1}^{n} |\zeta_i|^p \leq n^{1-p}\left(\sum_{i=1}^{n} |\zeta_i|\right)^p \tag{3}$$

Lemma 3 **([42]).** *The following Young's inequality is true for $x, y \geq 0$ and $p, q > 1$ such that $\frac{1}{p} + \frac{1}{q} = 1$:*

$$xy \leq \frac{x^p}{p} + \frac{y^q}{q} \tag{4}$$

Lemma 4 **([42]).** *The following inequality holds where τ, v are real variables and $a, b, c > 0$:*

$$|\tau|^a |v|^b \leq \frac{a}{a+b} c |\tau|^{a+b} + \frac{b}{a+b} c^{\frac{-a}{b}} |v|^{a+b} \tag{5}$$

2.2. Quadrotor Dynamics

The dynamic equations of a quadrotor are established in this section with the following assumptions [43]:

1. The quadrotor's construction is rigid and symmetrical.
2. The center of gravity of the quadrotor corresponds with the origin of the body axis.
3. The propellers are rigid.
4. Thrust and drag forces are related to the squares of the propeller's speed.
5. The UAV's roll and pitch angles are presumed to be operated within a range of $(-\pi/2, \pi/2)$, and yaw angle within $(-\pi, \pi)$.

A rotor on each arm of the quadrotor controls four fundamental movements that allow for it to achieve the required position and attitude. Four input signals are specified for the quadrotor's four motions. The initial input is u_1, which changes the speed of all propellers in the same proportion. u_1 controls the vertical movement of the quadrotor in the z axis. The second input signal, u_2, causes a torque along the body's x axis by changing the speed of the left propeller while also changing the speed of the right propeller by an equal amount. As a result, roll movement along the x axis obtained. The third input, u_3, which uses the same concept as u_2, adjusts the speed of the propellers on the front and rear rotors at the same time, leading in pitching motion along the y axis. The fourth input, u_4, simultaneously decreases or raises the speed of the left and right propellers while raising or reducing the speed of the rear and front propellers. As a result, yaw movement is achieved, which is the quadrotor turning around its own axis.

Because the rotors were symmetrically positioned, gyroscopic effects and aerodynamic torques balance each other out in flight, as seen in Figure 1. Two frames of reference are specified to express the quadrotor's position and attitude. The first frame of reference is the Earth frame, which is indicated by $E\{x, y, z\}$ and represents the position of the vehicle's center of gravity. The other is the fixed body frame, which is represented by $B\{x, y, z\}$.

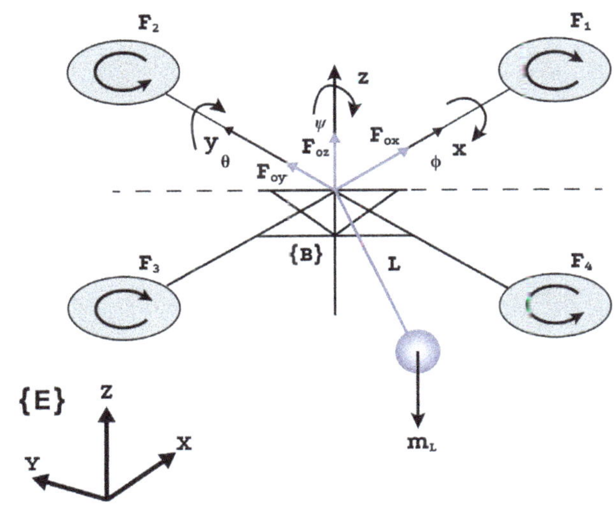

Figure 1. Quadrotor model with payload.

The dynamic quadrotor UAV model may be formed with parametric uncertainties and unknown external disturbances, and the model is as follows [44,45]:

$$\begin{aligned}
\ddot{x} &= \frac{1}{m}(\cos\phi\sin\theta\cos\psi+\sin\phi\sin\psi)u_1-\frac{K_1\dot{x}}{m+\Delta m}+D_x\\
\ddot{y} &= \frac{1}{m}(\cos\phi\sin\theta\sin\psi+\sin\phi\cos\psi)u_1-\frac{K_2\dot{y}}{m+\Delta m}+D_y\\
\ddot{z} &= \frac{1}{m}(\cos\phi\cos\theta)u_1-g-\frac{K_3\dot{z}}{m+\Delta m}+D_z\\
\ddot{\phi} &= \dot{\theta}\dot{\psi}\frac{I_y+\Delta I_y-I_z-\Delta I_z}{I_x-\Delta I_x}+\frac{J_r}{I_x-\Delta I_x}\Omega_r\dot{\theta}+\frac{1}{I_x-\Delta I_x}u_2-\frac{K_4 l}{I_x-\Delta I_x}\dot{\phi}+D_\phi\\
\ddot{\theta} &= \dot{\psi}\dot{\phi}\frac{I_z+\Delta I_z-I_x-\Delta I_x}{I_y-\Delta I_y}+\frac{J_r}{I_y-\Delta I_y}\Omega_r\dot{\phi}+\frac{1}{I_y-\Delta I_y}u_3-\frac{K_5 l}{I_y-\Delta I_y}\dot{\theta}+D_\theta\\
\ddot{\psi} &= \dot{\phi}\dot{\theta}\frac{I_x+\Delta I_x-I_y-\Delta I_y}{I_z-\Delta I_z}+\frac{1}{I_z-\Delta I_z}u_4-\frac{K_6}{I_z-\Delta I_z}\dot{\psi}+D_\psi
\end{aligned} \tag{6}$$

where (ϕ,θ,ψ) are Euler angles indicating roll, pitch and yaw angles; (x,y,z) are positions of the UAV in the space; m is the UAV's total mass; g is gravity acceleration; l is the space from the center of the UAV to the propellers; (I_x, I_y, I_z) are inertias with respect to axes; Jr is the inertia of the propeller; $K_i(i=1,2\ldots,6)$ are the drag coefficients; $\Omega_r=\Omega_1-\Omega_2+\Omega_3-\Omega_4$, Ω_i are the angular velocities of the propellers, $u_i(i=1,2,3,4)$ are the virtual inputs given in (7): $D_i(i=x,y,z,\phi,\theta,\psi)$ are external disturbances; Δ denotes parametric uncertainties.

$$\begin{aligned}
u_1 &= (F_1+F_2+F_3+F_4)\\
u_2 &= (-F_2+F_4)\\
u_3 &= (-F_1+F_3)\\
u_4 &= \frac{d(-F_1+F_2-F_3+F_4)}{b}
\end{aligned} \tag{7}$$

where $F_i(i=1,2,3,4)$ are rotor thrusts, b is the lift coefficient, d is the force-to-moment scaling vector.

2.3. Payload Dynamics

In this study, the effects of the load on the quadrotor in three dimensions are dynamically modeled and added to the quadrotor's own dynamics, and the total dynamic equations of the system are obtained. The payload model was developed on the following assumptions:

1. The payload may not rotate along the cable axis; therefore, payloads have only two degrees of freedom in tilting directions.
2. The load is suspended from a weightless, nonelastic cable, which is rigid.
3. The cable junction was placed at the quadrotor's centre of mass, and the tensile force of the cable does not directly impact the quadrotor's rotating motion.

As seen in Figure 2, where the payload was modeled as a three-dimensional point pendulum mass, the equations defining the dynamics of the load were derived by taking into account the α_x longitudinal suspension angles in the $x-z$ plane and the α_y longitudinal angles and forces in the $y-z$ plane. The following are the forces generated by the load on the respective axes [46]:

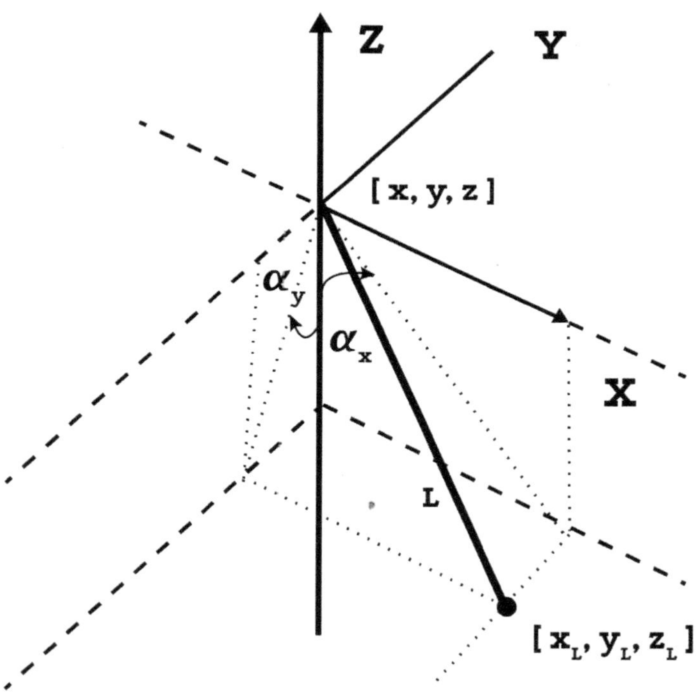

Figure 2. Point mass payload model.

$$
\begin{aligned}
F_{ox} &= m_L L\left(\ddot{\alpha}_x \cos\alpha_y \cos\alpha_x - \dot{\alpha}_x^2 \cos\alpha_y \sin\alpha_x\right) \\
F_{oy} &= m_L L\left(\ddot{\alpha}_y \cos\alpha_x \cos\alpha_y - \dot{\alpha}_y^2 \cos\alpha_x \sin\alpha_y\right) \\
F_{oz} &= m_L L\ddot{\alpha}_x \cos\alpha_y \sin\alpha_x + m_L L\dot{\alpha}_x^2 \cos\alpha_y \cos\alpha_x \\
&\quad + m_L L\ddot{\alpha}_y \cos\alpha_x \sin\alpha_y + m_L L\dot{\alpha}_y^2 \cos\alpha_x \cos\alpha_y - m_L g
\end{aligned}
\tag{8}
$$

where m_L is the payload's mass; F_{ox}, F_{oy}, F_{oz} are the payload's forces effect on quadrotor in three-dimensional space shown in Figure 2.

Hence, the entire system dynamics are as follows:

$$\begin{aligned}
\ddot{x} &= \frac{1}{(m+m_L)}(\cos\phi\sin\theta\cos\psi+\sin\phi\sin\psi)u_1 - \frac{K_1\dot{x}}{(m+m_L+\Delta m)} - \frac{F_{ox}}{(m+m_L)} + D_x \\
\ddot{y} &= \frac{1}{(m+m_L)}(\cos\phi\sin\theta\sin\psi+\sin\phi\cos\psi)u_1 - \frac{K_2\dot{y}}{(m+m_L+\Delta m)} - \frac{F_{oy}}{(m+m_L)} + D_y \\
\ddot{z} &= \frac{1}{(m+m_L)}(\cos\phi\cos\theta)u_1 - \frac{mg}{(m+m_L)} - \frac{K_3\dot{z}}{(m+m_L+\Delta m)} + \frac{F_{oz}}{(m+m_L)} + D_z \\
\ddot{\phi} &= \dot{\theta}\dot{\psi}\frac{I_y+\Delta I_y-I_z-\Delta I_z}{I_x-\Delta I_x} + \frac{J_r}{I_x-\Delta I_x}\Omega_r\dot{\theta} + \frac{l}{I_x-\Delta I_x}u_2 - \frac{K_4 l}{I_x-\Delta I_x}\dot{\phi} + D_\phi \\
\ddot{\theta} &= \dot{\psi}\dot{\phi}\frac{I_z+\Delta I_z-I_x-\Delta I_x}{I_y-\Delta I_y} - \frac{J_r}{I_y-\Delta I_y}\Omega_r\dot{\phi} + \frac{l}{I_y-\Delta I_y}u_3 - \frac{K_5 l}{I_y-\Delta I_y}\dot{\theta} + D_\theta \\
\ddot{\psi} &= \dot{\phi}\dot{\theta}\frac{I_x+\Delta I_x-I_y-\Delta I_y}{I_z-\Delta I_z} + \frac{1}{I_z-\Delta I_z}u_4 - \frac{K_6}{I_z-\Delta I_z}\dot{\psi} + D_\psi
\end{aligned} \tag{9}$$

2.4. Neural Network

In this work, a two-layer neural network (NN) structure was used to estimate uncertain dynamics as shown in Figure 3.

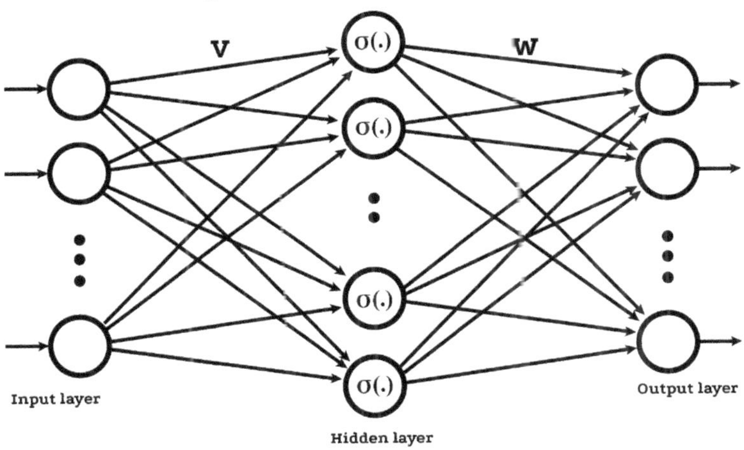

Figure 3. Neural-network structure.

The first layer is hidden and contains adjustable hidden weights, $W \in \mathbb{R}^{(Vh\times ko)}$, and the second layer consists of randomly determined constants, $\hat{V} \in \mathbb{R}^{(ki\times Vh)}$ where ki is the input count and ko is the output count. Vh represents the number of neurons in the hidden layer. Estimation function $f(x)$ can be given as $f(x) = W^T\sigma(\hat{V}^T\bar{x}) + \varepsilon$ where ε is the bounded NN estimation error that satisfies $\|\varepsilon\| < \varepsilon_M$, and $\sigma(\bullet) : \mathbb{R}^a \rightarrow \mathbb{R}^L$ is the activation function for the hidden layer. Because input-layer weights $\hat{V} \in \mathbb{R}^{ki\times Vh}$ are randomly chosen, for any input x, the estimation is viable; hence, activation function, $\sigma(x) = \sigma(\hat{V}^T\bar{x})$, establishes a stochastic base in compact set S [47]. $\bar{x}^T = [b_i\ x^T]$, where $b_i > 0$ is a bias term of input that lets threshold value to be in weight matrix $\hat{V}$. As an activation function, a tangent hyperbolic function in the form of $\sigma(x) = \frac{1-e^{-2x}}{1+e^{-2x}}$, was chosen in this study for all neurons. It was also assumed that target weights were limited to a known positive value W_M satisfying $\|W\|_F < W_M$ on any compact subset of $\mathbb{R}^n$ [48]. Moreover, $\|\bullet\|$ and $\|\bullet\|_F$ are considered as the vector and Frobenius norm, respectively [47].

Next, the controller design is given.

3. Controller Design

The suggested controller is outlined here. The proposed controller takes into consideration quadrotor dynamics, and the unpredictable dynamics caused by external disturbances and parametric uncertainties. To reduce the chattering influence of the SMC, the unknown dynamics are learned using two-layer NN. Because of the neural-network component, the controller can successfully govern the system without understanding the quadrotor dynamics. Furthermore, utilizing the updated learning structure, system dynamics may be appropriately updated even when there are uncertainties that change over time. With the assistance of a finite-time control system, the trajectory errors converge to zero in a finite time.

To enable effective controller design, the dynamic equations obtained in (9) are subjected to the following transformation [45]:

$$\begin{cases} \ddot{z} = f_1 + g_1 u_1 + \Delta f_1 \\ \begin{cases} \ddot{x} = f_2 + g_2 u_2 + \Delta f_2 \\ \ddot{\phi} = f_3 + g_3 u_2 + \Delta f_3 \end{cases} \\ \begin{cases} \ddot{y} = f_4 + g_4 u_3 + \Delta f_4 \\ \ddot{\theta} = f_5 + g_5 u_3 + \Delta f_5 \end{cases} \\ \ddot{\psi} = f_6 + g_6 u_4 + \Delta f_6 \end{cases} \tag{10}$$

where

$$\begin{aligned}
f_1 &= -\frac{mg}{(m+m_L)} - \frac{K_3\dot{z}}{(m+m_L)} + \frac{F_{oz}}{(m+m_L)} + D_z \\
\Delta f_1 &= \frac{\Delta m K_3 \dot{z}}{(m+m_L)(m+\Delta m+m_L)}, g_1 = \frac{\cos\phi\cos\theta}{(m+m_L)} \\
f_2 &= \frac{(\cos\phi\sin\theta\cos\psi + \sin\phi\sin\psi)u_1}{(m+m_L)} - \frac{K_1\dot{x}}{(m+m_L)} - \frac{F_{ox}}{(m+m_L)} + D_x \\
\Delta f_2 &= \frac{\Delta m K_1 \dot{x}}{(m+m_L)(m-\Delta m+m_L)}, g_2 = 0 \\
f_3 &= \frac{(\dot{\theta}\dot{\psi}(I_y - I_z) + J_r\dot{\theta}\Omega_r - K_4 l\dot{\phi})}{I_x} + D_\phi, g_3 = \frac{l}{I_x} \\
\Delta f_3 &= \frac{-(\dot{\theta}\dot{\psi}(I_y - I_z) + J_r\dot{\theta}\Omega_r - K_4 l\dot{\phi})\Delta I_x}{I_x(\Delta I_x + I_x)} + \dot{\theta}\dot{\psi}\left(\frac{\Delta I_y - \Delta I_z}{\Delta I_x + I_x}\right) \\
f_4 &= \frac{(\cos\phi\sin\theta\sin\psi + \sin\phi\cos\psi)u_1}{(m+m_L)} - \frac{K_2\dot{y}}{m+m_L} - \frac{F_{oy}}{m+m_L} + D_y \\
\Delta f_4 &= \frac{\Delta m K_2 \dot{y}}{(m+m_L)(m+\Delta m+m_L)}, g_4 = 0 \\
f_5 &= \frac{(\dot{\psi}\dot{\phi}(I_z - I_x) + J_r\dot{\phi}\Omega_r - K_5 l\dot{\theta})}{I_y} + D_\theta, g_5 = \frac{l}{I_y} \\
\Delta f_5 &= \frac{-(\dot{\psi}\dot{\phi}(I_z - I_x) + J_r\dot{\phi}\Omega_r - K_5 l\dot{\theta})\Delta I_y}{I_y(\Delta I_y + I_y)} + \dot{\psi}\dot{\phi}\left(\frac{\Delta I_z - \Delta I_x}{\Delta I_y + I_y}\right) \\
f_6 &= \frac{(\dot{\phi}\dot{\theta}(I_x - I_y) - K_6\dot{\psi})}{I_z} + D_\psi, g_6 = \frac{1}{I_z} \\
\Delta f_6 &= \frac{-(\dot{\phi}\dot{\theta}(I_x - I_y) - K_6\dot{\psi})\Delta I_z}{I_z(\Delta I_z + I_z)} + \dot{\psi}\dot{\phi}\left(\frac{\Delta I_x - \Delta I_y}{\Delta I_z + I_z}\right)
\end{aligned} \tag{11}$$

Moreover, sliding surface functions are selected as follows [49,50]:

$$
\begin{aligned}
s_1 &= a_{11}(\dot{z}_d - \dot{z}) + \int_0^t (a_{12} sign^{\alpha_1}(z_d - z) + a_{11} sign^{\alpha_2}(\dot{z}_d - \dot{z}))dt \\
s_2 &= a_{21}(\dot{x}_d - \dot{x}) + a_{23}(\dot{\phi}_d - \dot{\phi}) + \int_0^t (a_{22} sign^{\alpha_1}(x_d - x) + a_{21} sign^{\alpha_2}(\dot{x}_d - \dot{x}))dt \\
&+ \int_0^t (a_{24} sign^{\alpha_1}(\phi_d - \phi) + a_{23} sign^{\alpha_2}(\dot{\phi}_d - \dot{\phi}))dt \\
s_3 &= a_{31}(\dot{y}_d - \dot{y}) + a_{33}(\dot{\theta}_d - \dot{\theta}) + \int_0^t (a_{32} sign^{\alpha_1}(y_d - y) + a_{31} sign^{\alpha_2}(\dot{y}_d - \dot{y}))dt \\
&+ \int_0^t (a_{34} sign^{\alpha_1}(\theta_d - \theta) + a_{33} sign^{\alpha_2}(\dot{\theta}_d - \dot{\theta}))dt \\
s_4 &= a_{41}(\dot{\psi}_d - \dot{\psi}) + \int_0^t (a_{42} sign^{\alpha_1}(\psi_d - \psi) + a_{41} sign^{\alpha_2}(\dot{\psi}_d - \dot{\psi}))dt
\end{aligned} \tag{12}
$$

where $\alpha_1 \in (0,1)$, $\alpha_2 = \frac{2\alpha_1}{\alpha_1+1}$, and $a_{i,j}(i,j = 1,2,3,4)$ are the coefficients of sliding surfaces that are described later. $sign^\rho$ is the representation of $sign^\rho(x) = |x|^\rho sign(x)$ where $\rho > 0$.

3.1. Coefficients of Sliding Surfaces

The coefficients of the sliding manifolds in (12) are derived using the same Hurwitz stability criterion as in [45,51], and they are provided as follows:

$$
\begin{aligned}
& a_{11} > 0, a_{12} > 0, a_{41} > 0, a_{42} > 0 \\
& a_{21} = -\frac{m}{u_1 \cos\psi}(r_{11}r_{12} + r_{21}r_{31} + r_{32}r_{11}) \\
& a_{22} = -\frac{m}{u_1 \cos\psi}(r_{11}r_{21}r_{31}) \\
& a_{23} = 1, a_{24} = r_{12} + r_{21} + r_{31} \\
& a_{31} = \frac{m}{u_1 \cos\phi \cos\psi}(r_{12}r_{22} + r_{22}r_{32} + r_{32}r_{12}) \\
& a_{32} = \frac{m}{u_1 \cos\phi \cos\psi}(r_{12}r_{22}r_{32}) \\
& a_{33} = 1, a_{34} = r_{12} + r_{22} + r_{32}
\end{aligned} \tag{13}
$$

where $r_{ij}(i,j = 1,2,3,4)$ are the design parameters that are constant.

3.2. Ftnsmc Controller Design

In real-world systems, neural networks (NNs) are always used as one of the online estimation methodologies for dealing with unknown nonlinear uncertainties. The nonlinear quadrotor dynamics in (10), including parametric uncertainties and external disturbances, are considered to be unknown in this section. Unknown nonlinear dynamics $f_i \ \forall i = 1,2,3,4,5,6$ are estimated as follows:

$$
f_i = \Theta_i^T \sigma\left(H_i^T \varpi_i\right) + \chi_i \tag{14}
$$

where $\Theta_i \in \Re^{2\times h}$ are the required limited NN weights meeting $\frac{k_n}{2} \sum_{i=1}^{4} \|\Theta_i\|^2 \le \Theta_M$ and Θ_M is a positive constant, $k_n > 0$ represents the positive NN learning rate that is specified later, $h = 8$ is the number of neurons in the hidden layer, $\sigma(H_i\varpi_i)$ represents the basis function, and $H_i^T \in \Re^{h\times n}$ represents the mapping between the inputs and the hidden-layer neurons, where $n = 7$ is the number of inputs to the NN, $\frac{1}{2}\sum_{i=1}^{4} a_{i1}\chi_i^2 \leq \chi_M$ with χ_M being a positive constant, and χ_i being the limited NN reconstruction error fulfilling.

$\hat{\Theta}_i$ represents the unknown NN weights, and the approximated uncertain dynamics are given by:

$$\hat{f}_i = \hat{\Theta}_i^T \sigma\left(H_i^T \varpi_i\right) \tag{15}$$

The NN weight estimation error is defined as $\tilde{\Theta}_i = \Theta_i - \hat{\Theta}_i$, and the estimation error dynamics are $\dot{\tilde{\Theta}}_i = -\dot{\hat{\Theta}}_i$. The error in dynamics estimation can, therefore, be described as follows:

$$\begin{aligned} \tilde{f}_i &= f_i - \hat{f}_i \\ &= \Theta_i^T \sigma\left(H_i^T \varpi_i\right) + \chi_i - \hat{\Theta}_i^T \sigma\left(H_i^T \varpi_i\right) \\ &= \tilde{\Theta}_i^T \sigma\left(H_i^T \varpi_i\right) + \chi_i \end{aligned} \tag{16}$$

The estimated NN weights were adjusted using the adaption law shown below:

$$\dot{\hat{\Theta}}_i = -k_n \hat{\Theta}_i + s_i a_{i1} \sigma(H_i \overline{W_i}), \quad \forall i = 1,2,3,4. \tag{17}$$

where $k_n > 0$ is the learning rate.

An exponential reaching law is provided to achieve appropriate steady-state behavior [52]:

$$\dot{s}_i = -\eta_1 s_i - \eta_2 sign^{\beta}(s_i) \tag{18}$$

where $\eta_1 > 0, \eta_2 > 0$ and $\beta \in (0,1)$.

Taking the derivative of sliding surface s_i defined in (12) yields:

$$\begin{aligned} \dot{s}_1 &= a_{11}(\ddot{z}_d - \ddot{z}) + a_{12} sign^{\alpha_1}(z_d - z) + a_{11} sign^{\alpha_2}(\dot{z}_d - \dot{z}) \\ \dot{s}_2 &= a_{21}(\ddot{x}_d - \ddot{x}) + a_{22} sign^{\alpha_1}(x_d - x) + a_{21} sign^{\alpha_2}(\dot{x}_d - \dot{x}) \\ &\quad + a_{23}(\ddot{\phi}_d - \ddot{\phi}) + a_{24} sign^{\alpha_1}(\phi_d - \phi) + a_{23} sign^{\alpha_2}(\dot{\phi}_d - \dot{\phi}) \\ \dot{s}_3 &= a_{31}(\ddot{y}_d - \ddot{y}) + a_{32} sign^{\alpha_1}(y_d - y) + a_{31} sign^{\alpha_2}(\dot{y}_d - \dot{y}) \\ &\quad + a_{33}(\ddot{\theta}_d - \ddot{\theta}) + a_{34} sign^{\alpha_1}(\theta_d - \theta) + a_{33} sign^{\alpha_2}(\dot{\theta}_d - \dot{\theta}) \\ \dot{s}_4 &= a_{41}(\ddot{\psi}_d - \ddot{\psi}) + a_{42} sign^{\alpha_1}(\psi_d - \psi) + a_{41} sign^{\alpha_2}(\dot{\psi}_d - \dot{\psi}) \end{aligned} \tag{19}$$

The dynamics of the system, including parametric uncertainties and external disturbances, were considered to be unknown and they were estimated with the NN in this work. Therefore, by inserting the estimation function in (16) for all the dynamics, and taking the reaching law into consideration in (18) for $\dot{s}_i = 0$ control signals can be written as follows:

$$\begin{aligned} u_1 &= \frac{1}{a_{11}}\left\{a_{11}\left(\ddot{z}_d - \hat{f}_1\right) + a_{12} sign^{\alpha_1}(z_d - z) + a_{11} sign^{\alpha_2}(\dot{z}_d - \dot{z}) + \eta_1 s_1 + \eta_2 sign^{\beta}(s_1)\right\} \\ u_2 &= \frac{1}{a_{21} + a_{23}}\left\{\begin{array}{l} a_{21}\left(\ddot{x}_d - \hat{f}_2\right) + a_{22} sign^{\alpha_1}(x_d - x) + a_{21} sign^{\alpha_2}(\dot{x}_d - \dot{x}) \\ + a_{23}\left(\ddot{\phi}_d - \hat{f}_3\right) + a_{24} sign^{\alpha_1}(\phi_d - \phi) + a_{23} sign^{\alpha_2}(\dot{\phi}_d - \dot{\phi}) + \eta_1 s_2 + \eta_2 sign^{\beta}(s_2) \end{array}\right\} \\ u_3 &= \frac{1}{a_{31} + a_{33}}\left\{\begin{array}{l} a_{31}\left(\ddot{y}_d - \hat{f}_4\right) + a_{32} sign^{\alpha_1}(y_d - y) + a_{31} sign^{\alpha_2}(\dot{y}_d - \dot{y}) + a_{33}\left(\ddot{\theta}_d - \hat{f}_5\right) \\ + a_{34} sign^{\alpha_1}(\theta_d - \theta) + a_{33} sign^{\alpha_2}(\dot{\theta}_d - \dot{\theta}) + \eta_1 s_3 + \eta_2 sign^{\beta}(s_3) \end{array}\right\} \\ u_4 &= \frac{1}{a_{41}}\left\{a_{41}\left(\ddot{\psi}_d - \hat{f}_1\right) + a_{42} sign^{\alpha_1}(\psi_d - \psi) + a_{41} sign^{\alpha_2}(\dot{\psi}_d - \dot{\psi}) + \eta_1 s_4 + \eta_2 sign^{\beta}(s_4)\right\} \end{aligned} \tag{20}$$

3.3. Stability Analysis

Theorem 1. *Consider nonlinear System (10) with the sliding surface dynamics in (19), the controller inputs as in (20), and the estimated dynamics applied to the system, and the estimated NN weights tuned with the adaptation law (17), All NN weight estimation errors and sliding surfaces were SGPFS, and the tracking error converged in a finite time to a small neighborhood of the origin.*

Proof of Theorem 1. Define the Lyapunov candidate function of all sliding surfaces, s_i and NN weight estimation errors $\tilde{\Theta}_i$ as follows:

$$V = \sum_{i=1}^{4}\left(\frac{1}{2}{s_i}^2 + \frac{1}{2}\left(\tilde{\Theta}_i^T\tilde{\Theta}_i\right)\right) \tag{21}$$

by taking Derivative (21) and inserting the dynamics in (10), and considering that $\dot{\tilde{\Theta}}_i = -\dot{\hat{\Theta}}_i$ to obtain:

$$\begin{aligned}
\dot{V} &= \sum_{i=1}^{4}\left(s_i\dot{s}_i + \tilde{\Theta}_i^T\dot{\tilde{\Theta}}_i\right)\\
&= s_1\dot{s}_1 + s_2\dot{s}_2 + s_3\dot{s}_3 + s_4\dot{s}_4 + \tilde{\Theta}_1^T\dot{\tilde{\Theta}}_1 + \tilde{\Theta}_2^T\dot{\tilde{\Theta}}_2 + \tilde{\Theta}_3^T\dot{\tilde{\Theta}}_3 + \tilde{\Theta}_4^T\dot{\tilde{\Theta}}_4\\
&= s_1(a_{11}(\ddot{z}_d - f_1 + g_1u_1 + \Delta f_1) + a_{12}sign^{\alpha_1}(z_d - z) + a_{11}sign^{\alpha_2}(\dot{z}_d - \dot{z}))\\
&\quad + s_2\begin{pmatrix} a_{21}(\ddot{x}_d - f_2 + g_2u_2 + \Delta f_2) + a_{22}sign^{\alpha_1}(x_d - x) + a_{21}sign^{\alpha_2}(\dot{x}_d - \dot{x})\\ + a_{23}(\ddot{\phi}_d - f_3 + g_3u_2 + \Delta f_3) + a_{24}sign^{\alpha_1}(\phi_d - \phi) + a_{23}sign^{\alpha_2}(\dot{\phi}_d - \dot{\phi})\end{pmatrix}\\
&\quad + s_3\begin{pmatrix} a_{31}(\ddot{y}_d - f_4 + g_4u_3 + \Delta f_4) + a_{32}sign^{\alpha_1}(y_d - y) + a_{31}sign^{\alpha_2}(\dot{y}_d - \dot{y})\\ + a_{33}(\ddot{\theta}_d - f_5 + g_5u_3 + \Delta f_5) + a_{34}sign^{\alpha_1}(\theta_d - \theta) + a_{33}sign^{\alpha_2}(\dot{\theta}_d - \dot{\theta})\end{pmatrix}\\
&\quad + s_4(a_{41}(\ddot{\psi}_d - f_6 + g_6u_4 + \Delta f_6) + a_{42}sign^{\alpha_1}(\psi_d - \psi) + a_{41}sign^{\alpha_2}(\dot{\psi}_d - \dot{\psi}))\\
&\quad - \tilde{\Theta}_1^T\dot{\hat{\Theta}}_1 - \tilde{\Theta}_2^T\dot{\hat{\Theta}}_2 - \tilde{\Theta}_3^T\dot{\hat{\Theta}}_3 - \tilde{\Theta}_4^T\dot{\hat{\Theta}}_4
\end{aligned} \tag{22}$$

Using the controller inputs in (20), dynamic estimation error function in (16) and reaching law in (18), (22) can be rewritten in a compact form as:

$$\begin{aligned}
\dot{V} &= -\sum_{i=1}^{4}\eta_1 s_i^2 - \sum_{i=1}^{4}\eta_2|s_i|^{\beta+1} + \sum_{i=1}^{4}s_i a_{i1}\tilde{f}_i - \sum_{i=1}^{4}\tilde{\Theta}_i^T\dot{\hat{\Theta}}_i\\
&= -\sum_{i=1}^{4}\eta_1 s_i^2 - \sum_{i=1}^{4}\eta_2|s_i|^{\beta+1} + \sum_{i=1}^{4}s_i a_{i1}\left(\tilde{\Theta}_i^T\sigma\left(H_i^T\varpi_i\right) + \chi_i\right) - \sum_{i=1}^{4}\tilde{\Theta}_i^T\dot{\hat{\Theta}}_i\\
&= -\sum_{i=1}^{4}\eta_1 s_i^2 - \sum_{i=1}^{4}\eta_2|s_i|^{\beta+1} + \sum_{i=1}^{4}s_i a_{i1}\left(\tilde{\Theta}_i^T\sigma\left(H_i^T\varpi_i\right)\right) + \sum_{i=1}^{4}s_i a_{i1}\chi_i - \sum_{i=1}^{4}\tilde{\Theta}_i^T\left(-k_n\hat{\Theta}_i + s_i a_{i1}\sigma\left(H_i\overline{W}_i\right)\right)\\
&= -\sum_{i=1}^{4}\eta_1 s_i^2 - \sum_{i=1}^{4}\eta_2|s_i|^{\beta+1} + \sum_{i=1}^{4}s_i a_{i1}\left(\tilde{\Theta}_i^T\sigma\left(H_i^T\varpi_i\right)\right) + \sum_{i=1}^{4}s_i a_{i1}\chi_i + k_n\sum_{i=1}^{4}\tilde{\Theta}_i^T\hat{\Theta}_i - \sum_{i=1}^{4}s_i a_{i1}\tilde{\Theta}_i^T\left(\sigma\left(H_i\overline{W}_i\right)\right)\\
&= -\sum_{i=1}^{4}\eta_1 s_i^2 - \sum_{i=1}^{4}\eta_2|s_i|^{\beta+1} + \sum_{i=1}^{4}s_i a_{i1}\chi_i + k_n\sum_{i=1}^{4}\tilde{\Theta}_i^T\left(\Theta_i - \tilde{\Theta}_i\right)\\
&= -\sum_{i=1}^{4}\eta_1 s_i^2 - \sum_{i=1}^{4}\eta_2|s_i|^{\beta+1} + \sum_{i=1}^{4}s_i a_{i1}\chi_i - k_n\sum_{i=1}^{4}\left\|\tilde{\Theta}_i\right\|^2 + k_n\sum_{i=1}^{4}\tilde{\Theta}_i^T\Theta_i
\end{aligned} \tag{23}$$

The following inequalities hold from Young's inequality in Lemma 3:

$$
\begin{aligned}
k_n \sum_{i=1}^{4} \tilde{\Theta}_i^T \Theta_i &\leq \frac{k_n}{2} \sum_{i=1}^{4} \|\tilde{\Theta}_i\|^2 + \frac{k_n}{2} \sum_{i=1}^{4} \|\Theta_i\|^2 \\
\sum_{i=1}^{4} s_i a_{i1} \chi_i &\leq \frac{1}{2} \sum_{i=1}^{4} s_i^2 a_{i1} + \frac{1}{2} \sum_{i=1}^{4} a_{i1} \chi_i^2
\end{aligned} \tag{24}
$$

Upper bounds Θ_M and χ_M; inserting (24) into (23), we obtain:

$$
\begin{aligned}
\dot{V} &\leq -\sum_{i=1}^{4} \eta_1 s_i^2 - \sum_{i=1}^{4} \eta_2 |s_i|^{\beta+1} + \frac{1}{2} \sum_{i=1}^{4} s_i^2 a_{i1} + \frac{1}{2} \sum_{i=1}^{4} a_{i1} \chi_i^2 - k_n \sum_{i=1}^{4} \|\tilde{\Theta}_i\|^2 + \frac{k_n}{2} \sum_{i=1}^{4} \|\tilde{\Theta}_i\|^2 + \frac{k_n}{2} \sum_{i=1}^{4} \|\Theta_i\|^2 \\
&\leq -\sum_{i=1}^{4} \left(\eta_1 - \frac{a_{i1}}{2}\right) s_i^2 - \sum_{i=1}^{4} \eta_2 |s_i|^{\beta+1} - \frac{k_n}{2} \sum_{i=1}^{4} \|\tilde{\Theta}_i\|^2 + 4\chi_M + 4\Theta_M
\end{aligned} \tag{25}
$$

Here, by using Lemma 2, the inequality below holds:

$$
\left(\sum_{i=1}^{4} \eta_2 |s_i|^2 \right)^{\frac{\beta+1}{2}} \leq \sum_{i=1}^{4} \eta_2 |s_i|^{\beta+1} \tag{26}
$$

Moreover, according to Lemma 4, the inequalities. below exist:

$$
\begin{aligned}
\left(\|\tilde{\Theta}_i\|^2\right)^{\xi} &\leq (1-\xi)\xi^{\frac{\xi}{1-\xi}} + \|\tilde{\Theta}_i\|^2 \\
\left(\|s\|^2\right)^{\xi} &\leq (1-\xi)\xi^{\frac{\xi}{1-\xi}} + \|s\|^2
\end{aligned} \tag{27}
$$

where $0 < \xi = \frac{\beta+1}{2} < 1$.

By using (26) and (27), and Lemma 2, (25) can be rewritten as:

$$
\begin{aligned}
\dot{V} &\leq -\left(\eta_1 + \eta_2 - \frac{a_{i1}}{2}\right) \left(\sum_{i=1}^{4} \frac{1}{2} |s_i|^2 \right)^{\xi} - k_n \left(\sum_{i=1}^{4} \frac{1}{2} \|\tilde{\Theta}_i\|^2 \right)^{\xi} + 4\chi_M + 4\Theta_M + \left(\frac{k_n}{2} + \eta_1 - \frac{a_{i1}}{2}\right) \left((1-\xi)\xi^{\frac{\xi}{1-\xi}} \right) \\
&\leq -\bar{c} V_i^{\xi} + \bar{\delta}
\end{aligned} \tag{28}
$$

where $\bar{c} = \min\{-(\eta_1 + \eta_2 - \frac{a_{i1}}{2}), k_n\}$,
$\bar{\delta} = 4\chi_M + 4\Theta_M + \left(\frac{k_n}{2} + \eta_1 - \frac{a_{i1}}{2}\right)\left((1-\xi)\xi^{\frac{\xi}{1-\xi}}\right)$, with $\eta_1 \geq \frac{a_{i1}}{2}$.

Now let

$$
\bar{T} = \frac{1}{(1-\xi)\gamma\bar{c}} \left(V^{1-\xi}(s_i(0), \tilde{\Theta}_i(0)) - \left(\frac{\bar{\delta}}{(1-\gamma)\bar{c}} \right)^{\frac{1-\xi}{\xi}} \right) \tag{29}
$$

With Lemma 1, all signals are SGPFS, $\forall t \geq \bar{T}$, $V^{\xi}(s_i, \tilde{\Theta}_i) \leq \frac{\bar{\delta}}{(1-\xi)\bar{c}}$.

Moreover, from the definition of V, $\forall t \geq \bar{T}$, the following can be obtained:

$$
|e_i| \leq \left(\frac{\bar{\delta}}{(1-\xi)\bar{c}} \right)^{\frac{1}{2\xi}} \tag{30}
$$

After a finite time $\bar{T}$, the tracking errors of all states e_i converge to a small neighborhood of the origin and stay there. This concludes the proof. □

Remark 1. *According to stability analysis, design parameters* $\xi, \gamma, \bar{c}$, *and* $\bar{\delta}$ *significantly impacted the proposed controller's capacity for stability. High control precision and a quick convergence rate can result from large* $\xi, \gamma, \bar{c}$. *They could, however, provide comparatively huge control torques that would not be practical. High control precision and a quick convergence rate can also come from large* $\bar{\delta}$. *This might, however, potentially cause an infinite convergence. Therefore, it is important to carefully choose the design parameters of the suggested controller so that it can reach a sufficient level of control performance.*

Remark 2. *Due to their fractional power structure, finite-time-based controllers are more reliable than traditional controllers, as is demonstrated in the simulation results in the next section. However, because initial conditions impact the convergence time of the finite-time controller, this is not always possible in applications. Therefore, it is not always possible to compute the settling time of the finite-time controller. Since the estimated function's initial values are included in this study, it is also not possible to accurately compute the settling time. A fixed-time stability structure was suggested in the literature [53] as a solution to this issue, and this structure may serve as the subject of our next research.*

Next, numerical simulations are given in detail.

4. Numerical Simulations

To test the efficiency of the suggested controller, a nonlinear model of a quadrotor UAV was built and simulations in MATLAB were run; ode45 was used as the solver. Two different trajectory tracking simulations were performed on quadrotor UAV transporting a suspended payload system to demonstrate the effectiveness of the developed controller. Furthermore, the simulations were enlarged with two more control approaches, demonstrating the superiority of the suggested control over other control structures. First, a classical SMC was used to control the system. Then, the system was controlled with a finite-time SMC (FTSMC) control structure by adding a finite-time control structure to the classical SMC structure. Thus, before adding the neural-network component, the advantages of the finite-time control structure over the classical SMC were observed. Lastly, we clearly demonstrate the superiority of the results obtained using the proposed controller, finite-time-based neuro-sliding mode controller (FTNSMC) over the results obtained in the previous simulations.

The parameters of the quadrotor dynamical model are given in Table 1. The UAV's mass and inertia were also subject to a %20 uncertainty as $\Delta m = 0.20m$, $\Delta I_x = 0.20I_x$, $\Delta I_y = 0.20I_y$, $\Delta I_z = 0.20I_z$. Moreover, the quadrotor was assumed to be subjected to constant time-varying disturbances as $D_x = 0.02sin(t+5)$, $D_y = 0.02cos(2t+3)$, $D_z = 0.02sin(3t+2)$, $D_z = 0.02sin(3t+2)$, $D_\theta = 0.02sin(0.2t+5)$, $D_\phi = 0.02cos(0.2t-3)$, $D_\psi = 0.02sin(0.3t+2)$. Moreover, payload's desired angles are chosen as, $\alpha_x = A_{\alpha_x}cost$, $\alpha_y = A_{\alpha_y}sint$, where $A_{\alpha_x} = A_{\alpha_y} = 0.5$. Controller parameters were selected by trial and error as in Table 2. The hidden-layer neurons were employed at random, while the neural-network weights were initially set to zero during the simulations. The simulations were run with the following assumptions in mind.

Assumption 1. *The parametric uncertainties and external disturbances that were defined in (9) were assumed to be bound by a known bound.*

$$sat(x) = \begin{cases} x/\kappa, & |x| \le \kappa \\ sign(x), & |x| > \kappa \end{cases} \tag{31}$$

where $\kappa > 0$, *and it was taken to be* $\kappa = 0.5$ *in the simulations.*

Assumption 2. *The parametric uncertainties and external disturbances that were defined in (9) were assumed to be bounded by a known bound.*

Table 1. Model parameters.

Parameter	Value	Parameter	Value
m	1.1 kg	g	9.8 m/s^2
m_L	0.5 kg	L	1 m
I_x	1.22 Ns^2/rad	l	0.21 m
I_y	1.22 Ns^2/rad	b	5 Ns
I_z	2.22 Ns^2/rad	d	2 N/ms^2
K_1	0.1 Ns/m	K_4	0.12 Ns/rad
K_2	0.1 Ns/m	K_5	0.12 Ns/rad
K_3	0.1 Ns/m	K_6	0.12 Ns/rad
J_r	0.2 Ns^2/rad		

Table 2. Controller parameters.

Parameter	Value	Parameter	Value
a_{11}	0.01	r_{22}	0.3
a_{12}	0.05	r_{32}	0.03
a_{41}	0.01	η_1	2
a_{42}	0.02	η_2	5
r_{11}	0.03	β	1/3
r_{12}	0.3	α_1	83/97
r_{31}	0.03	k_n	0.1
r_{21}	0.03		

Two different scenarios were considered in the simulations for the trajectory tracking problem of quadrotor carrying a suspended payload.

4.1. Scenario 1

In the context of model uncertainties and external disturbances, the control goal was to allow the quadrotor to track a square desired trajectory. For simulation experiments, the quadrotor's beginning position and angle parameters were $[0, 0, 0]$m and $[0, 0, 0]$rad, and the chosen desired quadrotor trajectories are listed in Table 3 . The simulation time for all runs was 80 s.

Table 3. Desired trajectories for the first scenario.

Parameter	Trajectories	Time (s)
$[x_d \ y_d \ z_d]$	$[6, 6, 6]$ m	10
	$[3, 6, 6]$ m	20
	$[3, 3, 6]$ m	30
	$[6, 3, 6]$ m	40
	$[6, 6, 6]$ m	50
	$[6, 6, 0]$ m	80
$[\phi \ \theta_d \ \psi_d]$	$[0, 0, 0.5]$ rad	60
	$[0, 0, 0]$ rad	80

The x position of the quadrotor for three controllers is given in Figure 4. For a classical SMC, the quadrotor reached the first desired x value at around 7.89 s. The addition of a finite-time structure decreased the reaching time to around 6.08 s, and with the proposed method, the quadrotor reached the first desired x value at around 6.16 s.

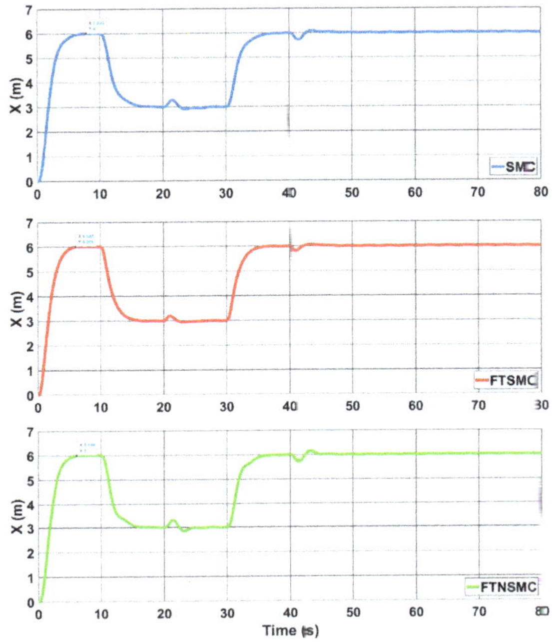

Figure 4. x positions of quadrotor for Scenario 1.

The y position of the quadrotor for three controllers is given in Figure 5. For a classical SMC, the quadrotor could not reach the first desired y value but it reaches about 5.97 m at 10 s. With FTSMC, the quadrotor reached the first desired y value at around 8.87 s, and with the addition of the neural-network component, this value decreased to around 8.55 s.

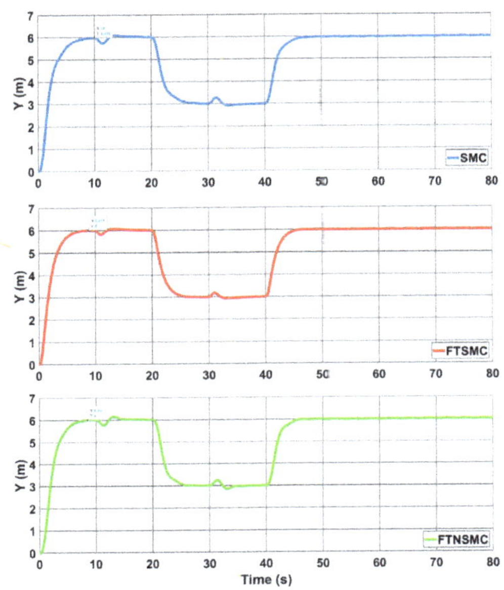

Figure 5. y positions of quadrotor for Scenario 1.

The z position of the quadrotor for three controllers is given in Figure 6. The quadrotor arrived at the desired 6 m value at around 12.68 s, whereas for FTSMC, this value decreased to 10.53 s; with the suggested controller, it settled to the desired value in less than 8.46 s.

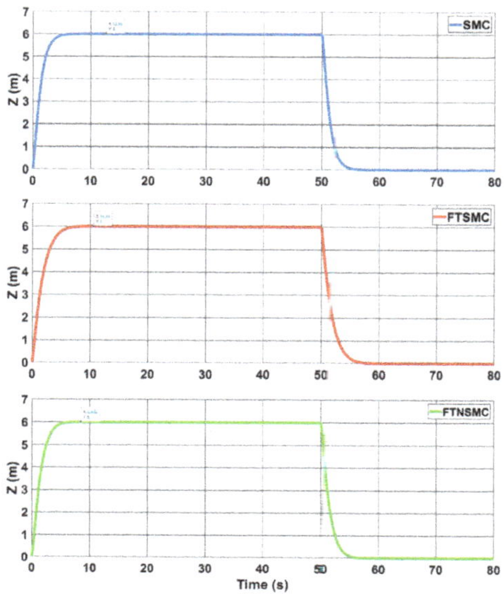

Figure 6. z positions of quadrotor for Scenario 1.

In Figure 7, the change in controller input u_1 is given. Although the initial values of the control input were similar in all three controllers, the chattering effect was clearly observed for the classical SMC. With FTSMC, this chattering decreased but remained. Using the proposed controller, the chattering effect completely disappeared. Moreover, the proposed controller exceptionally decreased the input value compared to SMC and FTSMC.

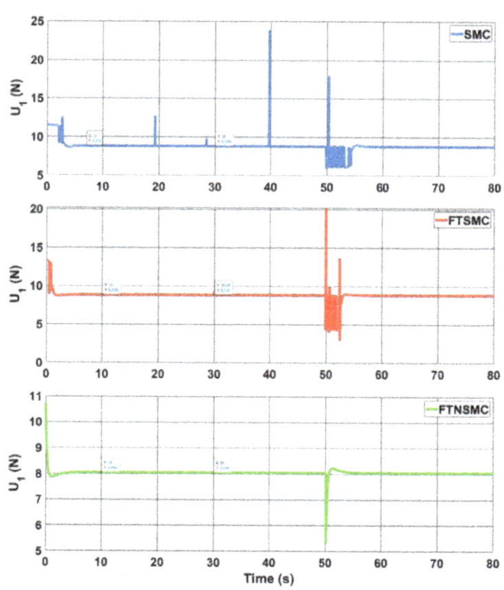

Figure 7. u_1 control inputs for Scenario 1.

Figure 8 displays the change in sliding variable s_1. It behaved as expected when the variables approached its sliding surfaces. Additionally, the oscillations of the sliding variable completely accounted for the position and velocity tracking inaccuracies of the system state variables.

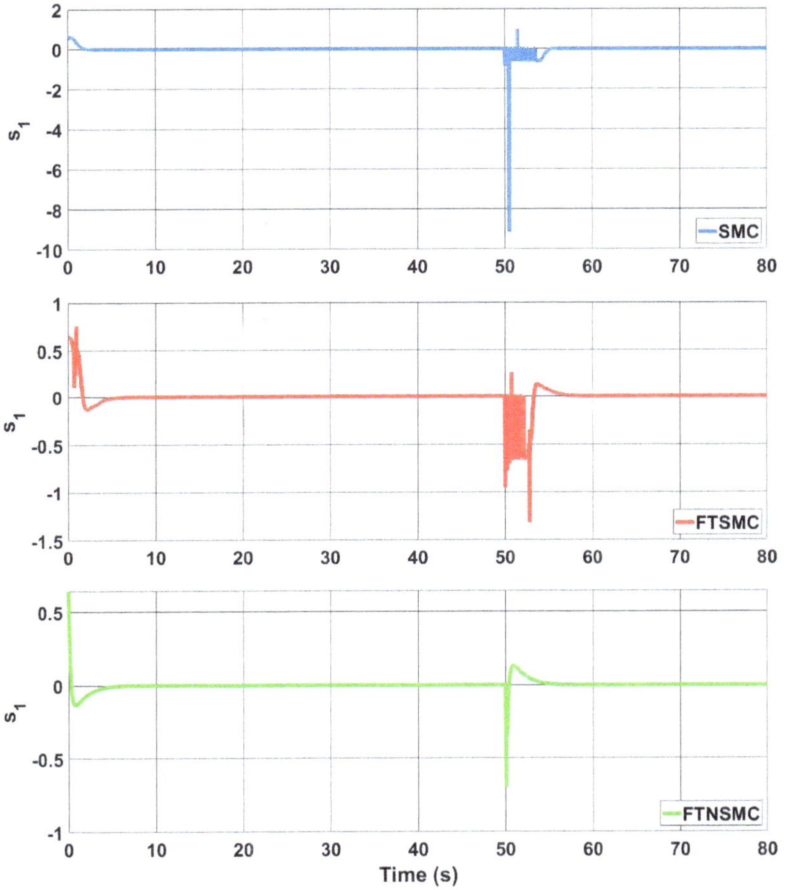

Figure 8. s_1 sliding variables for Scenario 1.

Variation in controller inputs u_2 and u_3 is seen in Figures 9 and 10 respectively. When SMC was used, both a chattering effect and a very high control input value were observed. By adding finite time to SMC, the chattering decreased as the value of the control input decreased with FTSMC. However, in both cases, high sparks and chattering could be seen clearly. Using the proposed controller not only significantly reduced the value of the control input, but also completely eliminated the chattering effect.

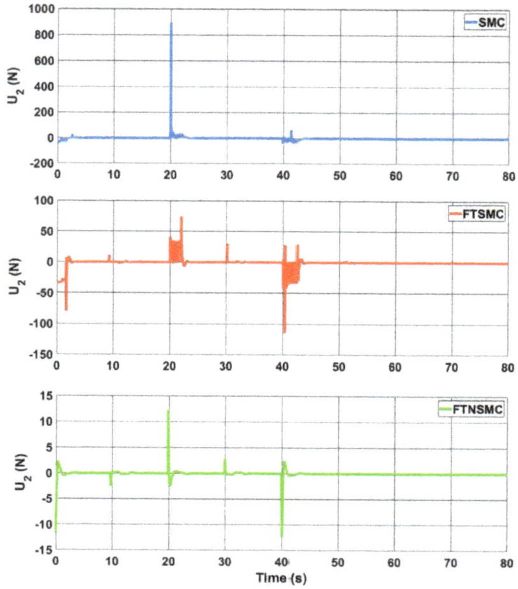

Figure 9. u_2 control inputs for Scenario 1.

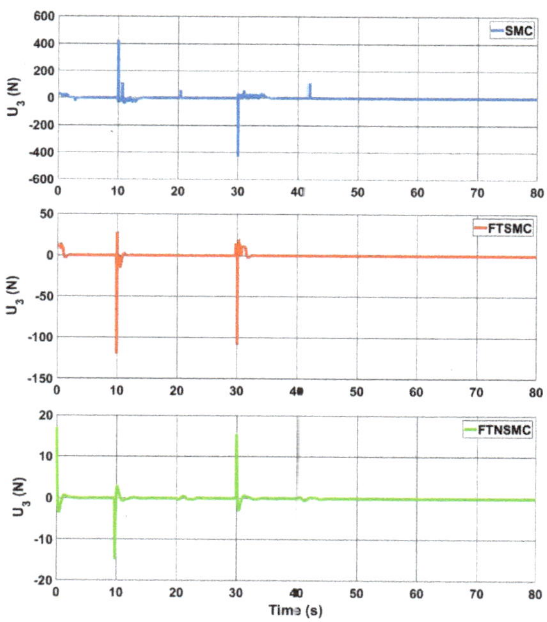

Figure 10. u_3 control inputs for Scenario 1.

In Figure 11, the change in controller input u_4 is given. Despite the fact that the starting values of the control input were comparable in all three controllers, the chattering effect was plainly seen in the conventional SMC. This chattering was reduced by FTSMC, but it still existed. When the proposed controller was used, the chattering effect was fully eliminated.

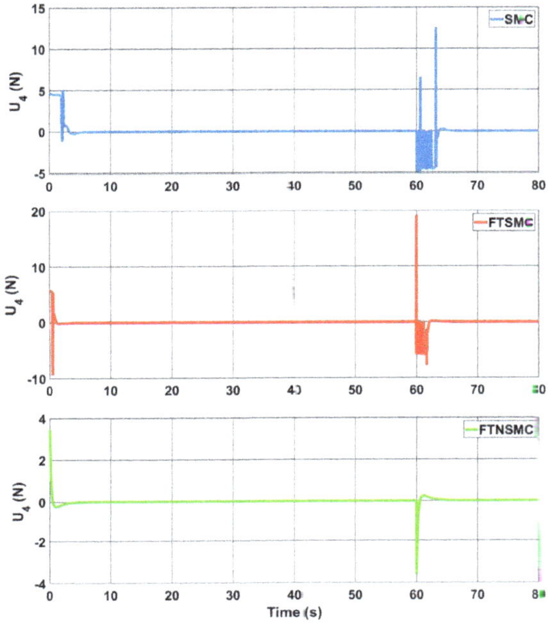

Figure 11. u_4 control inputs for Scenario 1.

Variations in quadrotor angles for proposed controller are given in Figure 12.

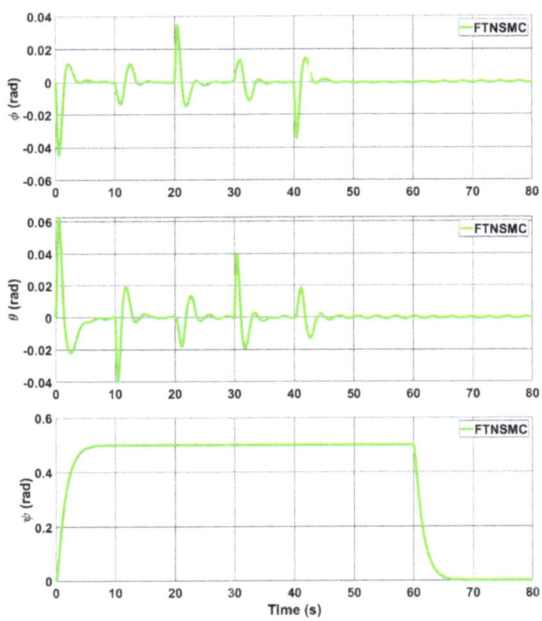

Figure 12. Quadrotor angles for proposed controller for Scenario 1.

The trajectory tracking of the UAV in 3D is given in Figure 13 for all three controllers. Adding a finite-time structure to the conventional SMC obviously sped up the system responsiveness. Additionally, the NN improved the system response even more.

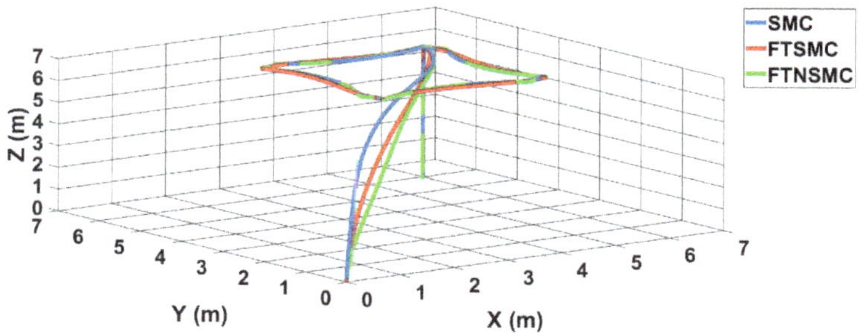

Figure 13. Trajectory of quadrotor in 3D space for Scenario 1.

Figure 14 shows the variation in the neural-network weights. As expected, the neural-network weights rapidly learnt the unknown dynamics after each movement of the UAV and converged to a value close to zero.

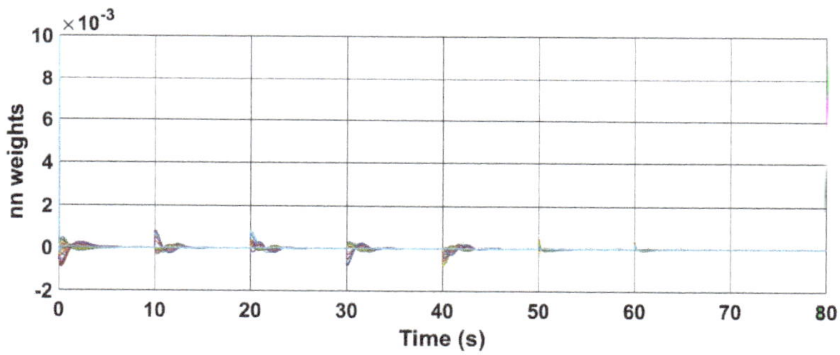

Figure 14. Change in neural-network weights for Scenario 1.

4.2. Scenario 2

In the second simulation, the control goal was to allow the quadrotor to track a circular desired trajectory and also ensures that the system state errors reach zero or to a close neighbourhood of zero in a finite time. The chosen desired quadrotor trajectories are listed in Table 4. The simulation time for all runs was 320 s.

Table 4. Desired trajectories for the second scenario.

Parameter	Trajectories
$[x_d\ y_d\ z_d]$	$50sin(0.05t)$ $50cos(0.05t)$ t
$[\theta_d\ \phi_d\ \psi_d]$	$[0, 0, 0.5]$ rad

The errors in the x axis for three controllers are given in Figure 15. Compared to a classical SMC, the proposed controller provided a fast convergence rate as accepted. The error in the x axis reached zero for the first time at around 4.3 s with the proposed controller and oscillated in the neighborhood of zero with a small margin because of the nature of desired trajectory, whereas it reached zero at around 6.6 s with FTSMC and 14 s with classical SMC.

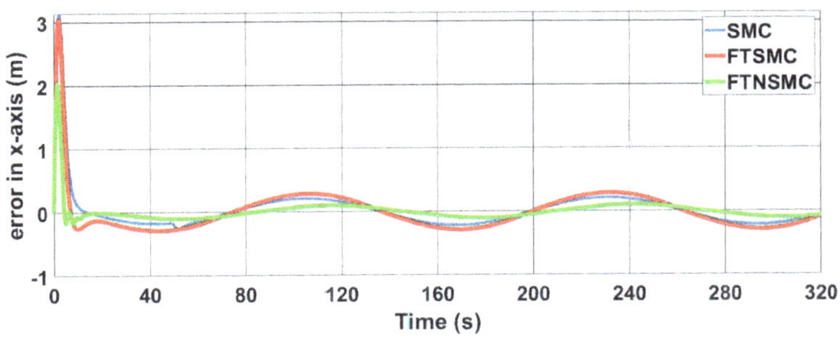

Figure 15. Errors in x-axis for Scenario 2.

Figure 16 displays the y-axis errors for three controllers. With the SMC, error convergence was very slow and it converged to zero after 50 s, as seen in the figure. The addition of finite-time structure decreased this value to around 13 s.

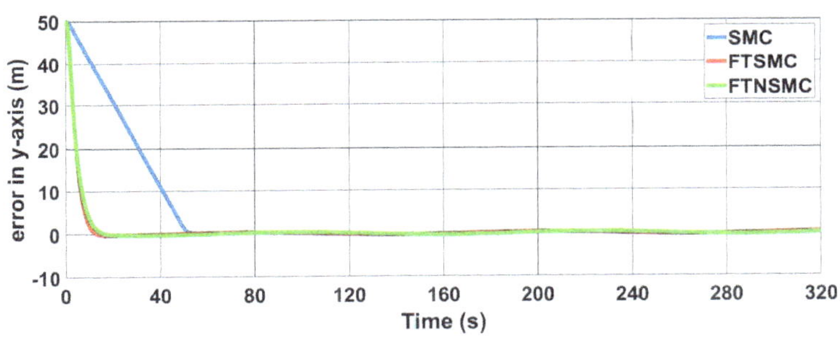

Figure 16. Errors in y-axis for Scenario 2.

In Figure 17 errors in the z axis are shown. For SMC controller error reached the neighborhood of zero within around 6.5 s. With FTSMC, this value was around 5.5 s, and with FTNSMC, it was around 4.5 s.

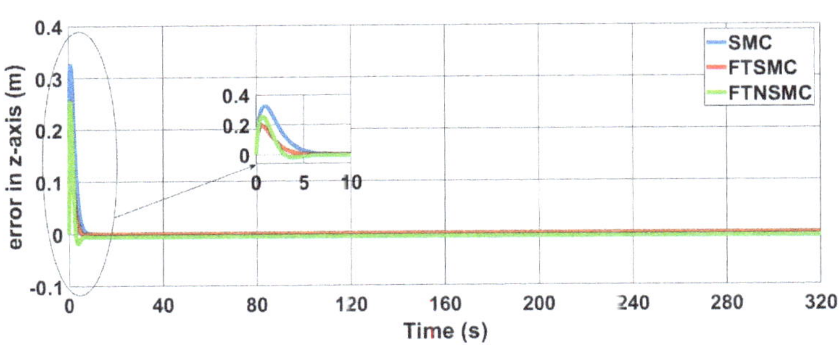

Figure 17. Errors in z-axis for Scenario 2.

u_1 controller inputs for three controllers are given in Figure 18. Since the desired trajectory was orbital, the progress of the motion in the z axis was expected to be slow. Changes in the control inputs for this axis were, therefore, close to each other within the three controllers. On the other hand, there was a difference between the convergence of the control inputs to nominal value. This time was around 4.5 s for SMC and FTSMC, and this time decreased to 2.8 s for the proposed controller. Moreover, small chattering could be detected with FTSMC, but this effect was also eliminated with FTNSMC.

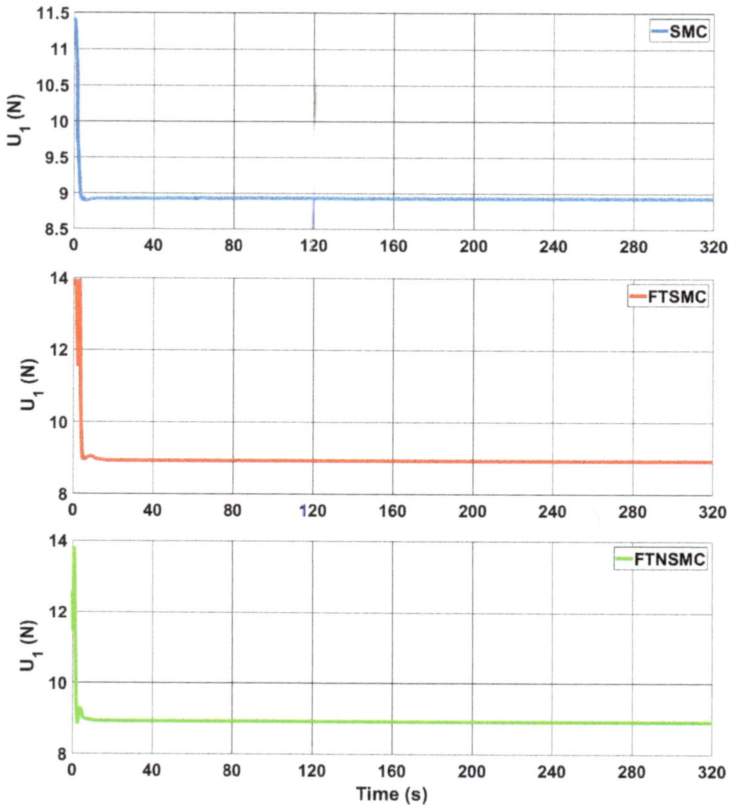

Figure 18. u_1 control inputs for Scenario 2.

Variations in control inputs u_2 and u_3 are given for the three controllers in Figures 19 and 20, respectively. The control signal for SMC exhibited undesirable sparks at the beginning with some chattering effect and, since it could only enter the desired trajectory after 50 s, high chattering occurred at these time values for input u_2. FTSMC reduced this chattering effect while increasing the initial value of control input; lastly, by using NNFTSMC, a chattering-free control input could be obtained while decreasing the initial value of control input of FTSMC. For the u_3 control input, a chattering effect could also be seen for SMC and FTSMC. However, the neural-network component in the proposed controller significantly reduced this effect.

Figure 21 demonstrates the change of control inputs u_4 for the three controllers. Although the initial values were close to each other for all controllers, with the proposed controller, the control input converged to zero more quickly after the initial movement and it was a smoother control input than the SMC and FTSMC methods. Furthermore, FTSMC showed more chattering-free control input behavior than the classical SMC did.

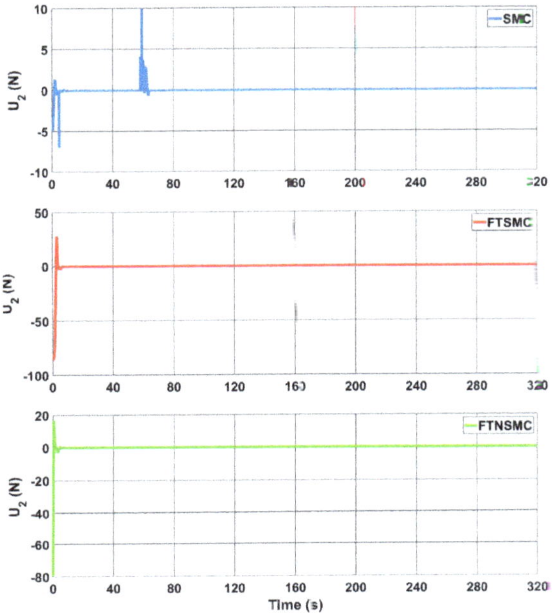

Figure 19. u_2 control inputs for Scenario 2.

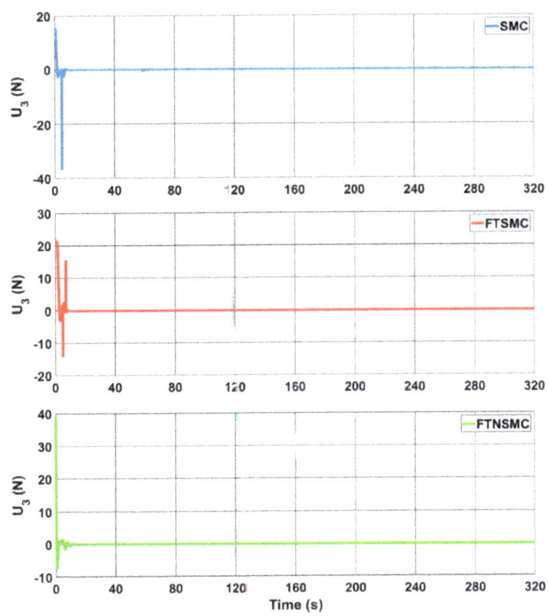

Figure 20. u_3 control inputs for Scenario 2.

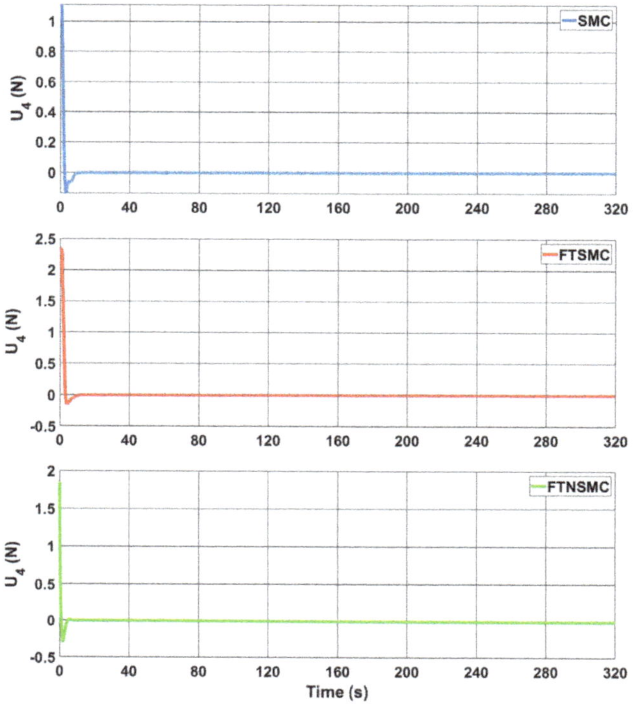

Figure 21. u_4 control inputs for Scenario 2.

In Figure 22, the trajectory of the quadrotor in 3D space is given for three controllers. The suggested controller tracking performance of the desired trajectory was superior to that of classical SMC and FTSMC.

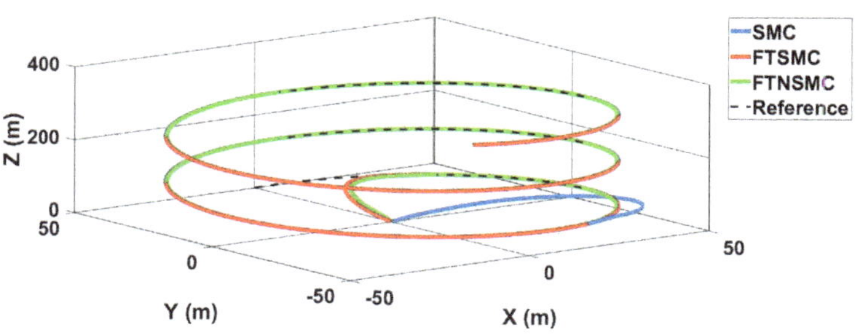

Figure 22. Trajectory of quadrotor in 3D space for Scenario 2.

The change in neural-network weights is presented in Figure 23. As expected, the neural-network weights rapidly learnt the unknown dynamics after each movement of the UAV and converged to a value close to zero.

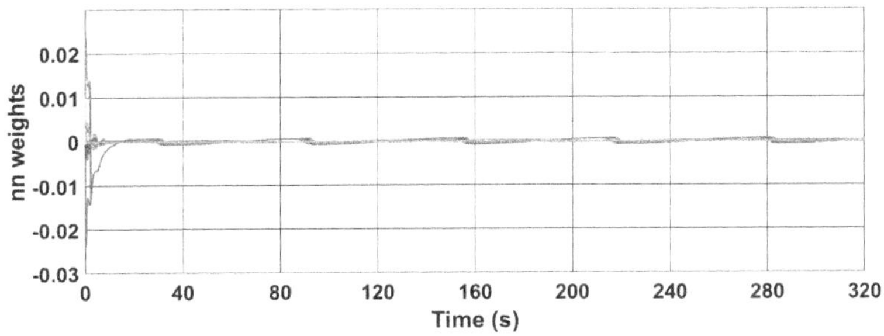

Figure 23. Change in neural-network weights for Scenario 2.

5. Conclusions

A neural-network-based finite-time sliding-mode controller was proposed to control a quadrotor UAV carrying a suspended payload with parametric uncertainties and external disturbances. After deriving the full nonlinear mathematical model of system, the proposed controller was developed, and three different simulations were performed to demonstrate the effectiveness of the proposed controller. First, the system was controlled with a classical SMC. In this simulation, errors in the trajectory tracking in all three axes due to the effects of unknown dynamics and serious chattering effects were observed in the control signals. Very high control signals were also needed to control the system. In the second simulation, a finite-time control structure was added to the SMC to reduce the values of the control signals and the trajectory tracking errors to converge to zero in a finite time. Although FTSMC improved the settling times and the values of the controller signals, it was not effective in eliminating the chattering effect. A neural-network structure was added to FTSMC to construct the proposed controller and effectively handle chattering. With the use of the neural network, chattering disappeared completely, the control signs were significantly reduced, and the settling times were further improved. Future work is to improve the finite-time stability to fixed-time stability to be able to free it of initial conditions to calculate the reaching time, demonstrating the effectiveness of the proposed controller with a real-time experiment.

Author Contributions: The article was written as part of the doctoral thesis of Ö.B. under the supervision of H.M.G. All authors have read and agreed to the published version of the manuscript.

Funding: This research received no external funding.

Institutional Review Board Statement: Not applicable

Informed Consent Statement: Not applicable.

Data Availability Statement: Not applicable.

Conflicts of Interest: The authors declare no conflict of interest.

References

1. Ding, Y.; Xin, B.; Chen, J. A Review of Recent Advances in Coordination between Unmanned Aerial and Ground Vehicles. *Unmanned Syst.* **2021**, *9*, 97–117. [CrossRef]
2. Shakhatreh, H.; Sawalmeh, A.H.; Al-Fuqaha, A.; Dou, Z.; Almaita, E.; Khalil, I.; Othman, N.S.; Khreishah, A.; Guizani, M. Unmanned Aerial Vehicles (UAVs): A Survey on Civil Applications and Key Research Challenges. *IEEE Access* **2019**, *7*, 48572–48634. [CrossRef]
3. Cai, G.; Dias, J.; Seneviratne, L. A Survey of Small-Scale Unmanned Aerial Vehicles: Recent Advances and Future Development Trends. *Unmanned Syst.* **2014**, *2*, 175–199. [CrossRef]
4. Shraim, H.; Awada, A.; Youness, R. A survey on quadrotors: Configurations, modeling and identification, control, collision avoidance, fault diagnosis and tolerant control. *IEEE Aerosp. Electron. Syst. Mag.* **2018**, *33*, 14–33. [CrossRef]
5. Faust, A.; Palunko, I.; Cruz, P.; Fierro, R.; Tapia, L Automated aerial suspended cargo delivery through reinforcement learning. *Artif. Intell.* **2017**, *247* [CrossRef]

6. Kim, S.; Choi, S.; Kim, H.J. Aerial manipulation using a quadrotor with a two DOF robotic arm. In Proceedings of the 2013 IEEE/RSJ International Conference on Intelligent Robots and Systems, Tokyo, Japan, 3–7 November 2013; pp. 4990–4995. [CrossRef]
7. Vahdanipour, M.; Khodabandeh, M. Adaptive fractional order sliding mode control for a quadrotor with a varying load. *Aerosp. Sci. Technol.* **2019**, *86*, 737–747. [CrossRef]
8. Sreenath, K.; Taeyoung Lee.; Kumar, V. Geometric control and differential flatness of a quadrotor UAV with a cable-suspended load. In Proceedings of the IEEE 52nd Annual Conference on Decision and Control (CDC 2013), Firenze, Italy, 10–13 December 2013; Volume 2019, pp. 2269–2274. [CrossRef]
9. Pizetta, I.H.B.; Brandao, A.S.; Sarcinelli-Filho, M. Modelling and control of a PVTOL quadrotor carrying a suspended load. In Proceedings of the ICUAS 2015—International Conference on Unmanned Aircraft Systems, Denver, CO, USA, 9–12 June 2015, [CrossRef]
10. Tang, S.; Kumar, V. Mixed Integer Quadratic Program trajectory generation for a quadrotor with a cable-suspended payload. In Proceedings of the 2015 IEEE International Conference on Robotics and Automation (ICRA), Seattle, WA, USA, 26–30 May 2015; pp. 2216–2222. [CrossRef]
11. Pounds, P.E.; Bersak, D.R.; Dollar, A.M. Stability of small-scale UAV helicopters and quadrotors with added payload mass under PID control. *Auton. Robots* **2012**, *33*, 129–142. [CrossRef]
12. Klausen, K.; Fossen, T.I.; Johansen, T.A. Nonlinear control of a multirotor UAV with suspended load. In Proceedings of the 2015 International Conference on Unmanned Aircraft Systems (ICUAS), Denver, CO, USA , 9–12 June 2015. [CrossRef]
13. Lee, T. Geometric Control of Quadrotor UAVs Transporting a Cable-Suspended Rigid Body. *IEEE Trans. Control. Syst. Technol.* **2018**, *26*, 255–264. [CrossRef]
14. Yang, S.; Xian, B. Energy-Based Nonlinear Adaptive Control Design for the Quadrotor UAV System With a Suspended Payload. *IEEE Trans. Ind. Electron.* **2020**, *67*, 2054–2064. [CrossRef]
15. Qian, L.; Liu, H.H. Path-Following Control of A Quadrotor UAV with A Cable-Suspended Payload Under Wind Disturbances. *IEEE Trans. Ind. Electron.* **2020**, *67*, 2021–2029. [CrossRef]
16. Wang, C.; Song, B.; Huang, P.; Tang, C.; Wang, C.; Song, B.; Huang, P.; Tang, C. Trajectory Tracking Control for Quadrotor Robot Subject to Payload Variation and Wind Gust Disturbance. *J. Intell. Robot Syst.* **2016**, *83*, 315–333. [CrossRef]
17. Bhat, S.P.; Bernstein, D.S. Finite-Time Stability of Continuous Autonomous Systems. *SIAM J. Control Optim.* **2000**, *38*, 751–766. [CrossRef]
18. Sun, K.; Qiu, J.; Karimi, H.R.; Gao, H. A Novel Finite-Time Control for Nonstrict Feedback Saturated Nonlinear Systems With Tracking Error Constraint. *IEEE Trans. Syst. Man, Cybern. Syst.* **2021**, *51*, 3968–3979. [CrossRef]
19. Sui, S.; Chen, C.L.P.; Tong, S. Fuzzy Adaptive Finite-Time Control Design for Nontriangular Stochastic Nonlinear Systems. *IEEE Trans. Fuzzy Syst.* **2019**, *27*, 172–184. [CrossRef]
20. Wang, N.; Karimi, H.R.; Li, H.; Su, S.F. Accurate Trajectory Tracking of Disturbed Surface Vehicles: A Finite-Time Control Approach. *IEEE/ASME Trans. Mechatronics* **2019**, *24*, 1064–1074. [CrossRef]
21. Du, P.; Liang, H.; Zhao, S.; Ahn, C.K. Neural-Based Decentralized Adaptive Finite-Time Control for Nonlinear Large-Scale Systems With Time-Varying Output Constraints. *IEEE Trans. Syst. Man Cybern. Syst.* **2021**, *51*, 3136–3147. [CrossRef]
22. Utkin, V.I. Survey Paper: Variable Structure Systems with Sliding Modes. *IEEE Trans. Automat. Contr.* **1977**. [CrossRef]
23. Muliadi, J.; Kusumoputro, B. Neural Network Control System of UAV Altitude Dynamics and Its Comparison with the PID Control System. *J. Adv. Transp.* **2018**, *2018*, 3823201. [CrossRef]
24. Slotine, J.J.; Sastry, S.S. Tracking Control of Nonlinear Systems using Sliding Surfaces. *Int. J. Control* **1983**, *38*, 465–492. [CrossRef]
25. Levant, A. Principles of 2-sliding mode design. *Automatica* **2007**, *43*, 576–586. [CrossRef]
26. Plestan, F.; Glumineau, A.; Laghrouche, S. A new algorithm for high-order sliding mode control. *Int. J. Robust Nonlinear Control.* **2008**, *18*, 441–453. [CrossRef]
27. Frikha, S.; Djemel, M.; Derbel, N. A New Adaptive Neuro-sliding Mode Control for Gantry Crane. *Int. J. Control. Autom. Syst.* **2018**, *16*, 559–565. [CrossRef]
28. Guzey, H.M.; Dierks, T.; Jagannathan, S.; Acar, L. Modified Consensus-based Output Feedback Control of Quadrotor UAV Formations Using Neural Networks. *J. Intell. Robot. Syst. Theory Appl.* **2019**. [CrossRef]
29. Razmi, H.; Afshinfar, S. Neural network-based adaptive sliding mode control design for position and attitude control of a quadrotor UAV. *Aerosp. Sci. Technol.* **2019**, *91*, 12–27. [CrossRef]
30. Bansal, S.; Akametalu, A.K.; Jiang, F.J.; Laine, F.; Tomlin, C.J. Learning quadrotor dynamics using neural network for flight control. In Proceedings of the IEEE 55th Conference on Decision and Control (CDC), Las Vegas, NV, USA, 12–14 December 2016, [CrossRef]
31. Li, S.; Wang, Y.; Tan, J.; Zheng, Y. Adaptive RBFNNs/integral sliding mode control for a quadrotor aircraft. *Neurocomputing* **2016**, *216*, 126–134. [CrossRef]
32. Lin, X.; Wang, Y.; Liu, Y. Neural-network-based robust terminal sliding-mode control of quadrotor. *Asian J. Control* **2022**, *24*, 427–438. [CrossRef]
33. Wang, L.; Chai, T.; Zhai, L. Neural-network-based terminal sliding-mode control of robotic manipulators including actuator dynamics. *IEEE Trans. Ind. Electron.* **2009**, *56*, 3296–3304. [CrossRef]

34. Bingöl, Ö.; Güzey, H.M. Neuro sliding mode control of quadrotor UAVs carrying suspended payload. *Adv. Robot.* **2021**, *35*, 255–266. [CrossRef]
35. Cabecinhas, D.; Cunha, R.; Silvestre, C. A trajectory tracking control law for a quadrotor with slung load. *Automatica* **2019**, *106*, 384–389. [CrossRef]
36. Yu, G.; Cabecinhas, D.; Cunha, R.; Silvestre, C. Nonlinear Backstepping Control of a Quadrotor-Slung Load System. *IEEE/ASME Trans. Mechatronics* **2019**, *24*, 2304–2315. [CrossRef]
37. Mofid, O.; Mobayen, S.; Wong, W.K. Adaptive terminal sliding mode control for attitude and position tracking control of quadrotor UAVs in the existence of external disturbance. *IEEE Access* **2020**, *9*, 3428–3440. [CrossRef]
38. Lv, Z.; Zhao, Q.; Li, S.; Wu, Y. Finite-time control design for a quadrotor transporting a slung load. *Control. Eng. Pract.* **2022**, *122*, 105082. [CrossRef]
39. Kang, B.; Miao, Y.; Liu, F.; Duan, J.; Wang, K. Jiang, S. A second-order sliding mode controller of quad-rotor UAV based on PID sliding mode surface with unbalanced load. *J. Syst. Sci. Complex.* **2021**, *34*, 520–536. [CrossRef]
40. Zhu, Z.; Xia, Y.; Fu, M. Attitude stabilization of rigid spacecraft with finite-time convergence. *Int. J. Robust Nonlinear Control* **2011**, *21*, 686–702. [CrossRef]
41. Lv, W.; Wang, F. Finite-time adaptive fuzzy tracking control for a class of nonlinear systems with unknown hysteresis. *Int. J. Fuzzy Syst.* **2018**, *20*, 782–790. [CrossRef]
42. Hardy, G.H.; Littlewood, J.E.; Pólya, G.; Pólya, G. *Inequalities*; Cambridge University Press: Cambridge, UK, 1952.
43. Xu, R.; Özgüner, Ü. Sliding mode control of a class of underactuated systems. *Automatica* **2008**, *44*, 233–241. [CrossRef]
44. Liu, J.; Gai, W.; Zhang, J.; Li, Y. Nonlinear Adaptive Backstepping with ESO for the Quadrotor Trajectory Tracking Control in the Multiple Disturbances. *Int. J. Control. Autom. Syst.* **2019**, *17*, 2754–2768. [CrossRef]
45. Xiong, J.J.; Zhang, G. Sliding mode control for a quadrotor UAV with parameter uncertainties. In Proceedings of the 2nd International Conference on Control, Automation and Robotics (ICCAR), Hong Kong, 28–30 April 2016. [CrossRef]
46. Feng, Y.; Rabbath, C.A.; Su, C.Y. Modeling of a Micro UAV with Slung Payload. In *Handbook of Unmanned Aerial Vehicles*; Springer: Berlin/Heidelberg, Germany, 2015. [CrossRef]
47. Lewis, F.L. Neural network control of robot manipulators. *IEEE Expert. Syst. their Appl.* **1996**, *11*, 64–75. [CrossRef]
48. Dierks, T.; Jagannathan, S. Output feedback control of a quadrotor UAV using neural networks. *IEEE Trans. Neural Netw.* **2010**, *21*, 50–56. [CrossRef]
49. Riani, A.; Madani, T.; Benallegue, A.; Djouani, K. Adaptive integral terminal sliding mode control for upper-limb rehabilitation exoskeleton. *Control Eng. Pract.* **2018**, *75*, 108–117. [CrossRef]
50. Lian, S.; Meng, W.; Shao, K.; Zheng, J.; Zhu, S.; Li, H. Full Attitude Control of a Quadrotor Using Fast Non-singular Terminal Sliding Mode with Angular Velocity Planning. *IEEE Trans. Ind. Electron.* **2022**, *69*, 1597–1607. [CrossRef]
51. Zheng, E.H.; Xiong, J.J.; Luo, J.L. Second order sliding mode control for a quadrotor UAV. *ISA Trans.* **2014**, *53*, 1350–1356. [CrossRef] [PubMed]
52. Fallaha, C.J.; Saad, M.; Kanaan, H.Y.; Al-Haddad, K. Sliding-mode robot control with exponential reaching law. *IEEE Trans. Ind. Electron.* **2011**, *58*, 600–610. [CrossRef]
53. Polyakov, A. Nonlinear Feedback Design for Fixed-Time Stabilization of Linear Control Systems. *IEEE Trans. Autom. Control.* **2012**, *57*, 2106–2110. [CrossRef]

drones

MDPI

Article

Deep Reinforcement Learning for the Visual Servoing Control of UAVs with FOV Constraint

Gui Fu [1,2], Hongyu Chu [1,*], Liwen Liu [2], Linyi Fang [2] and Xinyu Zhu [2]

[1] School of Information Engineering, Southwest University of Science and Technology, Mianyang 621010, China; abyfugui@cafuc.edu.cn

[2] Institute of Electronic and Electrical Engineering, Civil Aviation Flight University of China, Guanghan 618307, China; llw@cafuc.edu.cn (L.L.); brwn6g78q@126.com (L.F.); cafuczxy@dingtalk.com (X.Z.)

* Correspondence: chuhongyu@swust.edu.cn; Tel.:+86-187-8109-9091

Abstract: Visual servoing is a control method that utilizes image feedback to control robot motion, and it has been widely applied in unmanned aerial vehicle (UAV) motion control. However, due to field-of-view (FOV) constraints, visual servoing still faces challenges, such as easy target loss and low control efficiency. To address these issues, visual servoing control for UAVs based on the deep reinforcement learning (DRL) method is proposed, which dynamically adjusts the servo gain in real time to avoid target loss and improve control efficiency. Firstly, a Markov model of visual servoing control for a UAV under field-of-view constraints is established, which consists of quintuplet and considers the improvement of the control efficiency. Secondly, an improved deep Q-network (DQN) algorithm with a target network and experience replay is designed to solve the Markov model. In addition, two independent agents are designed to adjust the linear and angular velocity servo gains in order to enhance the control performance, respectively. In the simulation environment, the effectiveness of the proposed method was verified using a monocular camera.

Keywords: field-of-view (FOV) constraint; visual servoing; deep reinforcement learning; UAV

1. Introduction

An unmanned aerial vehicle (UAV) is an integrated agent capable of functions including environmental perception, motion planning, and servoing control among others [1]. Meanwhile, it also represents an integration of sensing technology, electronic engineering, communication engineering, computer engineering, automation control, and artificial intelligence [2,3]. Quadrotor UAVs can take off and land vertically, hover in the air, fly at low altitudes, cruise slowly, and so on. Hence, UAVs represent a focus of current research [4]. Quadrotors are widely used in daily life, e.g., in drones, small logistics transportation, pesticide spraying, and high-altitude auxiliary fire extinguishing [5].

Visual servoing adopts the image information of the target position obtained from airborne cameras to control the relative position and altitude of a quadrotor UAV [6]. Generally, visual servoing can be divided into position-based visual servoing (PBVS) and image-based visual servoing (IBVS) [7]. Errors in PBVS originate from the three-dimensional Cartesian space; thus, it is sensitive to noise, camera parameter errors, and the estimation accuracy of the target position [8]. IBVS uses the image feature error to calculate the input of the control robot directly, which enhances the robustness of the camera parameters. In contrast to PBVS, an accurate geometric model of the visual target is no longer required [9]. Thus, IBVS is widely applied in the control of mobile robots [10] and UAVs [11–13]. Servo gain is critical for IBVS because it affects the output and convergence speed of the whole system [14].

In various applications, such as cluster control based on vision [15,16], target tracking [17], and visual navigation [18], the target may be lost because of the limited FOV. Once the target is lost, these tasks cannot be performed properly. Therefore, it is important to consider the FOV constraint in the design stage of the servo controller.

To avoid feature loss, several methods have been proposed in recent decades. Hajiloo et al. [19] optimized the trajectory of features in image frames using the MPC method to address the FOV constraints. Chesi et al. [20] used the in-plane trajectory of a camera in three-dimensional Cartesian space but did not optimize the trajectory of features in the image frame [20]. Huang et al. [21] designed a constraint controller using the control obstacle function and quadratic programming to prevent the target on the ship exceeding the FOV of the camera. Zhang et al. [22] presented a new real-time optimal trajectory planning method for a rotorcraft with the constraint of speed and inputs of thrust and altitude. Their framework ensures the field-of-view constraints on the visual servoing control of the rotorcraft by transforming the FOV constraint into an altitude constraint. Zheng et al. [23] established a control barrier function to ensure the FOV constraint to maintain visibility. The barrier function is used to create a visible range that sets the maximum distance between the center of the image plane and the visual feature coordinates in the image plane. This visible range can be defined as the upper limit of the distance between these two points.

In recent years, with the rapid development of artificial intelligence, machine learning has been widely applied in various fields. Meanwhile, formal methods for the verification and validation of machine learning systems have been proposed in [24,25]. In the field of robot control, the reinforcement learning (RL) algorithm is used to complete the visual servoing task [26]. Wang et al. [27] introduced controllers to addresses the retention of visual features in the field of view of the camera for wheeled mobile robots (WMRs). When the feature is located in a dangerous area, a controller based on Q-learning is activated to avoid feature loss. This method uses Q-learning to directly control the motion of a WMR, but these controllable actions are rare and only applicable to some scenarios. In contrast to directly using the RL algorithm for the input of the control system, Shi et al. [28,29] used the Q-learning algorithm to choose the adaptive law to update the gain of IBVS, and the stability and convergence of the control system were improved. However, these methods did not consider the issue of field-of-view constraints. As we know, there is little research on learning-based FOV constraint control for UAVs. A comparison of our work with related works is provided in Table 1.

Table 1. Comparison of current related work.

Reference	Object	Method
[19]	Six-DOF robot manipulator	The constraints due to actuator limitations and visibility constraints can be taken into account using MPC strategy and computational complexity.
[20]	Six-DOF articulated arm	Trajectory in the 3D space satisfying FOV constraints. It depends on accurate environmental and system models.
[21]	VTOL UAVs	Generate the constrained control inputs to ensure the nonsingular attitude extraction and FOV.
[22]	Quadrotor	FOV is indirectly guaranteed by attitude constraints.
[23]	Quadrotor	The system is bounded by a visible set. Control barrier function
[27]	WMRs	Using Q-learning to design a controller, the action is simple.
[28,29]	Quadrotor	Using Q-learning to design adaptive laws, the control effectiveness is improved without considering the FOV
Ours	Quadrotor	Using DQN to design adaptive laws, the control effectiveness is improved considering the FOV

In order to solve the problem of feature loss and improve the efficiency of IBVS, we propose a method of IBVS control of UAVs with visibility constraints based on the deep reinforcement learning method, which dynamically adjusts the servo gain in real time to avoid target loss and improve control efficiency. The Markov model of visual servoing control for a UAV under field-of-view constraints is established, and an improved DQN algorithm with a target network and experience replay is designed to solve the Markov model. Then, to enhance the control performance of the control system, we designed two independent servo gains for linear and angular velocities to replace the single servo gain in the traditional method.

The remainder of this paper is organized as follows. In Section 2, the works related to the proposed method are presented, which include a model of a quadrotor UAV and the method of classical IBVS. Section 3 describes the proposed algorithm for adjusting the adaptive servo gain using a DQN. To demonstrate the effectiveness of the proposed method, simulations are presented in Section 4. Finally, Section 5 summarizes the conclusions of the study.

2. Preliminaries

2.1. Quadrotor Model Description

The coordinate frames of a quadrotor are shown in Figure 1, where the body coordinate frame and world coordinate frame are denoted as $X_b - Y_b - Z_b$ and $X_i - Y_i - Z_i$, respectively. The coordinate of the quadrotor's centroid $\boldsymbol{\xi} = (X, Y, Z)$:

$$\begin{cases} \dot{\boldsymbol{\xi}} = \boldsymbol{R}\boldsymbol{v} \\ m\dot{\boldsymbol{v}} = -m\boldsymbol{\omega} \times \boldsymbol{v} + \boldsymbol{F} \\ \dot{\boldsymbol{R}} = \boldsymbol{R} \cdot sk(\boldsymbol{\omega}) \\ \boldsymbol{I}\dot{\boldsymbol{\omega}} = -\boldsymbol{\omega} \times \boldsymbol{I}\boldsymbol{\omega} + \boldsymbol{\Gamma} \\ \boldsymbol{F} = mg\boldsymbol{R}^T\boldsymbol{e} - u_1\boldsymbol{e_z} \\ \boldsymbol{\Gamma} = [u_2, u_3, u_4] \end{cases} \tag{1}$$

where " $\times$ " represents the cross multiplication operation; $\boldsymbol{R} \in SO(3)$ is the rotation matrix from the body coordinate to world coordinate; $\boldsymbol{I} \in \Re^{3\times 3}$ is the UAVs constant inertia matrix around the centroid; $sk()$ is a skew-symmetric matrix with $sk(a)b = a \times b$; u_1 is the total lift force along the axis of the coordinate system; and u_1, u_2, u_3 are the moment of rotation for the x, y, and z axes of the body coordinate system. u_1, u_2, u_3, u_4 in the body coordinates can be calculated as follows:

$$\begin{bmatrix} u_1 \\ u_2 \\ u_3 \\ u_4 \end{bmatrix} = \begin{bmatrix} K_F & K_F & K_F & K_F \\ -K_F L & 0 & K_F L & 0 \\ 0 & -K_F L & 0 & K_F L \\ K_M & -K_M & K_M & -K_M \end{bmatrix} \begin{bmatrix} w_1^2 \\ w_2^2 \\ w_3^2 \\ w_4^2 \end{bmatrix} \tag{2}$$

where K_F and K_M are the lift coefficient and moment coefficient, respectively. The parameter L is the arm of force, indicating the displacement of the motor relative to the center of mass of the aircraft. w_i represents the rotational speed of each motor.

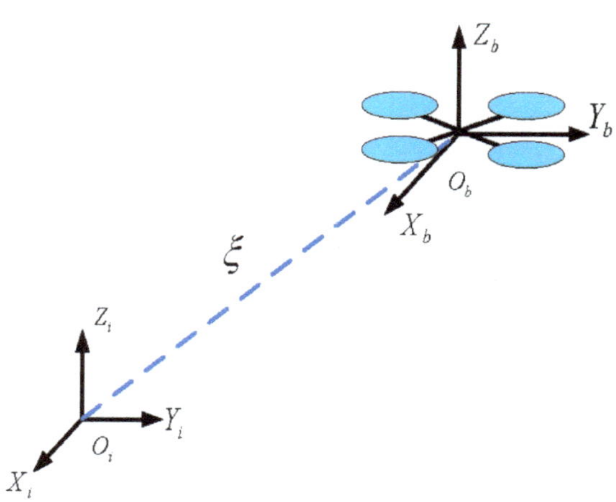

Figure 1. The coordinate frames of a quadrotor.

2.2. The Classical IBVS Method

Assuming that $P_i^C = (X_i^C, Y_i^C, Z_i^C)^T$ is a point in the camera coordinate and $p_i = (x_i, y_i)$ is the observed target point in the image, the relationship between P_i^C and p_i is illustrated in Figure 2. The camera coordinate and image plane are represented by X-Y-Z and x-y, respectively. The principal point is the intersection of the Z-axis and image plane; its coordinates can be expressed as (u_0, v_0).

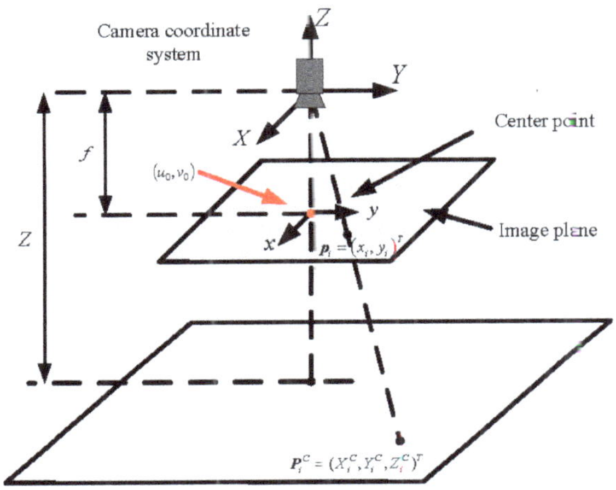

Figure 2. The perspective imaging mode of the camera.

The conversion between a point in the camera coordinate system and the image coordinate system refers to the mapping of a 3D point in the world captured with a camera to a 2D point on the camera sensor. The conversion can be described as

$$\begin{cases} x_i = f\frac{X_i^C}{Z_i^C} \\ y_i = f\frac{Y_i^C}{Z_i^C} \end{cases} \quad (3)$$

The derivative of Equation (3) with respect to time is expressed as

$$\begin{bmatrix} \dot{x}_i \\ \dot{y}_i \end{bmatrix} = \frac{-1}{Z_i^C}\begin{bmatrix} f_1 & 0 \\ 0 & f_2 \end{bmatrix}\begin{bmatrix} 1 & 1 & -X_i^C/Z_i^C \\ 0 & 1 & -Y_i^C/Z_i^C \end{bmatrix}\begin{bmatrix} \dot{X}_i^C \\ \dot{Y}_i^C \\ \dot{Z}_i^C \end{bmatrix} \quad (4)$$

Substituting Equation (3) into Equation (4) yields

$$\begin{cases} \dot{x}_i = \frac{\left(\dot{X}_i^C - x\dot{Z}_i^C\right)}{Z_i^C} \\ \dot{y}_i = \frac{\left(\dot{Y}_i^C - y\dot{Z}_i^C\right)}{Z_i^C} \end{cases} \quad (5)$$

The relationship between the velocity of the three-dimensional points and the spatial velocity of the camera is established using the spatial kinematic equation:

$$\begin{cases} \dot{X}_i^C &= -v_x - \omega_y Z_i^C + \omega_z Y_i^C \\ \dot{Y}_i^C &= -v_y - \omega_z X_i^C + \omega_x Z_i^C \\ \dot{Z}_i^C &= -v_z - \omega_x Y_i^C + \omega_y X_i^C \end{cases} \tag{6}$$

Substituting Equation (6) into Equation (5) yields

$$\begin{cases} \dot{x}_i = -\frac{v_x}{Z} + \frac{xv_z}{Z} + xy\omega_x - (1+x^2)\omega_y + y\omega_z \\ \dot{y}_i = -\frac{v_y}{Z} + \frac{yv_z}{Z} - xy\omega_y - (1+x^2)\omega_x - x\omega_z \end{cases} \tag{7}$$

Equation (7) can be rewritten in matrix form:

$$\begin{bmatrix} \dot{x}_i \\ \dot{y}_i \end{bmatrix} = J_i \begin{bmatrix} v \\ \omega \end{bmatrix} \tag{8}$$

where $v = (v_x, v_y, v_z)^T$ and $\omega = (\omega_x, \omega_y, \omega_z)^T$. J_i is the Jacobian matrix for each target point and it satisfies:

$$J_i = \begin{bmatrix} -\frac{1}{Z} & 0 & \frac{x_i}{Z} & x_i y_i & -(1+x_i^2) & y_i \\ 0 & -\frac{1}{Z} & \frac{y_i}{Z} & 1+y_i^2 & -x_i y_i & -x_i \end{bmatrix} \tag{9}$$

The crucial aspect of IBVS control is the discrepancy between the current coordinates of the feature target and the desired coordinates in the image. The feature error can be expressed as follows:

$$e = s - s^* \tag{10}$$

where $s \in \Re^{2M\times 1}$ are current positions of the target in the image and $s^* \in \Re^{2M\times 1}$ are the desired positions of the target in the image. Meanwhile, the motion of the target position in the image is directly related to the movement of the drone. The derivative of Equation (9) can be expressed as

$$\dot{e} = \dot{s} = J \cdot V \tag{11}$$

where $V = (v_x, v_y, v_z, \omega_x, \omega_y, \omega_z)^T$ is the velocity vector of the UAV in the body frame and $J \in \Re^{2M\times 6}$ is the image Jacobian matrix.

To ensure the exponential decline in the characteristic error, the relationship between the UAV speed and the characteristic error vector can be obtained using the following method:

$$V^T = -\lambda J^+ e \tag{12}$$

where $J^+ = (J^T J)^{-1} J^T$ is the pseudo-inverse matrix of J and λ is the servo gain vector of IBVS.

3. IBVS with Deep Reinforcement Learning

This section presents an IBVS method based on a DQN. As shown in Figure 3, the MDP model is first established. Next, the error e and state are fed into two different agents, respectively. Then, the DQN is applied to solve the MDP problems and automatic servo gain. The servo gains and depth Z are input into IBVS controller. Finally, the speed control quantity $(v, \omega)^T$ is obtained.

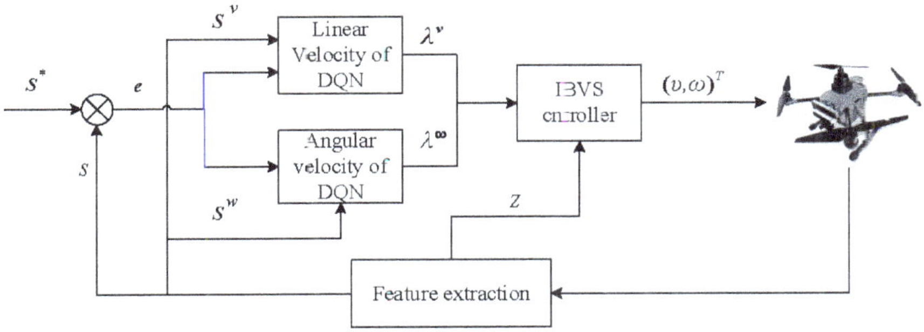

Figure 3. The scheme of IBVS based on the DQN method.

3.1. The Markov Decision Process Model

To design an adaptive law based on the DQN algorithm, this task needs to be formulated as an MDP model consisting of quintuplet (S, A, P, γ, r). The model of state transition P is determined by the environment, and γ is generally set according to experience. Thus, this paper focuses on the design of S, A, and r.

- State Space

The image plane is divided into $c \times v$ by pixel. A coordinate system is established in the current image plane, and the coordinates of the desired position in the image plane can be set as the origin. In the process of discretizing the image plane space, it is necessary to divide the pixel plane into a certain number of grids. The image plane can be divided into n_x segments along the x-axis and n_y segments along the y-axis, and it needs to meet the complete division, i.e.,

$$c\%n_x = 0; v\%n_y = 0 \tag{13}$$

From the above formula, it can be deduced that the size of the divided state space is:

$$K = \sum_{i=1}^{\frac{n_x}{2}} \left(\frac{n_y}{2} - i + 1 \right) \tag{14}$$

where K is the size of the state space.

If the pixel coordinate of the current position of the UAV in the image plane is (c_i, v_i), and the expected position is (c_a, v_a), then the formula can be obtained:

$$\hat{S} = \mathrm{sgn}(c_i - c_a)^2 + \mathrm{sgn}(v_i - v_a)^2 \tag{15}$$

where $\hat{S}$ is the pixel distance of the image plane coordinates and sgn() is the sign function.

Figure 4 shows that the pixel coordinates are divided into state spaces. When the agent is positioned in the black circle shown in Figure 4, it can be expressed as state 1; when the agent is in the blue circle, it is state 3; and when the agent is in the gray circle, it is state 5. Using this division method, a state space of size can be obtained, which is expressed as

$$S = \{s_i | i = 1, 2, 3, \cdots, K\} \tag{16}$$

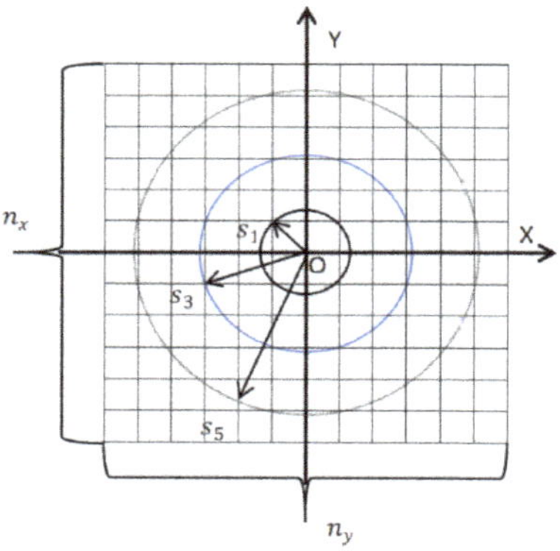

Figure 4. The state spaces in the discrete gird coordinate system.

- Action Selection

The servo gain is the key factor affecting the control efficiency of the visual servo control system. It is directly related to the control input; thus, this paper selects two different gains for angular and linear velocities, respectively. If the servo gain is directly used as the action of the agent, the action space will be too large and the phenomenon of dimension disaster will arise. Therefore, it is necessary to perform an initial assignment to the two servo gains, respectively, and then further design a limited regulation range for this initial assignment, i.e., the action range. The action interval is also discrete; thus, we can better carry out the learning process.

First, we design the action set according to the above idea: set an initial value as the initial servo gain and then set the size of the action set. The angular and linear velocity servo gains have the same size.

$$\begin{cases} \alpha^v = \{a_i^v | i = 1, 2, \cdots, 2n+1\} \\ \alpha^\omega = \{a_i^\omega | i = 1, 2, \cdots, 2n+1\} \end{cases} \tag{17}$$

The action space A^v, A^ω can be expressed as

$$\begin{cases} A^v = \{-n^v d^v, -(n^v - 1)d^v, \cdots, -d^v, 0, d^v \cdots, (n^v - 1)d^v, n^v d^v\} \\ A^\omega = \{-n^\omega d^\omega, -(n^\omega - 1)d^\omega, \cdots, -d^\omega, 0, d^\omega \cdots, (n^\omega - 1)d^\omega, n^\omega d^\omega\} \end{cases} \tag{18}$$

The update rules of the servo gain satisfy

$$\begin{cases} \lambda_{t+1}^v = \lambda_t^v (a_t^v + 1) \\ \lambda_{t+1}^\omega = \lambda_t^\omega (a_t^\omega + 1) \end{cases} \tag{19}$$

- Reward Function

The reward function is the agent's feedback with respect to the environment, which is crucial to the convergence and stability of the whole learning algorithm. Indeed, it determines the effectiveness of the reinforcement learning algorithm to some extent. The reward function is the standard used to evaluate whether agents can learn the optimal strategy in the process of continuous interaction with the environment. If we choose to give only a reward value to the final goal or expected position of reinforcement learning in the design of the reward function, the convergence time of the algorithm will be greatly increased in the environment because the agent will obtain the reward only upon achieving the final goal. In the state action value function, the reward feedback of other positions in the state space is zero, and the amount of reinforcement learning will also be greatly increased. If a positive value smaller than the reward value fed back by the final goal is randomly designed in the reward function of a certain area in the state space, it is likely that a phenomenon similar to the repeated movement of the agent in this area will occur because the agent constantly swipes the reward value in this area without choosing to move to other areas in order to find a better reward or the final reward by using the influence of strategy.

In the design of the abovementioned state space and action set, we can understand that the UAV is controlled by the image visual servo controller, and deep reinforcement learning is responsible for regulating the servo gain. Therefore, in reinforcement learning, the agent can also represent the motion of the target in the image plane, which has a good mapping relationship with the three-dimensional inertial space. On this basis, the feedback of agents in the environment can be divided into three situations.

When the distance between all the feature points of the target and the coordinates of the desired position feature points in the image plane is less than a fixed value, it can be considered that the rotor UAV has reached the desired position and the one-time learning algorithm of reinforcement learning is completed. In these circumstances, the agent can obtain a maximum reward value.

When the UAV cannot obtain the total number of feature points of the desired target in the state space, it means that the UAV has exceeded the observation range of the target position by the visual servoing system. In this case, the target is lost and the minimum reward value should be given to the agent.

When the rotor UAV can detect all the feature points but has not reached the desired position, a function of feature error can be set:

$$e = \sum_{i=1}^{N} |s_i - s_i^*| / N \tag{20}$$

where N is the number of characteristic points, s_i is the current location, and $s_i{}^*$ is the desired location.

The reward function can be set according to the above three situations as

$$r = \begin{cases} 100, \text{ arriving at desired position} \\ -100, \text{ feature loss} \\ -qe - pa_v, \text{ otherwise} \end{cases} \tag{21}$$

where q is the weight coefficient of the feature error vector and p is the weight coefficient of penalized large actions.

3.2. DQN Algorithm

The Q function of Q-learning represents the value of each pair of state–action in a specific configuration environment. In order to represent the Q function that may have a large number of state actions, the deep Q-Network (DQN) uses the neural network approach to approximate the characteristics, and the Q function is expressed as a weight parameter θ as $Q(s, a; \theta)$. s and a represent state and action, respectively. The next state s' is obtained after performing a.

There are two major improvement measures for DQNs:

- Target Network: With only one network, updating the Q function in real time can result in a chaotic trajectory and poor training. To avoid instability caused by updating the Q function while simultaneously acquiring the Q value, a target network is used. The target network provides a stable Q value for the Q function to be updated. The target network is updated with the new Q function to improve performance.
- Experience Replays: Experience replays will build a replay buffer $D = \{e_1, e_2, e_3 \cdots |D|\}$. The replay buffer is also called replay memory. Instead of using the samples in the standard sequence, small batches are randomly selected from the data set for training to diminish the relativity between training samples.

In each iteration, the state and action are sent to the DQN and the estimated Q value is generated.

$$L_i(\theta_i) = E_{s,a,r,s'}[\frac{1}{2}\left(y_i^{DQN} - Q(s,a;\theta_i)\right)^2] \quad (22)$$

The loss function expressed in the expected form is:

$$y_i{}^{DQN} = r + \gamma \max_{a'} Q(s',a';\theta^-) \quad (23)$$

where θ^- denotes the target network parameters.

Including the experience replays in Equation (22),

$$L_i(\theta_i) = E_{(s,a,r,s')\sim D}[\frac{1}{2}\left(y_i^{DQN} - Q(s,a;\theta_i)\right)^2] \quad (24)$$

The gradient of loss function is given by:

$$\nabla_{\theta i} L_i(\theta_i) = E_{(s,a,r,s')\sim D}\left[\left(y_i^{DQN} - Q(s,a;\theta_i)\right)\nabla_{\theta i} Q(s,a;\theta_i)\right] \quad (25)$$

The optimization algorithm is gradient descent, which iteratively updates the model's parameters to find the optimal set of values. The parameters of separate target networks are determined according to

$$\theta^- = \tau\theta + (1-\tau)\theta^- \quad (26)$$

The proposed iterative algorithm's process is provided in Algorithm 1, where the max episode n, T denotes the MaxStep.

Algorithm 1: DQN-based IBVS method

Initialization;
For episode = 1:n
  For t = 1:T
    If $s_i > K$ or $s_i \leq 1$ (Termination condition);
      Break
    End
    Generate random number: u $(0 < \mu < 1)$;
    If $\mu < \varepsilon$
      Random selection of action a_t^v, a_t^ω;
    Else
      Select the corresponding strategic actions a_t^v, a_t^ω;
      obtain the rewards r_t^v, r_t^ω and servo gains λ_t^v and λ_t^ω;
      λ_t^v and λ_t^ω are substituted to $\mathrm{V^T} = -\boldsymbol{\lambda}\,\mathbf{J}^{+}\mathrm{e}$;
      Observe the next State s_{i+1};
      Store the experience replay with $[s_i, a_t^v, a_t^w, r_t^v, r_t^\omega, s_{i+1}]$;
      $s_i = s_{i+1}$;
    End
    If Experience replay full
      Randomly selected datasets of buffer D;
      Train the network by gradient descent method and updating network parameters θ according to (24);
      After training a certain number of times, update the target network θ^- according to (25);
    End
  End
End

4. Simulations and Experiments

This section describes various simulations that were carried out to prove the effectiveness and performance of IBVS based on a DQN. Due to the concise language and high programming efficiency of MATLAB, the proposed control system was simulated using MATLAB R2021b.

The parameters of the quadrotor used in the simulation are m = 1 kg, g = 9.81 m/s^2. and J = diag{0.0081, 0.008, 0.0142} kg.m^2/rad^2. The focal length of the camera is 3.2 mm. The length and width of the camera pixels are 1.4×10^{-6} m. The targets are four points on the horizontal plane. Their coordinates relative to the inertial system are (3,2,3), (2,2,3), (2,3,3), and (3,3,3). The parameters of the DQN are shown in Table 2, where ε is the error threshold, γ is the discount parameter, and α is the learning rate, while $\lambda^{\omega}*$ and $\lambda^{\omega}*$ represent the initial servo gain values of angular velocity and linear velocity, respectively.

Table 2. Intrinsic parameters of the DQN.

Parameters	Value
ε	5 pixels
γ	0.8
α	0.9
$\lambda^{\omega*}$	0.04
λ^{v*}	0.6
episode	500

Figure 5 shows the two groups of servo gain obtained through deep reinforcement learning. In the beginning, small servo gains are selected to avoid target loss, because a large gain value will lead to an excessive control input and cause the UAV to move violently. In the later stage, a large gain is chosen to improve the efficiency of IBVS.

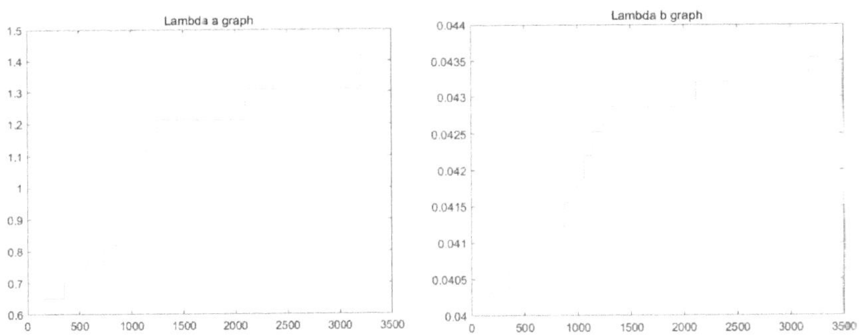

Figure 5. The final servo gains obtained using the DQN algorithm.

To prove the effectiveness and practicability of the new method, we compare C-IBVS [28] under different servo gains in the simulation platform. Figure 6 shows the position path trajectory diagram of the quadrotor UAV under various conditions.

Figure 7 shows the feature trajectories of the image, and four colored lines represent the trajectories of different feature points in the image. It can be seen that when a gain of $\lambda = 1.5$ is adopted in the C-IBVS method, the trajectories vary drastically and there is a risk of losing the target because the feature points have reached the image border. When $\lambda = 1$, there is a safe distance between the image trajectory and the edge. When $\lambda = 0.5$, the image trajectory is the gentlest and the trajectory is shortest. The DQN method is also relatively gentle, and the image trajectory is a safe distance from the edge, which ensures that the target features are not lost.

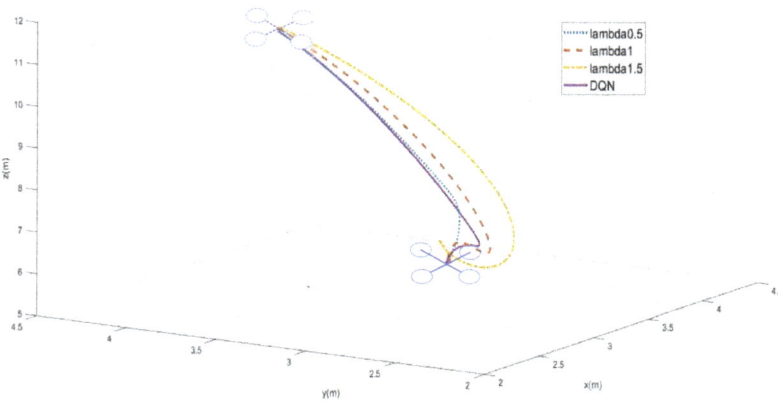

Figure 6. Position trajectory of the UAV.

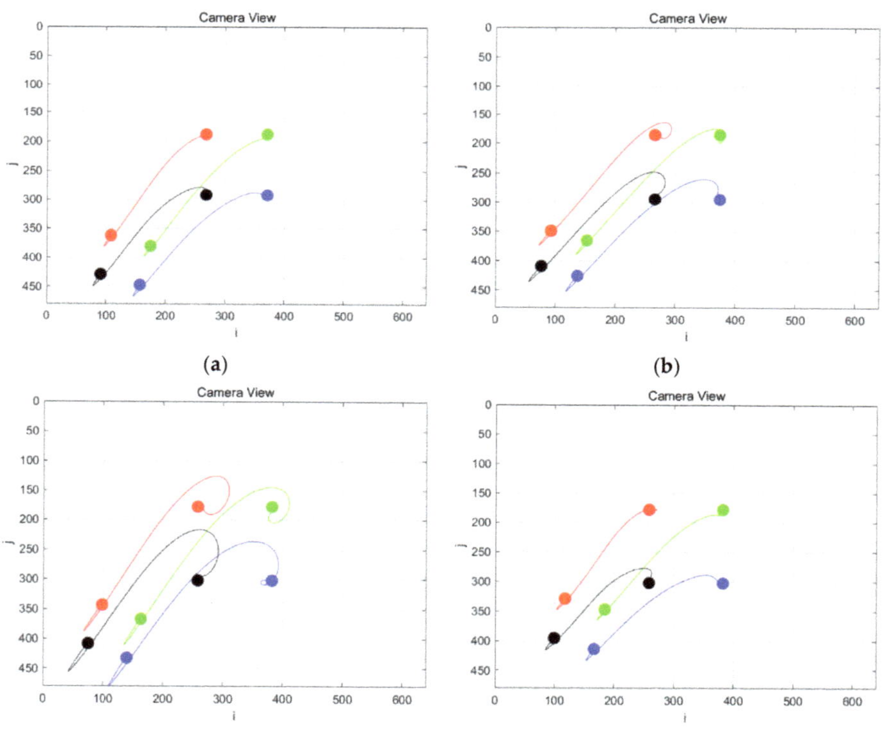

Figure 7. Feature trajectories in the image plane under different conditions: (**a**) C-IBVS (λ = 0.5); (**b**) C-IBVS (λ = 1); (**c**) C-IBVS (λ = 1.5); (**d**) IBVS based on a DQN.

Figure 8 shows the position movement trajectories of a UAV in four cases. It can be seen that when λ = 0.5, the convergence needs 25 time steps. When λ = 1, it needs 15 time steps, and when λ = 1.5, it is also needs 15 time steps because it involves overshoot and for other reasons. The convergence of the adopted DQN method takes 12 time steps, and the curve has fast convergence and no overshoot.

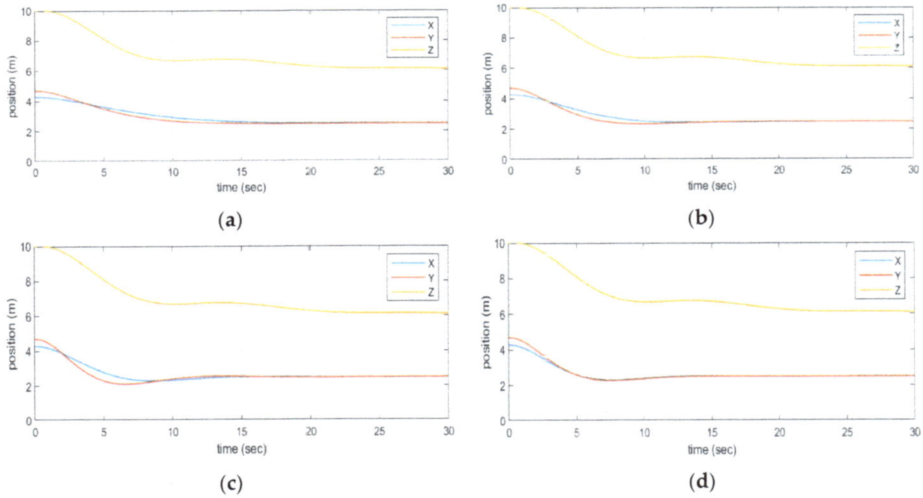

Figure 8. Position movement trajectory of the UAV under different conditions: (**a**) C-IBVS (λ = 0.5); (**b**) C-IBVS (λ = 1); (**c**) C-IBVS (λ = 1.5); (**d**) IBVS based on a DQN.

5. Conclusions

To address the problems of target loss and the low control efficiency in image-based visual servoing (IBVS), this paper proposes a DQN-based IBVS adaptive servo gain adjustment method. First, a Markov model of UAV visual servoing control under field-of-view constraints was established by designing the state space, action set, and reward function, which considers the improvement of the control efficiency. Second, an improved DQN algorithm with a target network and experience replay was designed to solve the Markov model. To enhance the control performance, two independent agents were designed, with each agent responsible for adjusting the linear and angular velocity servo gains, respectively. Through a simulation analysis of a quadrotor UAV equipped with a monocular camera, the proposed method can avoid target loss caused by FOV constraint and improve the efficiency of IBVS.

The process of action selection in this method is discrete and the number of actions is limited. In future work, a continuous action space will be designed to enhance the algorithm's performance. Additionally, the algorithm will be further validated through real flight experiments.

Author Contributions: Conceptualization, G.F. and L.L.; methodology, G.F. and X.Z.; software, L.F.; validation, G.F. and L.F.; formal analysis, G.F.; writing—original draft preparation, G.F.; writing—review and editing, H.C. and G.F. All authors have read and agreed to the published version of the manuscript.

Funding: This research was funded by the National Natural Science Foundation of China, grant number 61601382; Fund project for basic scientific research expenses of central universities, grant number J2022-024; Fund project for basic scientific research expenses of central universities, grant number J2022-07; and the Independent research project of the Key Laboratory of Flight Techniques and Flight Safety, grant number FZ2021ZZ04.

Institutional Review Board Statement: Not applicable.

Informed Consent Statement: Not applicable.

Data Availability Statement: The data used to support the findings of this study are available from the corresponding author upon request.

Conflicts of Interest: The authors declare no conflict of interest.

Abbreviations

The following abbreviations are used in this manuscript:

UAV	unmanned aerial vehicle
FOV	field of view
DRL	deep reinforcement learning
DQN	deep Q-network
PBVS	position-based visual servoing
IBVS	image-based visual servoing
WMRs	wheeled mobile robots
MPC	model predictive control

References

1. Alzahrani, B.; Oubbati, O.S.; Barnawi, A.; Atiquzzaman, M.; Alghazzawi, D. UAV assistance paradigm: State-of-the-art in applications and challenges. *J. Netw. Comput. Appl.* **2020**, *166*, 102706. [CrossRef]
2. Mahony, R.; Kumar, V.; Corke, P. Multirotor Aerial Vehicles: Modeling, Estimation, and Control of Quadrotor. *IEEE Robot. Autom. Mag.* **2012**, *19*, 20–32. [CrossRef]
3. Zhen, Z.; Chen, Y.; Wen, L.; Han, B. An intelligent cooperative mission planning scheme of UAV swarm in uncertain dynamic environment. *Aerosp. Sci. Technol.* **2020**, *100*, 105826. [CrossRef]
4. Liu, Y.; Liu, H.; Tian, Y.; Sun, C. Reinforcement learning based two-level control framework of UAV swarm for cooperative persistent surveillance in an unknown urban area. *Aerosp. Sci. Technol.* **2020**, *98*, 105671. [CrossRef]
5. Wu, Y.; Hu, X. An online route planning method for multi-rotor drone in urban environments. *Control. Decis.* **2021**, *36*, 2851–2860.
6. Zheng, D.; Wang, H.; Wang, J.; Chen, S.; Chen, W.; Liang, X. Image-Based Visual servoing of a quadrotor using virtual camera approach. *IEEE/ASME Trans. Mechatron.* **2017**, *22*, 972–982. [CrossRef]
7. Chaumette, F.; Hutchinson, S. Visual servo control. I. Basic approaches. *IEEE Robot. Autom. Mag.* **2006**, *13*, 82–90. [CrossRef]
8. Chen, C.; Tian, Y.; Lin, L.; Chen, S.; Li, H.; Wang, Y.; Su, K. Obtaining World Coordinate Information of UAV in GNSS Denied Environments. *Sensors* **2020**, *20*, 2241. [CrossRef]
9. Abdessameud, A.; Janabi-Sharifi, F. Image-based tracking control of VTOL unmanned aerial vehicles. *Automatica* **2015**, *53*, 111–119. [CrossRef]
10. Zhang, X.; Fang, Y.; Li, B.; Wang, J. Visual servoing of nonholonomic mobile robots with uncalibrated Camera-to-Robot parameters. *IEEE Trans. Ind. Electron.* **2017**, *64*, 390–400. [CrossRef]
11. Zhang, X.; Fang, Y.; Zhang, X.; Jiang, J.; Chen, X. Dynamic Image-Based Output Feedback Control for Visual Servoing of Multirotors. *IEEE Trans. Ind. Inform.* **2020**, *16*, 7624–7636. [CrossRef]
12. Ceren, Z.; Altuğ, E. Image based and hybrid visual servo control of an unmanned aerial vehicle. *J. Intell. Robot. Syst.* **2012**, *65*, 325–344. [CrossRef]
13. Liu, N.; Shao, X.; Yang, W. Desired compensation RISE-based IBVSS control of quadrotor for tracking a moving target. *Nonlinear Dyn.* **2019**, *95*, 2605–2624. [CrossRef]
14. Santamaria-Navarro, À.; Andrade-Cetto, J. Uncalibrated image based visual servoing. In Proceedings of the 2013 IEEE International Conference on Robotics and Automation, Karlsruhe, Germany, 6–10 May 2013; pp. 5247–5252.
15. Miao, Z.; Zhong, H.; Lin, J.; Wang, Y.; Chen, Y.; Fierro, R. Vision-Based Formation Control of Mobile Robots With FOV Constraints and Unknown Feature Depth. *IEEE Trans. Control. Syst. Technol.* **2021**, *29*, 2231–2238. [CrossRef]
16. Lopez-Nicolas, G.; Aranda, M.; Mezouar, Y. Formation of differential-drive vehicles with field-of-view constraints for enclosing a moving target. In Proceedings of the 2017 IEEE International Conference on Robotics and Automation (ICRA), Singapore, Singapore, 29 May–3 June 2017; IEEE: Piscataway, NJ, USA, 2017; pp. 261–266.
17. Bhagat, S.; Pb, S. UAV Target Tracking in Urban Environments Using Deep Reinforcement Learning. In Proceedings of the 2020 International Conference on Unmanned Aircraft Systems (ICUAS), Athens, Greece, 1–4 September 2020.
18. Bruno, H.M.S.; Colombini, E.L. LIFT-SLAM: A deep-learning feature-based monocular visual SLAM method. *Neurocomputing* **2021**, *455*, 97–110. [CrossRef]
19. Hajiloo, A.; Keshmiri, M.; Xie, W.F.; Wang, T.T. Robust Online Model Predictive Control for a Constrained Image-Based Visual Servoing. *IEEE Trans. Ind. Electron.* **2016**, *63*, 2242–2250.
20. Chesi, G. Visual Servoing Path Planning via Homogeneous Forms and LMI Optimizations. *IEEE Trans. Robot.* **2009**, *25*, 281–291. [CrossRef]
21. Huang, Y.; Zhu, M.; Zheng, Z.; Low, K.H. Homography-based visual servoing for underactuated VTOL UAVs tracking a 6-DOF moving ship. *IEEE Trans. Veh. Technol.* **2022**, *71*, 2385–2398. [CrossRef]

22. Zhang, X.; Fang, Y.; Zhang, X.; Shen, P.; Jiang, J.; Chen, X. Attitude-Constrained Time-Optimal Trajectory Planning for Rotorcrafts: Theory and Application to Visual Servoing. *IEEE/ASME Trans. Mechatron.* **2020**, *25*, 1912–1921. [CrossRef]
23. Zheng, D.; Wang, H.; Wang, J.; Zhang, X.; Chen, W. Toward Visibility Guaranteed Visual Servoing Control of Quadrotor UAVs. *IEEE/ASME Trans. Mechatron.* **2019**, *24*, 1087–1095. [CrossRef]
24. Krichen, M.; Mihoub, A.; Alzahrani, M.Y.; Adoni, W.Y.H.; Nahhal, T. Are Formal Methods Applicable To Machine Learning And Artificial Intelligence? In Proceedings of the 2022 2nd International Conference of Smart Systems and Emerging Technologies (SMARTTECH), Riyadh, Saudi Arabia, 9–11 May 2022; pp. 48–53.
25. Seshia, S.A.; Sadigh, D.; Sastry, S.S. Toward Verified Artificial Intelligence. *Commun. ACM* **2022**, *65*, 46–55. [CrossRef]
26. Sutton, R.S.; Barto, A.G. *Reinforcement Learning: An Introduction*; MIT Press: Cambridge, MA, USA, 2018.
27. Wang, Y.; Lang, H.; De Silva, C.W. A Hybrid Visual Servo Controller for Robust Grasping by Wheeled Mobile Robots. *IEEE/ASME Trans. Mechatron.* **2010**, *15*, 757–769. [CrossRef]
28. Shi, H.; Li, X.; Hwang, K.S.; Pan, W.; Xu, G. Decoupled visual servoing with fuzzy Q-learning. *IEEE Trans. Ind. Inform.* **2018**, *14*, 241–252. [CrossRef]
29. Shi, H.; Shi, L.; Sun, G.; Hwang, K.S. Adaptive Image-Based Visual Servoing for Hovering Control of Quad-rotor. *IEEE Trans. Cogn. Dev. Syst.* **2020**, *12*, 417–426. [CrossRef]

drones

MDPI

Article

Quadcopter Trajectory Tracking Based on Model Predictive Path Integral Control and Neural Network

Yong Li *, Qidan Zhu * and Ahsan Elahi

College of Intelligent Systems, Science and Engineering, Harbin Engineering University, Harbin 150001, China; ahsan.elahi@hrbeu.edu.cn

* Correspondence: llyy5927@hrbeu.edu.cn (Y.L.); zhuqidan@hrbeu.edu.cn (Q.Z.); Tel.: +86-188-4516-3353 (Y.L.)

Abstract: This paper aims to address the trajectory tracking problem of quadrotors under complex dynamic environments and significant fluctuations in system states. An adaptive trajectory tracking control method is proposed based on an improved Model Predictive Path Integral (MPPI) controller and a Multilayer Perceptron (MLP) neural network. The technique enhances control accuracy and robustness by adjusting control inputs in real time. The Multilayer Perceptron neural network can learn the dynamics of a quadrotor by its state parameter and then the Multilayer Perceptron sends the model to the Model Predictive Path Integral controller. The Model Predictive Path Integral controller uses the model to control the quadcopter following the desired trajectory. Experimental data show that the improved Model Predictive Path Integral–Multilayer Perceptron method reduces the trajectory tracking error by 23.7%, 34.7%, and 10.3% compared to the traditional Model Predictive Path Integral, MPC with MLP, and Model Predictive Path Integral with a two-layer network, respectively. These results demonstrate the potential application of the method in complex environments.

Keywords: quadcopter; model predictive path integral; neural network; multilayer perceptron

1. Introduction

Recent research on quadcopters has established them as a significant platform for the development of unmanned aerial vehicles (UAVs) owing to their simple construction and maintenance. These quadcopters demonstrate exceptional maneuverability and are capable of hovering, taking off, flying, and landing in confined spaces, owing to their vertical takeoff and landing (VTOL) capability [1]. A quadcopter comprises four rotors positioned at the corners of a cross-shaped frame, and it is controlled by adjusting the rotational speeds of these rotors [2,3]. Quadcopters are employed in various applications such as surveillance, search-and-rescue missions, and hazardous operations in congested and Global Positioning System (GPS)-denied settings, such as dense forests, crowded offices, corridors, and warehouses. To perform these tasks effectively, quadcopters must have robust navigation abilities and a thorough understanding of their operational environment. Substantial research work examined various aspects of quadcopter dynamics and control, motion planning, and trajectory generation in unstructured environments [4–8].

Sampling-based optimal control methods have attracted considerable interest in path planning and trajectory tracking applications. These control techniques function by generating multiple trajectories at each control step through input sampling (referred to as rollouts) and subsequently determining a sub-optimal sequence of control inputs based on the sampled trajectories. This process involves executing a finite time-horizon optimization,

https://doi.org/10.3390/drones9010009

utilizing the optimal input for the initial time step and then iteratively repeating the optimization for subsequent time steps. The path integral optimal control framework [9–11] provides a mathematical foundation for developing optimal control algorithms through stochastic trajectory sampling. The main principle is based on the transformation of the value function for the optimal control problem using the Feynman–Kac lemma [12,13] into an expectation over all possible trajectories, referred to as a path integral. This transformation enables the resolution of stochastic optimal control problems by a Monte Carlo approximation employing a forward sampling of stochastic diffusion processes. Furthermore, the utilization of path integral control theory has gained prominence in recent research.

Model Predictive Control (MPC) has been used on various robotic systems, such as self-driving cars [14] and quadcopters [15], and it is widely employed in industrial applications [16]. The path integral control framework can be integrated with the Model Predictive Control paradigm. In this scenario, an open-loop control sequence is continuously optimized in the background, whereas the system finds the current optimal estimate of the controller in parallel. One big problem with this approach is that it needs to sample a lot of different trajectories simultaneously, which is challenging when the systems being considered have complex dynamics. A possible solution to this issue involves simplifying the system model through a hierarchical scheme [17]. This employs path integral control to create trajectories for a simplified point mass model that a low-level controller can then follow. Model Predictive Path Integral (MPPI) control was introduced in [18], displaying its ability in aggressive autonomous driving. It explored information-theoretic dualities between free energy and relative entropy, which led to the formation of the Model Predictive Path Integral control logic. Williams developed the iterative path integral method, known as the Model Predictive Path Integral control framework [16]. That method iteratively modified the control sequence to find optimal solutions based on the importance of sampling trajectories. In [19], an alternative iterative method was proposed that removed the control and noise affine constraints inherent in the original Model Predictive Path Integral framework, which could accommodate non-affine dynamics. The central idea behind that extension was to use the information-theoretic interpretation of optimal control through the Kullback–Leibler (KL) divergence and free energy. It was different from the original method, which used a linearization of the Hamilton–Jacobi–Bellman (HJB) equation and applied the Feynman–Kac lemma. Despite these methods coming from different derivations, they were practically equivalent and theoretically linked. A learning model to generate informed sampling distributions was presented in [20], which was an extension of Williams's information-theoretic-based method. In [15], the generation of time-optimal trajectories through multiple waypoints for quadrotors was investigated.

The MPPI controller has been mainly utilized in aggressive driving and UAV navigation in congested conditions. However, UAV navigation in natural conditions often involves incomplete state information. In environments such as forests and urban canyons, the navigation system cannot directly measure all state variables due to sensor limitations or occlusion problems [1,6]. Although most control schemes depend on the accuracy of the underlying models, specific adaptive control, sliding mode control, and robust control methods can remain effective despite model inaccuracies [21–23]. A comprehensive analysis of advanced aerodynamic effects controlling quadrotor flight, such as blade flapping and airflow dynamics, is conducted in [24]. These effects are fundamentally complicated to model, posing considerable challenges in controller design. The standard Model Predictive Path Integral approach may necessitate substantial computational resources or manual parameter adjustment to maintain control performance during dynamic environmental variations or severe system state fluctuations. Therefore, relying exclusively on the Model Predictive Path Integral control mechanism is inadequate for responding to modern complicated missions. Incorporating an adaptive mechanism into the Model Predictive Path Integral framework allows for the dynamic modification of essential parameters according to system states, and control performance enhance the robustness and efficiency of the control strategy. Moreover, to overcome modeling problems, data-driven and learning-based control strategies have been proposed [25]. A promising approach employs neural networks (NNs) to model system dynamics. Neural networks are acknowledged as universal function approximators, capable of modeling highly nonlinear functions and latent states directly from observed data, which may be challenging to model otherwise explicitly [26]. The study in [27] demonstrates that simple feedforward networks can proficiently learn and generalize system dynamics with high accuracy.

This paper proposes an adaptive control method that integrates the enhanced Model Predictive Path Integral (MPPI) method and a Multilayer Perceptron (MLP) neural network, which can dynamically modify essential parameters according to system states or control performance. In experiments, the method achieves a root-mean-square error (RMSE) in trajectory tracking that is more than 10% lower than alternative methods, surpassing the MPPI control without learning dynamics and the Model Predictive Path Integral method employed with a two-layer neural network. The paper is organized as follows: Section 2 explains the dynamic model of quadcopter; Section 3 presents the training and validation of the MLP; and Section 4 propcses the adaptive Model Predictive Path Integral approach, which can adjust its sampling based on trajectory error. Section 5 presents a comparison of Model Predictive Path Integral control without learning, utilizing two hidden layers of neural network (NN), to adaptive Model Predictive Path Integral control employing a Multilayer Perceptron. Experimental results prove that the adaptive Model Predictive Path Integral method utilizing a Multilayer Perceptron outperforms alternative methodologies.

2. Material

In this section, we first present the nonlinear dynamic model of the quadcopter to be used and the associated model uncertainties and external disturbances.

A general discrete-time stochastic nonlinear system is of the form [1,9]:

$$x_{t+1} = f(x_t, u_t + \varepsilon_t) + \omega_t \tag{1}$$

where $x \in \mathbb{R}^N$ is the state, and $u \in \mathbb{R}^M$ is the control input. The term $\varepsilon_t \sim N(0, \sigma_\varepsilon^2)$ represents a noise disturbance with a normal distribution in the control signals generated in the lower-level controllers for real-world robot systems. The term ω_t represents the combined errors due to inaccurate model representations and purely external disturbances, such as wind in the case of quadcopters.

Quadcopter Model

Figure 1 shows the quadcopter model, with an inertial reference frame defined as W and a body-fixed frame defined as B, with the origin defined at the center of the quadcopter. The Euler angles to describe the quadrotor rotations in W about the body-fixed axes x_B, y_B, and z_B, are roll ϕ, pitch θ, and yaw ψ, respectively.

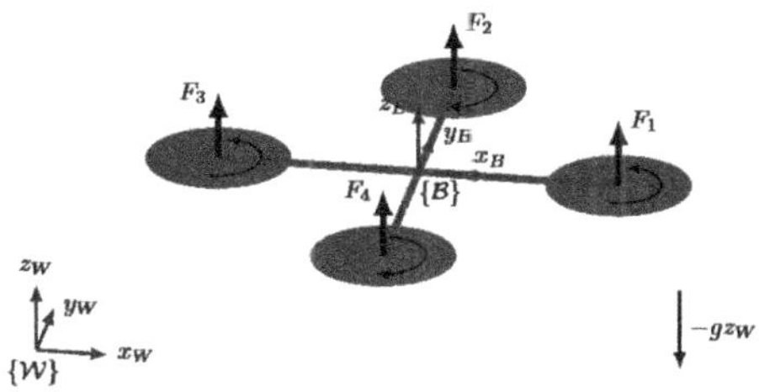

Figure 1. Quadcopter model.

The rotation from W to B with a ZYX transformation is expressed as:

$$ {}^{W}R_{\mathrm{B}} = \begin{bmatrix} c_{\psi}c_{\theta} & c_{\psi}s_{\phi}s_{\theta} - c_{\phi}s_{\psi} & s_{\psi}s_{\phi} + c_{\phi}c_{\psi}s_{\theta} \\ c_{\theta}s_{\psi} & c_{\phi}c_{\psi} + s_{\psi}s_{\phi}s_{\theta} & c_{\phi}s_{\psi}s_{\theta} - c_{\psi}s_{\phi} \\ -s_{\theta} & c_{\theta}s_{\phi} & c_{\phi}c_{\theta} \end{bmatrix} \tag{2} $$

where $s_x = \sin(x), c_x = \cos(x)$ for $\forall x \in \{\phi, \theta, \psi\}$.

The body angular velocity Ω represents the angular velocity vector relative to the body frame and is defined as:

$$ \Omega = \begin{bmatrix} p \\ q \\ r \end{bmatrix} \tag{3} $$

where p is the rotation rate around the x_B axis; q is the rotation rate around the y_B axis; r is the rotation rate around the z_B axis.

The Euler angles (ϕ, θ, ψ) define the orientation of the quadrotor by rotating from the inertial frame to the body frame. The relationship between the body angular velocity Ω and the Euler angle derivatives $\left(\dot{\phi}, \dot{\theta}, \dot{\psi}\right)$ is given by:

$$ \Omega = \mathbf{K}\dot{\Phi} \tag{4} $$

where $\dot{\Phi} = \begin{bmatrix} \dot{\phi} \\ \dot{\theta} \\ \dot{\psi} \end{bmatrix}$.

To derive the transformation matrix $\mathbf{K}$, we decompose the rotation in the inertial frame. After applying the sequence of rotations, we can express Ω in terms of $\dot{\phi}, \dot{\theta}, \dot{\psi}$ as:

$$ \Omega = \begin{bmatrix} 1 & 0 & -\sin(\theta) \\ 0 & \cos(\phi) & -\sin(\phi)\cos(\theta) \\ 0 & -\sin(\phi) & \cos(\phi)\cos(\theta) \end{bmatrix} \begin{bmatrix} \dot{\phi} \\ \dot{\theta} \\ \dot{\psi} \end{bmatrix} \tag{5} $$

Taking the inverse, we can express the Euler angle rates as:

$$ \dot{\Phi} = \mathbf{K}^{-1}\Omega \tag{6} $$

where $\mathbf{K}^{-1} = \mathbf{T}$ is the required transformation matrix.

$$ \mathbf{T} = \begin{bmatrix} 1 & s_{\phi}t_{\theta} & c_{\phi}t_{\theta} \\ 0 & c_{\phi} & -s_{\phi} \\ 0 & \frac{s_{\phi}}{c_{\theta}} & \frac{c_{\phi}}{c_{\theta}} \end{bmatrix} \tag{7} $$

The linear motion equations describe the translational motion in space. Using Newton's second law, we have:

$$m\ddot{\mathbf{r}} = \mathbf{F}_{total} \tag{8}$$

where m is the mass of the quadrotor; $\mathbf{r}$ is the position vector of the quadrotor; $\mathbf{F}_{total}$ is the total force acting on the quadrotor.

The total force consists of two components: the gravitational force $-mg\mathbf{e}_3$, where $\mathbf{e}_3 = [0,0,1]^T$ is the unit vector pointing upwards, and the thrust force $\mathbf{F} = {}^W R_B F\mathbf{e}_3$, where the thrust is always directed along the z_B-axis of the body frame. The rotation matrix ${}^W R_B$ converts that thrust into the inertial frame. The dynamics of the quadcopter are given by:

$$\dot{\mathbf{x}} = v \tag{9}$$

$$m\dot{\mathbf{v}} = -mg\mathbf{e}_3 + {}^W R_B F\mathbf{e}_3 + \mathbf{d}_F \tag{10}$$

$$\dot{\mathbf{\Phi}} = T^{-1}\mathbf{\Omega} \tag{11}$$

$$\mathbf{J}\dot{\mathbf{\Omega}} = \mathbf{\Gamma} + \mathbf{d}_\Gamma - \mathbf{\Omega} \times \mathbf{J}\mathbf{\Omega} \tag{12}$$

where m is the mass of the quadcopter, $\mathbf{J}$ is the inertia matrix expressed in B, $\mathbf{x} = [x,y,z]^T$ is the position vector, $\mathbf{v} = [v_x, v_y, v_z]^T$ is the velocity vector, $\mathbf{\Phi} = [\phi,\theta,\psi]^T$ is the Euler angle vector, $\mathbf{\Omega} = [p,q,r]^T$ is the body angular rate vector, $\mathbf{e}_3 = [0,0,1]^T \in \mathbb{R}^3, F \in \mathbb{R}^+$ is the body thrust input generated by the motors, $\mathbf{\Gamma} = (\tau_x, \tau_y, \tau_z)^T \in \mathbb{R}^3$ is the torque input to the vehicle in B , and $L \in \mathbb{R}^+$ is the arm length of the quadrotor. The disturbances ε and ω are explicitly included in the quadrotor dynamics equations, affecting both thrust $\mathbf{F}$ and torque $\mathbf{\Gamma}$. $\mathbf{d}_F$ represents the external force disturbance vector affecting the thrust $\mathbf{F}$. $\mathbf{d}_\Gamma$ represents the external torque disturbance vector affecting the torque $\mathbf{\Gamma}$. The state of the system is defined as $\mathbf{x} = [x,y,z,\phi,\theta,\psi,v_x,v_y,v_z,p,q,r]^T \in \mathbb{R}^{12}$. Each rotor produces a vertical force $F_i = k_F\omega_i^2$ and moment $M_i = k_M\omega_i^2$, where ω_i is the angular velocity of ith rotor. We map the control inputs $\mathbf{F}$ and $\mathbf{\Gamma}$ to the system inputs $\mathbf{u} = [\omega_1,\omega_2,\omega_3,\omega_4]$ as

$$\begin{bmatrix} F \\ \tau_x \\ \tau_y \\ \tau_z \end{bmatrix} = \begin{bmatrix} k_F & k_F & k_F & k_F \\ 0 & k_F L & 0 & -k_F L \\ -k_F L & 0 & k_F L & 0 \\ k_M & -k_M & k_M & -k_M \end{bmatrix} \begin{bmatrix} \omega_1^2 \\ \omega_2^2 \\ \omega_3^2 \\ \omega_4^2 \end{bmatrix} \tag{13}$$

where k_F is the thrust coefficient, k_M is the moment (torque) coefficient, L is the distance from the rotor to the center of mass.

We deal with the actuator dynamics by introducing a first-order lag in the control inputs. This accounts for the physical limitations and response times of the quadrotor's actuators, ensuring that rapid changes in control commands do not lead to unrealistic actuator behaviors. Specifically, the actuator dynamics are represented as follows.

The actuator is modeled as a first-order linear system characterized by its time constant τ:

$$\tau\dot{\mathbf{u}} + \mathbf{u} = \mathbf{u}_{cmd} \tag{14}$$

where τ is the actuator time constant, $\mathbf{u}$ is the actual control input, and $\mathbf{u}_{cmd}$ is the commanded control input. Rearranging the equation, we obtain:

$$\dot{\mathbf{u}} = -\frac{1}{\tau}\mathbf{u} + \frac{1}{\tau}\mathbf{u}_{cmd} \tag{15}$$

This first-order model captures the delayed and smoothed response of the actuators to control commands, reflecting real-world actuator behavior.

By simulating the actuator dynamics, the dataset more accurately reflects real-world quadcopter behavior, including the effects of rotor inertia and drag. In subsequent sections, we simplify the dynamics during control implementation because the neural network has already learned the influence of actuator dynamics. The inclusion of Equation (14) in the data generation process ensures that the learned model accurately captures the effects of actuators on the quadcopter's overall dynamics.

3. Methods

3.1. Neural Network

This section describes the neural network architecture employed to mimic the quadrotor dynamics. The improvement in computational capacity enables the execution of large-scale batch predictions within minimal timeframes. The adequately trained neural networks can effectively generalize the system's dynamic equations using supervised learning techniques. Subsequently, employing dynamic equations directly in iterative control schemes has disadvantages, as these methods frequently neglect model uncertainties and external disturbances. These constraints can be mitigated by replacing traditional dynamic models with neural networks, which can be continuously updated online as new data are acquired during quadrotor flights. This adaptability enables the neural networks to adjust changes in the model, hence improving control performance. This section provides a detailed explanation of the data collection process, neural network model design, training procedures, and the resulting training outcomes.

3.1.1. Data Generation

First, we created a dataset of input–output pairs, where the inputs were the current states and control inputs, and the outputs were the corresponding state derivatives. The input $\mathbf{x}_{input} = [x, y, z, v_x, v_y, v_z, \phi, \theta, \psi, p, q, r, \sigma_1, \sigma_2, \sigma_3, \sigma_4, \gamma]$ to the neural network was 17-dimensional data, and the output $\mathbf{y}_{output} = [\dot{x}, \dot{y}, \dot{z}, \dot{v}_x, \dot{v}_y, \dot{v}_z, \dot{\phi}, \dot{\theta}, \dot{\psi}, \dot{p}, \dot{q}, \dot{r}]$ was the 12-state derivatives. Multiple input combinations were sampled from uniform distributions within the operational range defined for the quadrotor. The state and their ranges are specified in Table 1.

Table 1. Inputs' sampling and ranges.

Input	Range	Description
x, y	$[-10, 10]$	X-axis position, Y-axis position, simulates variations in horizontal position
z	$[-10, 10]$	Z-axis position, considers flight requirements below the reference plane
(v_z, v_y, v_z)	$[0, 10]$	Velocity
(ϕ, θ)	$[-\pi/4, \pi/4]$	Roll and pitch, prevent drone instability
(ψ)	$[-\text{pi}, \text{pi}]$	Yaw, covers all possible directions
(r, p, q)	$[-2, 2]$	Angular velocity
$(\sigma_1, \sigma_2, \sigma_3, \sigma_4)$	$[0, 900]$	Rotor speed

The sampled variables $(\sigma_1, \sigma_2, \sigma_3, \sigma_4)$ were not used as inputs to the NN directly for training. For a quadcopter, the thrust T_i produced by each rotor can be represented as:

$$T_i = k_F \omega_i^2 \tag{16}$$

where k_F is the thrust coefficient, which represents the efficiency of thrust generation by the rotor, and ω_i is the angular velocity of the ith rotor.

The total thrust F is the sum of the thrusts produced by all four rotors:

$$F = \sum T_i = k_F\left(\omega_1^2 + \omega_2^2 + \omega_3^2 + \omega_4^2\right) \quad (17)$$

In addition to thrust, each rotor generates roll, pitch, and yaw moments. The moments consist of two parts: (1) roll and pitch moments and (2) yaw moments. These are moments caused by the eccentric position of the rotors. For roll and pitch, the moments are:

$$\tau_x = Lk_F\left(-\sigma_2^2 + \sigma_4^2\right), \tau_y = Lk_F\left(-\sigma_1^2 + \sigma_3^2\right) \quad (18)$$

where L is the distance from the rotor to the center of mass of the quadcopter.

The sign of the moment depends on the arrangement and rotation direction of the rotors.

The yaw moment: this moment is generated by the reaction torques of the rotors and can be expressed as:

$$\tau_z = k_M\left(-\sigma_1^2 + \sigma_2^2 - \sigma_3^2 + \sigma_4^2\right) \quad (19)$$

where k_M is the coefficient representing the efficiency of moment generation.

Gyroscopic precession effect: differences in the speed of the rotors also lead to a net gyroscopic effect, which affects the system's dynamic response.

$$\gamma = \sigma_1 - \sigma_2 + \sigma_3 - \sigma_4 \quad (20)$$

where γ is the net effect due to differences in motor speeds, contributing to gyroscopic precession.

We used these sampled variables to calculate the following additional inputs to the NN as

$$\begin{aligned} F &= k_F(\sigma_1^2 + \sigma_2^2 + \sigma_3^2 + \sigma_4^2) \\ \tau_x &= Lk_F(-\sigma_2^2 + \sigma_4^2) \\ \tau_y &= Lk_F(-\sigma_1^2 + \sigma_3^2) \\ \tau_z &= k_M(-\sigma_1^2 + \sigma_2^2 - \sigma_3^2 + \sigma_4^2) \\ \gamma &= \sigma_1 - \sigma_2 + \sigma_3 - \sigma_4 \end{aligned} \quad (21)$$

where F is the sum of thrusts produced by all four rotors, τ_x is the roll torque causing a rotation around the X-axis, τ_y is the pitch torque causing a rotation around the Y-axis, τ_z is the yaw torque causing rotation around the Z-axis, γ is the net effect due to differences in motor speeds, contributing to gyroscopic precession.

These inputs were passed through the quadrotor dynamics defined in Section 2. The training output for the NN was the state derivatives obtained for each input.

The use of neural networks allowed for the implicit approximation of disturbances in the dynamic model. During training, the neural network employed supervised learning, using input data, which included the effects of disturbances, and output data to learn the dynamic relationships of the entire system and establish an approximate dynamic model:

$$\hat{f}(x_t, u_t) \approx f(x_t, u_t + \varepsilon_t) + \omega_t$$

where $\hat{f}$ is the dynamic model learned by the neural network.

By approximating ε_t and ω_t through the neural network, the design phase of the controller did not need to explicitly introduce stochastic disturbances. Instead, these disturbances were considered as part of the learned deterministic model.

3.1.2. NN Model and Training

This section describes the selection of a Multilayer Perceptron (MLP) neural network architecture to imitate the quadrotor dynamics [27,28]. The network consisted of six hidden layers containing 256, 256, 128, 128, 64, and 64 neurons, respectively, as shown in Figure 2. Every hidden layer employed the Leaky Rectified Linear Unit (LeakyReLU) activation function to incorporate nonlinearity and mitigate the vanishing gradient problem. The training process utilized a dataset consisting of 15,000 samples, processed over 2000 epochs with a batch size of 64. A Graphics Processing Unit (GPU) was employed to accelerate training, facilitating fast computation and supporting a gradual decrease in the loss function through multiple iterations, thus improving the model's predictive performance. The optimization utilized the Adam optimizer in conjunction with a custom learning rate scheduler to enhance accuracy. The learning rate followed an exponential decay strategy, starting with an initial rate of 0.001 and reducing to 0.0001 over 1000 epochs. That method ensured stable convergence and prevented overshooting during training. The neural network model was developed with TensorFlow 2 and trained on a system comprising an Intel i9-13900 HX processor, 16 GB of RAM, and an NVIDIA GeForce RTX 3060 Ti GPU with 8 GB of memory. The model achieved a Mean Squared Error (MSE) of 0.013 and a Mean Absolute Error (MAE) of 0.047 on the test dataset on completion of training, which indicated high predictive accuracy and robustness.

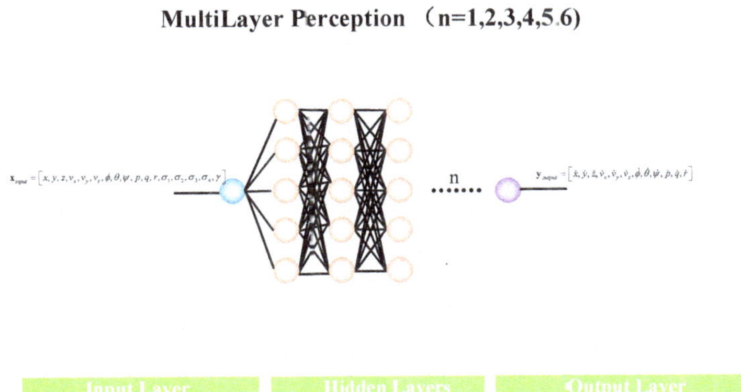

Figure 2. Multilayer Perceptron architecture for learning dynamics.

The LeakyReLU activation function is as follows:

$$f(x) = \begin{cases} x & if x > 0 \\ \alpha x & if x \leq 0 \end{cases} \tag{22}$$

where α is a small positive constant, typically set to $\alpha = 0.01$ or another small value. Specifically, when the input x is positive, Leaky ReLU outputs x; when the output is negative, it outputs αx, retaining a slight negative gradient.

The Mean Squared Error was used for the loss function; it can measure the average squared difference between the predicted and actual state derivatives:

$$L(\theta) = \frac{1}{N}\sum_{i=1}^{N}(\hat{y}_i - y_i)^2 \tag{23}$$

where $\hat{y}_i$ is the predicted state derivative from the MLP, y_i is the ground-truth state derivative.

On completion of training, the network approximated the function $f : \mathbf{x}_{input} \rightarrow \mathbf{y}_{output}$, essentially functioning as a surrogate model for the quadrotor dynamics.

3.1.3. Validation

A distinct validation set was used to assess the model's performance on unseen data after completing training, thereby preventing the model from overfitting and from memorizing the training data. The MSE and MAE outputs on the validation set indicated the model generalization capability. Figure 3 shows the desired value and actual value on validation set. Points situated along the diagonal line $y = x$ in Figure 3 indicate flawless predictions. Figure 4 displays the actual position and predicted position of (x, y, z) over time on the validation set. Figure 5 shows the actual angular and predicted angular velocity of (r, p, q) over time on the validation set.

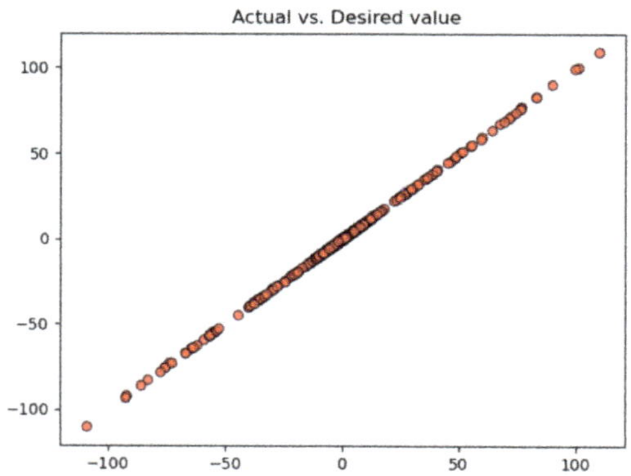

Figure 3. Desired value and actual value on the validation test.

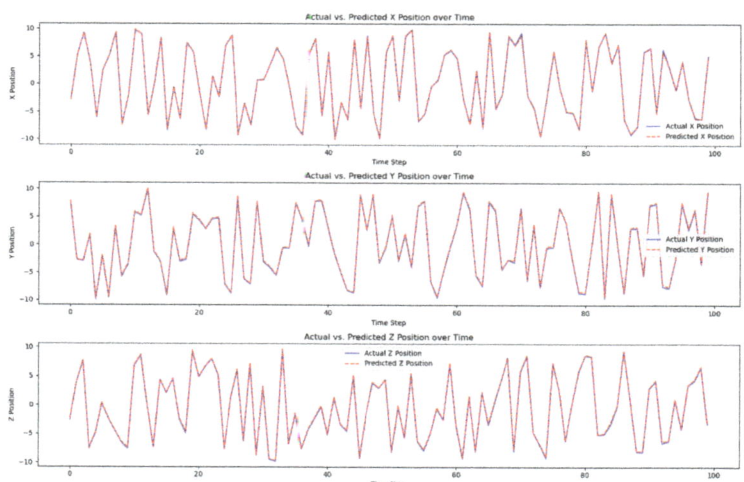

Figure 4. Actual position and predicted position of (x, y, z) over time on the validation set.

3.2. *Improved Model Predictive Path Integral*

For a discrete-time dynamical system with states $\mathbf{x}_t \in \aleph$, control inputs $\mathbf{u}_t \in U$, ε error in control input, ω as external disturbances, and mapping $f : \aleph \times U \rightarrow \aleph$ from the current state and input to the next state given by

$$x_{k+1} = f(x_k, u_k) \tag{24}$$

a general optimal control problem consists of finding a control policy $\pi(x)$, a map from the current state to the optimal input, $\pi : \aleph \to U$, such that a cost $J : \aleph \times U \to \mathbb{R}^{+}$ is minimized and is given by

$$\pi(x) = \underset{u}{\operatorname{argmin}} J(\mathbf{x}, \mathbf{u}) \tag{25}$$

$$J(\mathbf{x}, \mathbf{u}) = \sum_{t=0}^{T} \left(\left\| x_t - x_{ref} \right\|_Q^2 + \| u_t \|_R^2 \right) \tag{26}$$

where x_t is the current state, x_{ref} is the reference state, u_t is the control input, and Q and R are weighting matrices, subject to $\mathbf{x}_0 = \mathbf{x}$

$$\mathbf{x}_{k+1} = f(\mathbf{x}_k, \mathbf{u}_k) \tag{27}$$

$$\mathbf{x}_k \in \aleph, \mathbf{u} \in U \tag{28}$$

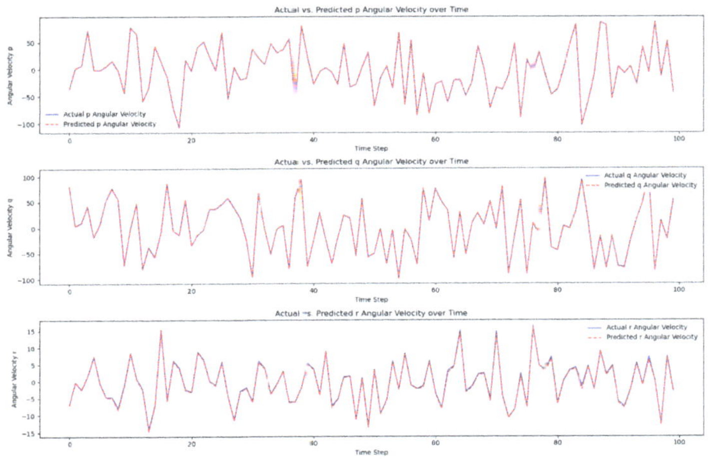

Figure 5. Actual angular and predicted angular velocity of (r, p, q) over time on the validation set.

The path integral control method for developing optimal control algorithms is based on stochastic trajectory samplings [29,30]. This method's benefit is in its avoidance of using derivatives for the system dynamics or the cost function, thus providing robustness in estimating the system dynamics and formulating the cost function. Given a transition model F, a sample size N, number of time steps (horizon) T, and an initial control sequence $\mathbf{u}_t^i = \left\{ \mathbf{u}_{t|t}^i, \mathbf{u}_{t+1|t}^i, \dots \mathbf{u}_{t+T-1|t}^i \right\}$, the algorithm calculates K rollouts for T time steps for the inputs and evaluates different possible trajectories by running them through the dynamic model F. The cost and weight for each of these rollouts are calculated using the provided cost function. The K weights are used to evaluate the contribution of each rollout to the final optimal input sequence, with only the first input being applied. The remainder of the sequence is used as the initial input for the subsequent iteration, and the process continues. The traditional MPPI method may require considerable computational resources or manual parameter adjustments to maintain control performance in response to dynamic environmental changes or significant swings in system states. Consequently, incorporating an adaptive mechanism into the MPPI control method enables the dynamic adjustment of its essential parameters according to system states or control performance, thereby enhancing the control's robustness and efficiency.

At each time step t, multiple control sequences were generated by randomly sampling the control inputs. The sampled control input sequences were denoted as follows:

$$\left\{ \mathbf{u}_t^i \right\}_{t=1}^N \tag{29}$$

Each control sequence $\mathbf{u}_t^i = \left\{ \mathbf{u}_{t|t}^i, \mathbf{u}_{t+1|t}^i, \ldots \mathbf{u}_{t+T-1|t}^i \right\}$ had the length of the prediction horizon T.

The real-time system error $e(t)$ as the distance between the current system state $\mathbf{x}$ and the target state $\mathbf{x}_d$ was defined as:

$$e(t) = \|\mathbf{x}_d - \mathbf{x}\| \tag{30}$$

The Euclidean distance was used to calculate the system error as follows:

$$e(t) = \sqrt{(\mathbf{x}_d - \mathbf{x})^T (\mathbf{x}_d - \mathbf{x})} \tag{31}$$

Based on the real-time error $e(t)$, we dynamically adjusted the sampling number N and the prediction horizon T:

$$N = N_o + k_N e(t) \tag{32}$$

$$T = T_o + k_T e(t) \tag{33}$$

where N_o and T_o are the baseline sampling number and baseline prediction horizon used when the error is small. k_N and k_T are gain coefficients that control the responsiveness of the sampling number and prediction horizon to the error.

When the error $e(t)$ increased, the sampling number N was increased to enhance sampling density and control precision. Conversely, when the error decreased, the sampling number was reduced to save computational resources. Similarly, the prediction horizon T was dynamically adjusted based on the error magnitude. An increased error led to a longer prediction horizon, allowing the controller to consider state changes over a more extended period. A decreased error shortened the prediction horizon, improving computational efficiency.

The cumulative cost for each trajectory was calculated as follows:

$$J^i = \sum_{k=0}^{T-1} \ell\left(\mathbf{x}_{t+k|t}^i, \mathbf{u}_{t+k|t}^i\right) \tag{34}$$

The weight for each trajectory was calculated as follows:

$$\omega^i = \frac{\exp(-J^i/\lambda)}{\sum_{j=1}^N \exp(-J^j/\lambda)} \tag{35}$$

Subsequently, the current control input using a weighted average of all sampled control inputs was updated as follows:

$$\mathbf{u}_t = \sum_{i=1}^N \omega^i \mathbf{u}_t^i \tag{36}$$

After applying the updated control input $\mathbf{u}_t$ to the system, we moved to the next time step t + 1 and repeated the above steps.

The MLP and improved MPPI control block diagram is shown in Figure 6:

Flight data and dynamic inputs: Provide a realistic dynamic description of the system, serving as the true data source in the simulation. These dynamics describe the motion behavior of the quadrotor UAV under external influences, including position, velocity, angular velocity, noise, and disturbance.

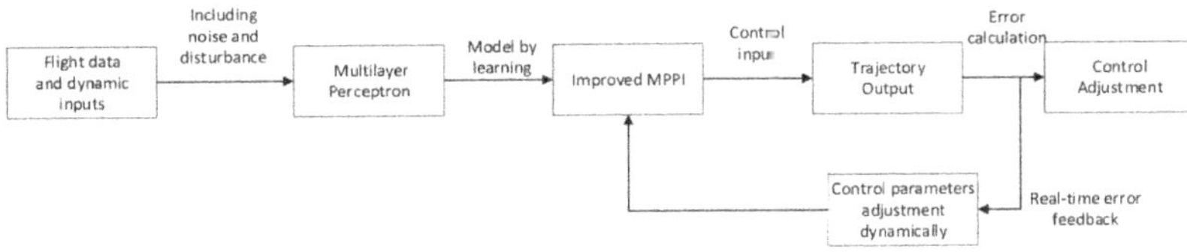

Figure 6. MLP and improved MPPI control block diagram.

Multilayer Perceptron: The MLP is used to learn the system dynamics, effectively replacing traditional dynamic models to some extent. This approach is particularly effective in complex dynamic environments or under external disturbances. The neural network's inputs are the current system state and control inputs, and the output is the derivative of the system state.

Improved MPPI controller: Based on the path integral method, the controller generates multiple possible trajectories through stochastic sampling and selects the optimal control input from them. By adaptively adjusting the sampling number and prediction horizon, the improved MPPI controller can better handle dynamic changes in the environment. When the error increases, the sampling number is increased to improve control accuracy; when the error decreases, the sampling number is reduced to save computational resources.

Trajectory output and control adjustment: The MPPI controller dynamically adjusts the control input based on real-time state errors, generating the trajectory output for the UAV. When there is an error between the actual trajectory and the target trajectory, the control adjustment module provides real-time feedback on this error and optimizes the process by dynamically adjusting the sampling number and prediction horizon, thereby enhancing control performance.

3.3. Stability Analysis

3.3.1. Neural Network Convergence Analysis

According to the universal approximation theorem, neural networks with appropriate structures and sufficient training data can approximate any continuous function with arbitrary precision. Therefore, we assumed that the MLP neural network satisfied the following after training:

$$\left\|\hat{f}(x,u) - f(x,u)\right\| \leq \varepsilon, \forall x, u \in D \tag{37}$$

where $\hat{f}(x,u)$ is the neural network-approximated dynamics model, and $f(x,u)$ is the true dynamics model; ε is a small positive constant, and D is the system's operating space.

During the training process, the neural network reduced the approximation error ε iteratively through learning, ultimately keeping it within a small bounded range. Through learning mechanisms and Figures 3–5, it ensured that the approximation error always remained within a controllable range. The above analysis demonstrates that the neural network reliably converged to an accurate approximation of the quadcopter dynamics.

3.3.2. Stability Analysis of the Closed-Loop System

To prove the stability of the closed-loop system, we adopted a control framework based on passivity, following the method proposed by Ruggiero et al. [31]. The passivity-based control method is suitable for nonlinear systems and effectively handles system uncertainties and external disturbances.

Choose the control law as:

$$\mathbf{u} = \mathbf{u}_{MPPI} - \mathbf{K}\mathbf{e} \tag{38}$$

where $\mathbf{u}_{MPPI}$ is the control input obtained from the Model Predictive Path Integral (MPPI) controller, and $\mathbf{K}$ is the gain matrix designed to stabilize the tracking error, $\mathbf{e}$ is the error.

Define a Lyapunov function V to evaluate the stability of the system. For the quadrotor system, we selected the following candidate Lyapunov function:

$$V(e) = \mathbf{e}^T\mathbf{P}\mathbf{e} \tag{39}$$

where $\mathbf{e} = \mathbf{x} - \mathbf{x}_{ref}$ is the tracking error of the system, and P is a symmetric positive definite matrix.

Compute the time derivative of the Lyapunov function:

$$\dot{V} = 2\mathbf{e}^T\mathbf{P}\dot{\mathbf{e}} \tag{40}$$

Based on the system dynamics:

$$\dot{\mathbf{x}} = f(\mathbf{x}, \mathbf{u}) + \mathbf{d} \tag{41}$$

where $\mathbf{d}$ represents all external disturbances and model uncertainties.
Based on $\mathbf{e} = \mathbf{x} - \mathbf{x}_{ref}$ and Equation (40), we obtained:

$$\dot{\mathbf{e}} = f(\mathbf{x}, \mathbf{u}) + \mathbf{d} - \dot{\mathbf{x}}_{ref} \tag{42}$$

The neural network approximation becomes:

$$f(\mathbf{x}, \mathbf{u}) \approx \hat{f}(\mathbf{x}, \mathbf{u}) - \varepsilon(\mathbf{x}, \mathbf{u}) \tag{43}$$

Thus,

$$\dot{\mathbf{e}} = \hat{f}(\mathbf{x}, \mathbf{u}) - \varepsilon(\mathbf{x}, \mathbf{u}) + \mathbf{d} - \dot{\mathbf{x}}_{ref} \tag{44}$$

Assuming that the MPPI controller ensures that in the absence of errors and disturbances:

$$\hat{f}\left(\mathbf{x}_{ref}, \mathbf{u}_{MPPI}\right) = \dot{\mathbf{x}}_{ref} \tag{45}$$

we can linearize the tracking error dynamics around $\mathbf{x}_{ref}$:

$$\dot{\mathbf{e}} \approx (\mathbf{A} - \mathbf{B}\mathbf{K})\mathbf{e} - \varepsilon(\mathbf{x}, \mathbf{u}) + \mathbf{d} \tag{46}$$

Substituting $\dot{\mathbf{e}}$ into $\dot{V}$:

$$\dot{V} = 2\mathbf{e}^T\mathbf{P}((\mathbf{A} - \mathbf{B}\mathbf{K})\mathbf{e} - \varepsilon(\mathbf{x}, \mathbf{u}) + \mathbf{d}) \tag{47}$$

where $\mathbf{A}$ and $\mathbf{B}$ are the Jacobian matrices of the system dynamics with respect to $\mathbf{x}$ and $\mathbf{u}$, respectively.

Under the passivity-based control framework, the control law and Lyapunov function parameters are designed such that the Lyapunov derivative satisfies:

$$\dot{V} \le -\lambda||\mathbf{e}||^2 + \gamma||\mathbf{d}||^2 + \beta||\varepsilon||^2 \tag{48}$$

where $\lambda > 0, \gamma > 0, and \beta > 0$ are design parameters.

To ensure stability, we require the following:

$$\dot{V} \le 0 \tag{49}$$

To achieve this, the matrix $\mathbf{P}$ and the control gain matrix $\mathbf{K}$ are selected by solving the Lyapunov equation.

Choose $\mathbf{P}$ and $\mathbf{K}$ so that they satisfy the following equation:

$$(\mathbf{A} - \mathbf{B}\mathbf{K})^T\mathbf{P} + \mathbf{P}(\mathbf{A} - \mathbf{B}\mathbf{K}) = -\mathbf{Q} \tag{50}$$

where $\mathbf{Q} > 0$ is a positive definite matrix. This condition ensures that the linear part of $\dot{V}$ is negative definite:

$$2\mathbf{e}^T\mathbf{P}(\mathbf{A}-\mathbf{BK})\mathbf{e} \leq -\lambda||\mathbf{e}||^2 \tag{51}$$

where $\lambda > 0$ is the smallest eigenvalue of $\mathbf{Q}$.

Apply the Cauchy–Schwarz inequality to the remaining terms:

$$-2\mathbf{e}^T\mathbf{P}\varepsilon + 2\mathbf{e}^T\mathbf{P}\mathbf{d} \leq 2||\mathbf{P}||\,||\varepsilon||\,||\mathbf{e}|| - 2||\mathbf{P}||\,||\mathbf{d}||\,||\mathbf{e}|| \tag{52}$$

To maintain $\dot{V} \leq 0$, we require the following:

$$-\lambda||\mathbf{e}||^2 + 2||\mathbf{P}||\,||\varepsilon||\,||\mathbf{e}|| + 2||\mathbf{P}||\,||\mathbf{d}||\,||\mathbf{e}|| \leq 0 \tag{53}$$

To satisfy the above inequality, we can bound the terms involving $||\mathbf{e}||$ by ensuring that

$$\lambda||\mathbf{e}||^2 \geq 2||\mathbf{P}||\,||\varepsilon||\,||\mathbf{e}|| + 2||\mathbf{P}||\,||\mathbf{d}||\,||\mathbf{e}|| \tag{54}$$

Dividing both sides by $||\mathbf{e}||$ (assuming $||\mathbf{e}|| > 0$):

$$\lambda||\mathbf{e}|| \geq 2||\mathbf{P}||(||\varepsilon|| + ||\mathbf{d}||) \tag{55}$$

This condition holds if:

$$||\mathbf{e}|| \geq \frac{2||\mathbf{P}||(||\varepsilon|| + ||\mathbf{d}||)}{\lambda} \tag{56}$$

When the tracking error $||\mathbf{e}||$ is sufficiently large, $\dot{V} \leq 0$. For smaller errors, the bounded approximation error and disturbances are dominated by the negative definite term, ensuring that $\dot{V} \leq 0$ overall.

From the above analysis, we conclude that the tracking error $\mathbf{e}$ converges to a bounded region around zero, with the bound determined by the approximation error ε and disturbances $||\mathbf{d}||$. The tracking error remains bounded all the time, ensuring robust performance despite disturbances and model uncertainties. The control law satisfies the passivity condition, preventing energy injection that could destabilize the system.

Therefore, the closed-loop quadrotor system under the proposed MPPI controller and MLP neural network approximation achieves global asymptotic stability, ensuring reliable trajectory tracking in the presence of disturbances and model uncertainties.

4. Results

The neural network was trained using TensorFlow 2 on Google Colab, utilizing local system resources. Drone simulations were conducted in MATLAB, leveraging the MATLAB Deep Learning Toolbox, which facilitates the import and utilization of pre-trained TensorFlow models. We implemented three control methods: (1) Model Predictive Path Integral (MPPI) control without learning dynamics, (2) MPPI control with dynamics learned by a simple fully connected neural network, (3) MPC with dynamics learned by a Multilayer Perceptron (MLP) and (4) MPPI control with dynamics learned by a Multilayer Perceptron (MLP). Simulations were performed across two distinct trajectories: (1) waypoint navigation and (2) tracking a figure-eight trajectory. In all simulations, the plant was integrated at a frequency of 100 Hz (every 0.01 s), and the controller operated at 20 Hz (every 0.05 s). The initial position of the quadcopter was set to [0,10] with an initial control input of [620.6108, 620.6108, 620.6108, 620.6108]. For the waypoint navigation task, the desired position was [3,10]. The figure-eight trajectory was defined with a scale factor of 2, and an additional spiral path was utilized to evaluate the algorithm's performance further. All algorithms were evaluated using a comprehensive cost function that penalized deviations in position tracking, velocity magnitudes, angular rotations, deviations of control signals from the nominal hovering control (indirectly reflecting control effort), and changes in control signals. A critical penalty was imposed if the sampled trajectory descended below z = 0 (below the reference ground level). The Root-Mean-Square Error (RMSE) metric was employed to assess the experimental results across the three trajectory tasks:

$$RMSE = \sqrt{\frac{1}{N}\sum_{i=1}^{N} e_i^2}$$

4.1. MPPI Control Without Learning Dynamics

(1) Waypoint navigation: The number of rollouts per step was 800, with a prediction horizon of 25. At each step, the controller took an average of 0.12 s to execute. The program execution time was 618.76 s. The final tracking error was 0.1426 m. The results are shown in Figure 7a–f. The RMSE was 0.7124 m.

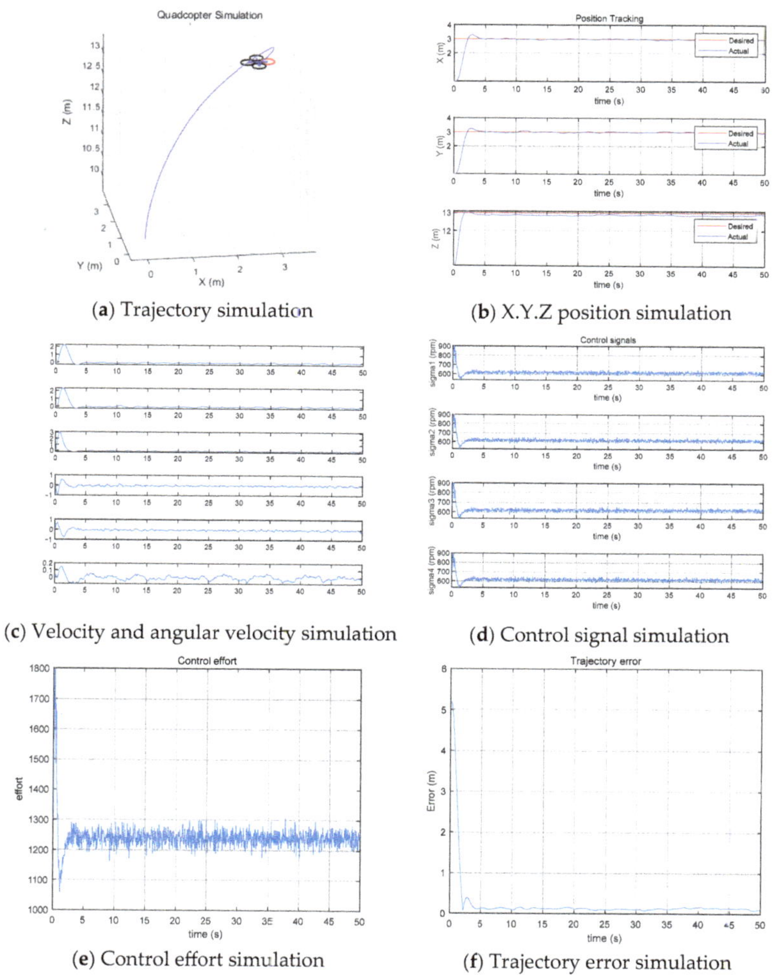

(a) Trajectory simulation (b) X.Y.Z position simulation

(c) Velocity and angular velocity simulation (d) Control signal simulation

(e) Control effort simulation (f) Trajectory error simulation

Figure 7. Dynamic simulation of MPPI control without learning on waypoint trajectory (the blue line represents the result from the MPPI control without learning; the red line represents the desired result); RMSE = 0.7124 m.

(2) Figure-eight trajectory tracking: The number of rollouts per step was 800, with a prediction horizon of 25. At each step, the controller took an average of 0.21 s to execute. The program execution time was 938.49 s. The results are shown in Figure 8a–f. The RMSE was 1.9756 m.

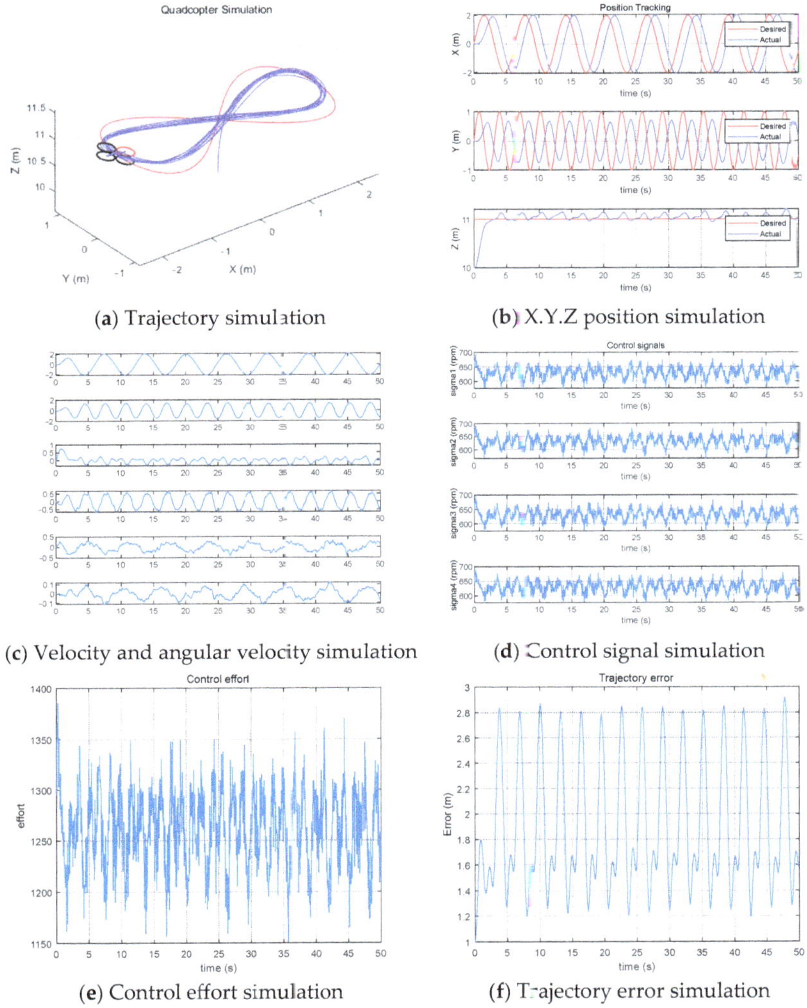

(**a**) Trajectory simulation

(**b**) X.Y.Z position simulation

(**c**) Velocity and angular velocity simulation

(**d**) Control signal simulation

(**e**) Control effort simulation

(**f**) Trajectory error simulation

Figure 8. Dynamic simulation of MPPI control without learning on figure-eight trajectory (blue line represents the result of the MPPI control without learning; the red line represents the desired result); RMSE = 1.9756 m.

4.2. MPPI Control with Learning Dynamics with Two Hidden Layers Connected to Neural Network

(1) Waypoint navigation: The number of rollouts per step was 800, with a prediction horizon of 25. At each step, the controller took an average of 0.48 s to execute. The program execution time was 723.15 s. The results are shown in Figure 9a–f. The RMSE was 0.7053 m.

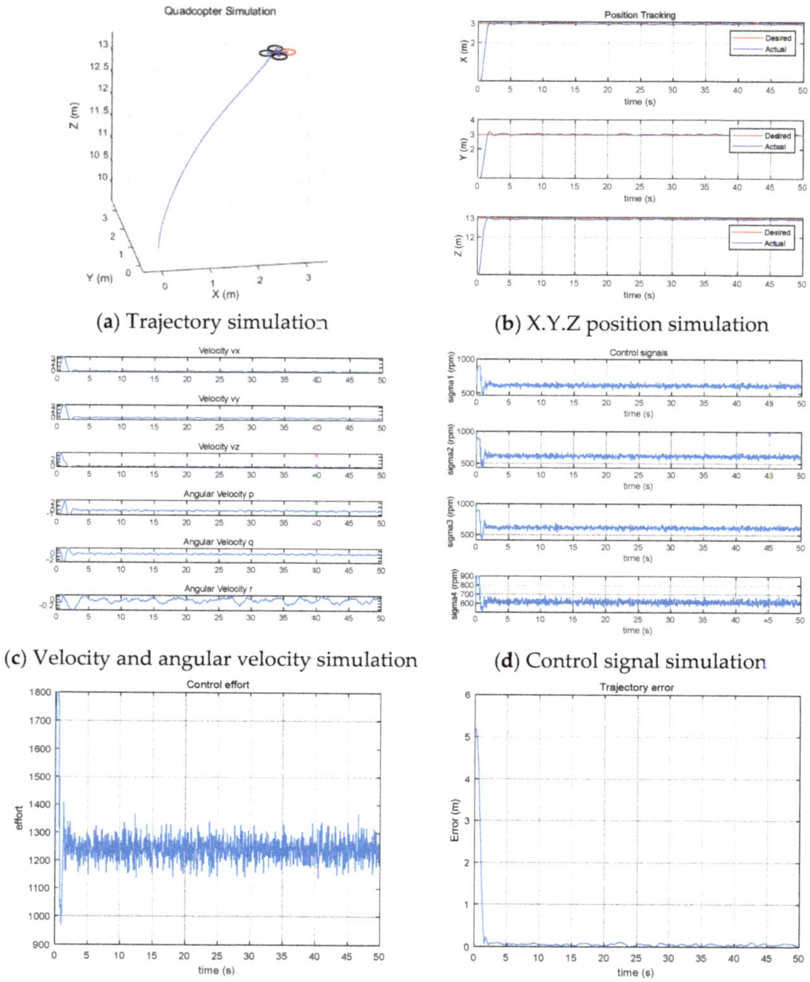

(**a**) Trajectory simulation (**b**) X.Y.Z position simulation (**c**) Velocity and angular velocity simulation (**d**) Control signal simulation (**e**) Control effort simulation (**f**) Control effort simulation

Figure 9. Simulation of MPPI control with 2-hidden-layer- network on waypoint trajectory (the blue line represents the result of MPPI control with 2 hidden layers; the red line represents the desired result); RMSE = 0. 7053 m.

(2) Figure-eight trajectory tracking: The number of rollouts per step was 800, with a prediction horizon of 25. At each step, the controller took an average of 0.5 s to execute. The program execution time was 986.37 s. The results are shown in Figure 10a–f. The RMSE was 1.3883 m.

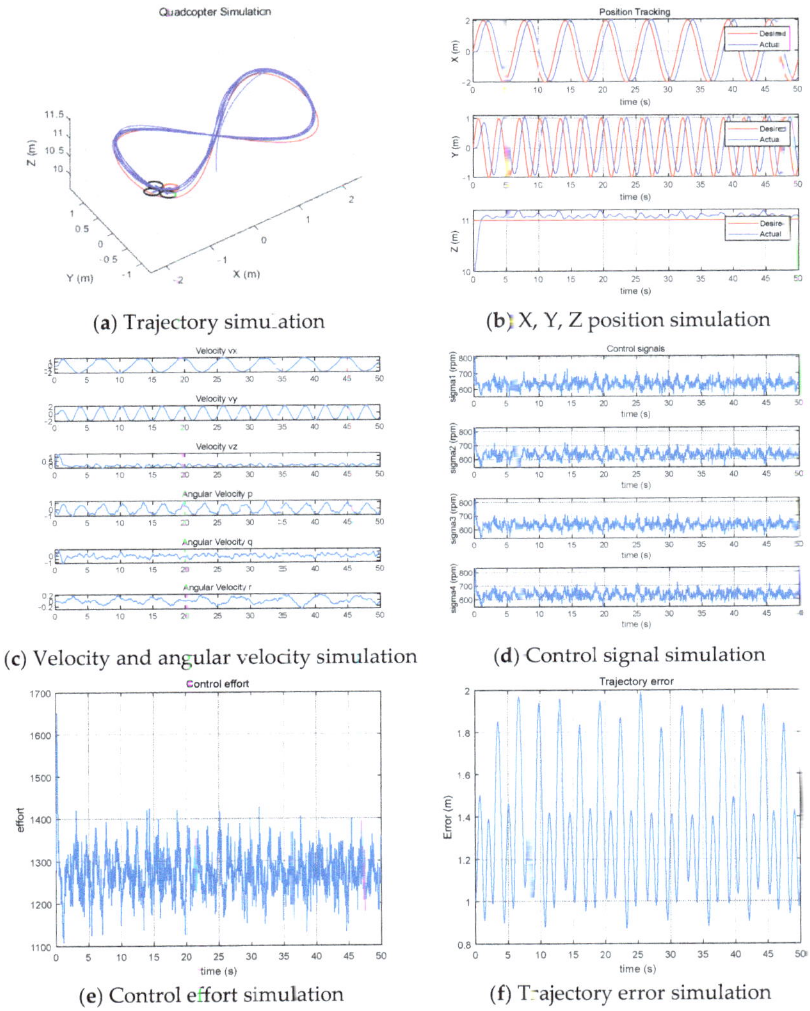

(a) Trajectory simulation

(b) X, Y, Z position simulation

(c) Velocity and angular velocity simulation

(d) Control signal simulation

(e) Control effort simulation

(f) Trajectory error simulation

Figure 10. Simulation of MPPI control with 2-hidden-layer network on figure-eight trajectory (the blue line represents the result of MPPI control with 2 hidden layers; the red line represents the desired result); RMSE = 1.3883 m.

4.3. *MPC Control with Learning Dynamics by Multilayer Perceptron*

(1) Waypoint navigation: The number of rollouts per step was 800, with a prediction horizon of 25. At each step, the controller took an average of 1.6 s to execute. The program execution time was 650.25 s. The results are shown in Figure 11a–f. The RMSE was 0.9747 m.

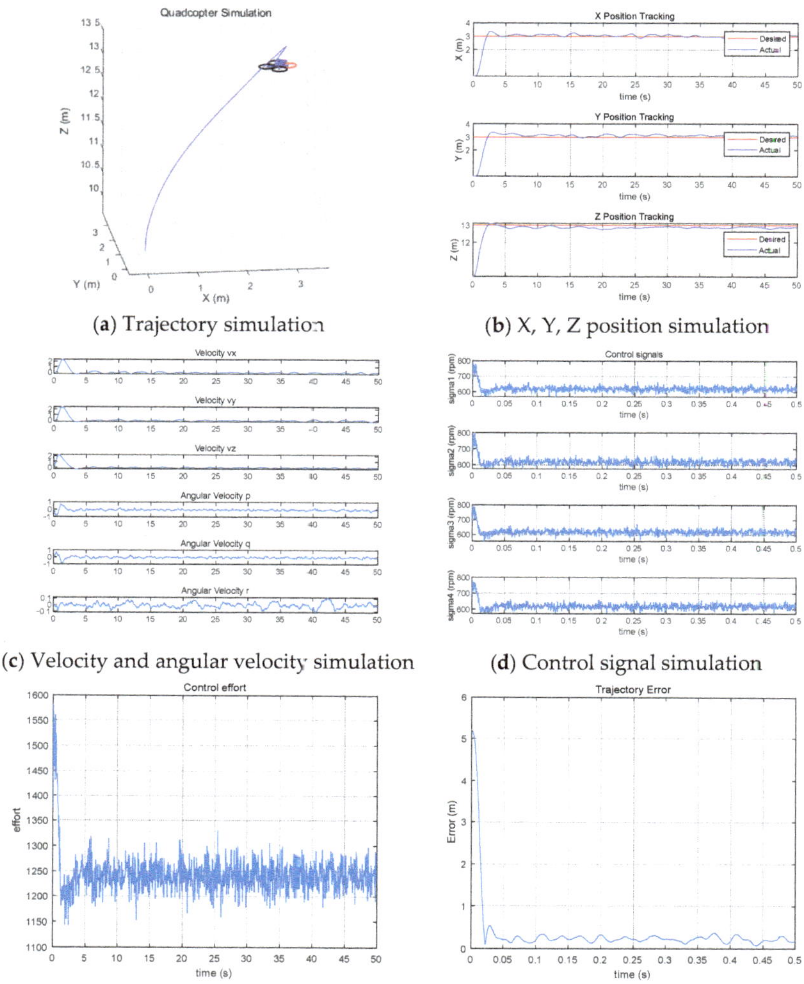

(**a**) Trajectory simulation (**b**) X, Y, Z position simulation

(**c**) Velocity and angular velocity simulation (**d**) Control signal simulation

(**e**) Control effort simulation (**f**) Trajectory error simulation

Figure 11. Simulation of MPC with MLP on waypoint trajectory (the blue line represents MPC with MLP; the red line represents the desired result); RMSE = 0.9747 m.

(2) Figure-eight trajectory tracking: The number of rollouts per step was 800, with a prediction horizon of 25. At each step, the controller took an average of 0.53 s to execute. The simulation time was 30 s, and the actual program execution time was 1253.27 s. The results are shown in Figure 12a–f. The RMSE was 1.8973 m.

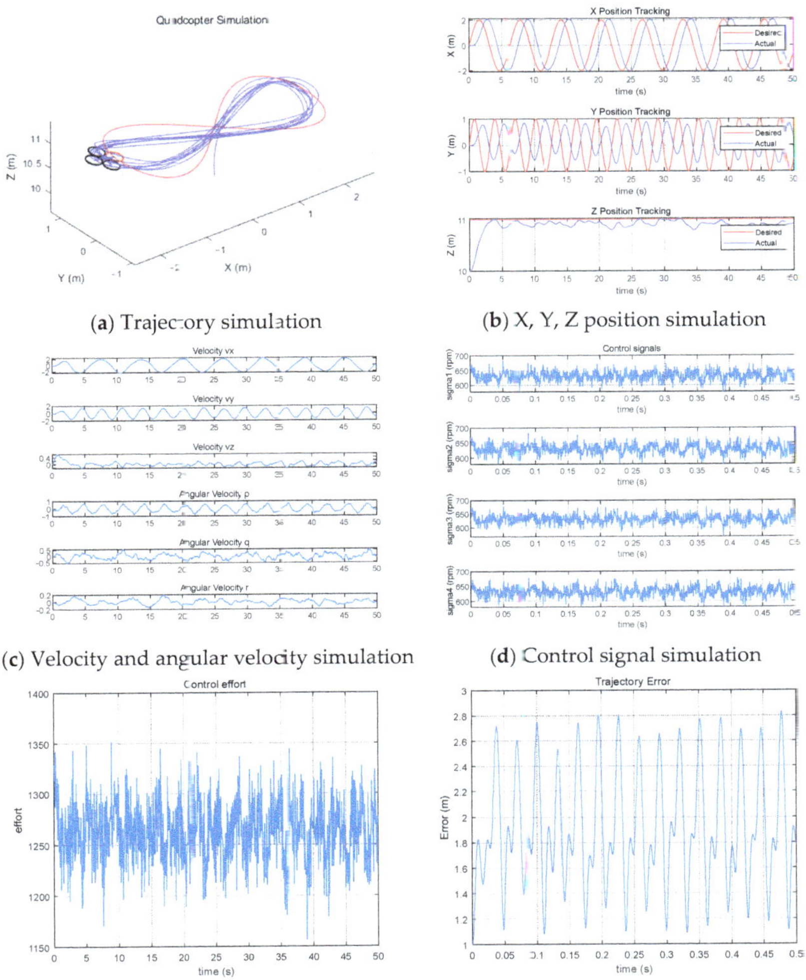

(**a**) Trajectory simulation (**b**) X, Y, Z position simulation

(**c**) Velocity and angular velocity simulation (**d**) Control signal simulation

(**e**) Control effort simulation (**f**) Trajectory error simulation

Figure 12. Simulation of MPC with MLP on figure-eight trajectory (the blue line represents the result of MPC with MLP; the red line represents the desired result); RMSE = 1.8973 m.

4.4. MPPI Control with Learning Dynamics by Multilayer Perceptron

(1) Waypoint navigation: The number of rollouts per step was 800, with a prediction horizon of 25. At each step, the controller took an average of 1.8 s to execute. The program execution time was 819.68 s. The results are shown in Figure 13a–f. The RMSE was 0.6515 m.

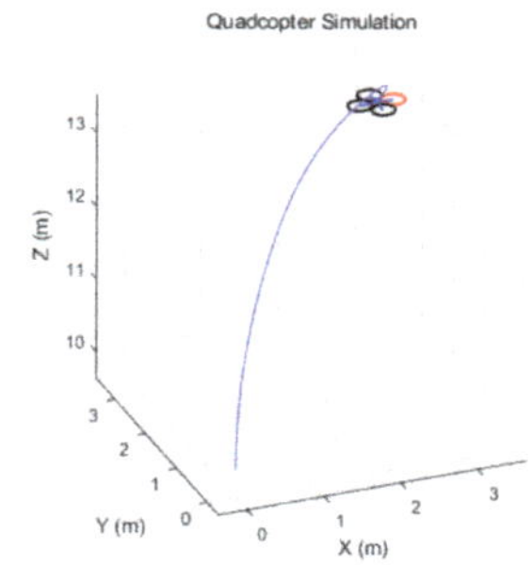

(**a**) Trajectory simulation

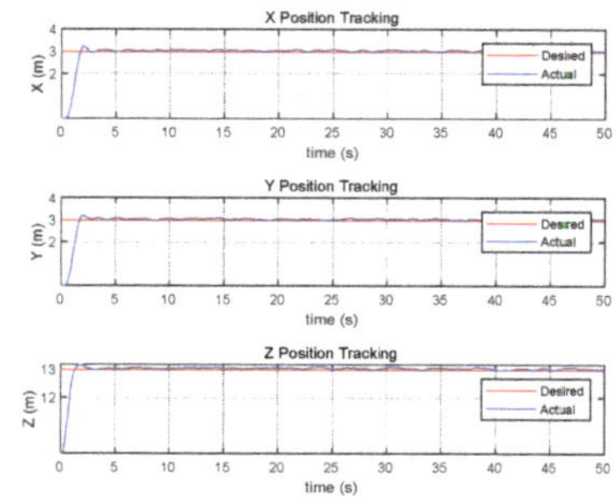

(**b**) X, Y, Z position simulation

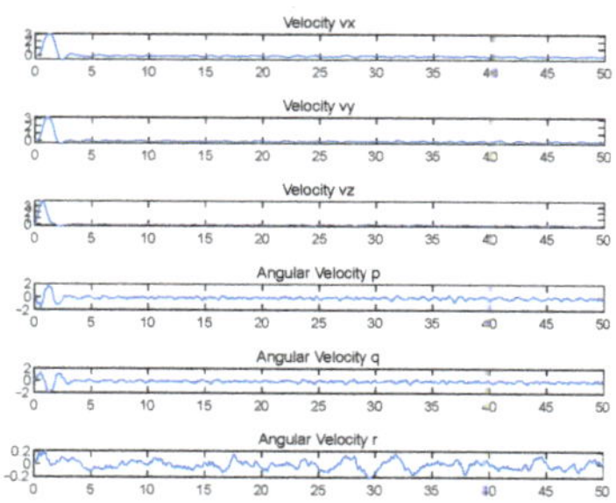

(**c**) Velocity and angular velocity simulation

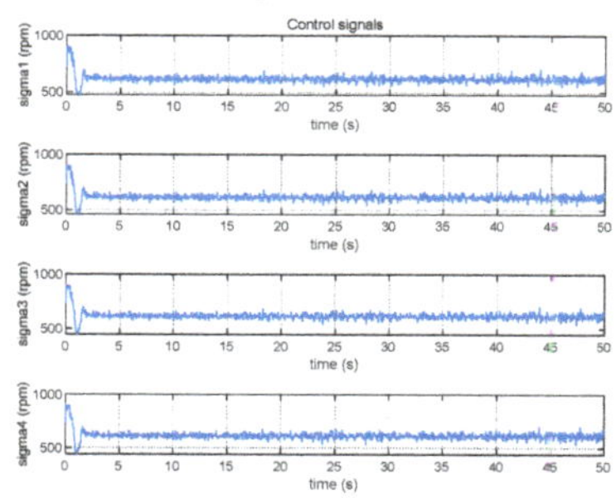

(**d**) Control signal simulation

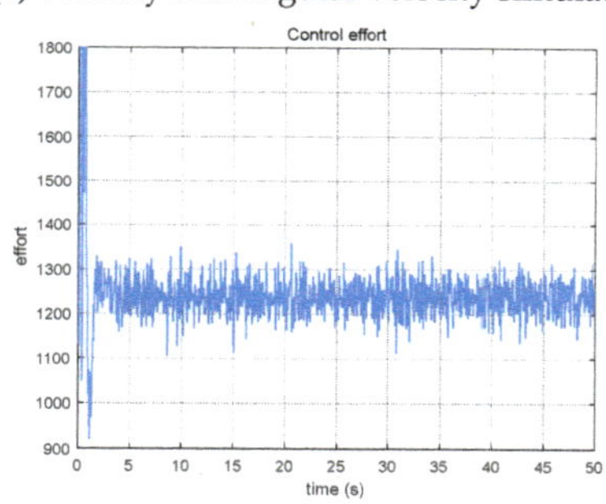

(**e**) Control effort simulation

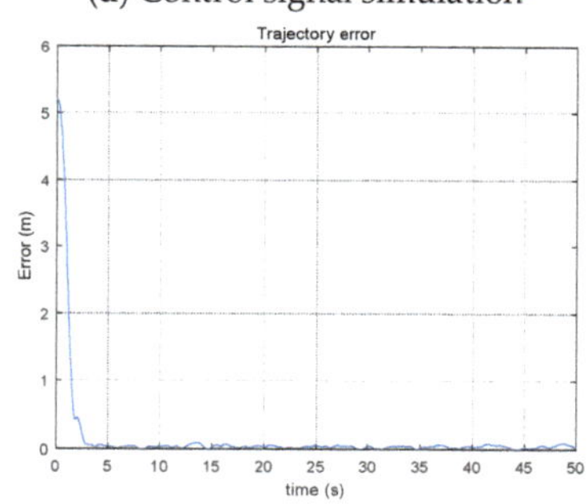

(**f**) Trajectory error simulation

Figure 13. Simulation of MPPI control with MLP on waypoint trajectory (the blue line represents the result of MPPI control with MLP; the red line represents the desired result); RMSE = 0.6515 m.

(2) Figure-eight trajectory tracking: The number of rollouts per step was 800, with a prediction horizon of 25. At each step, the controller took an average of 0.45 s to execute. The simulation time was 30 s, and the actual program execution time was 1142.27 s. The results for the same are shown in Figure 14a–f. The RMSE was 1.1093 m.

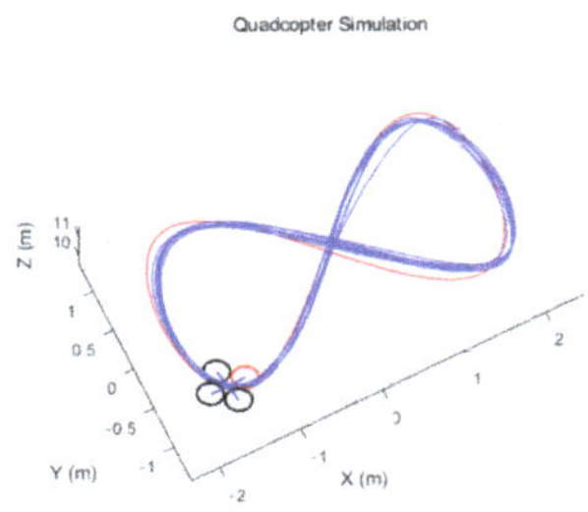

(**a**) Trajectory simulation

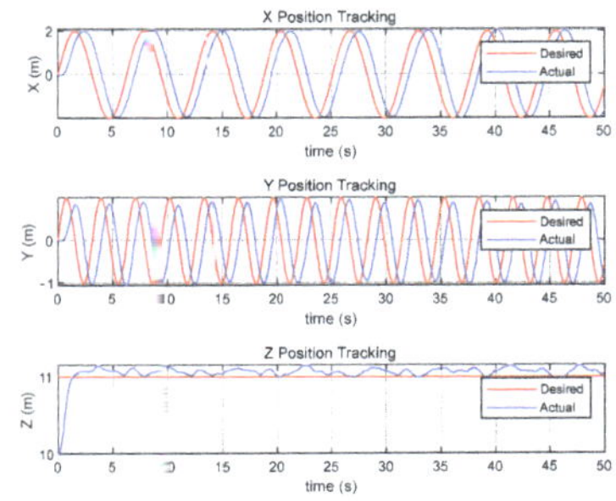

(**b**) X, Y, Z position simulation

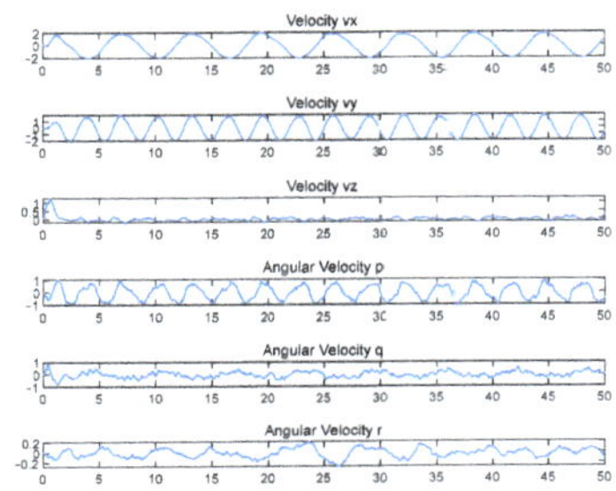

(**c**) Velocity and angular velocity simulation

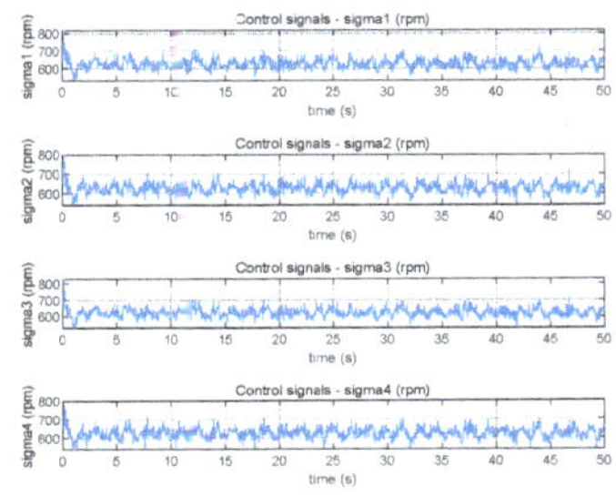

(**d**) Control signal simulation

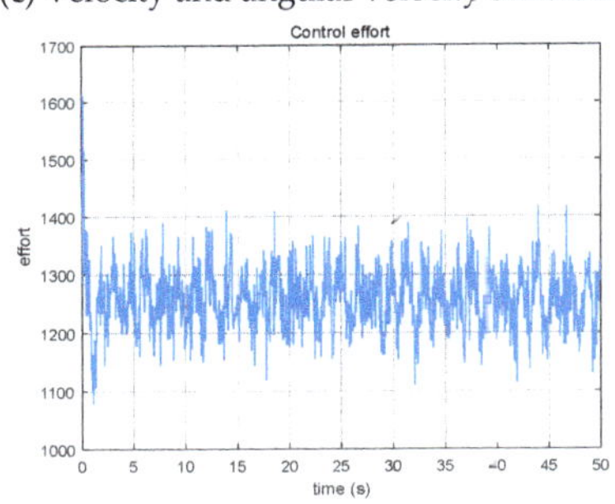

(**e**) Control effort simulation

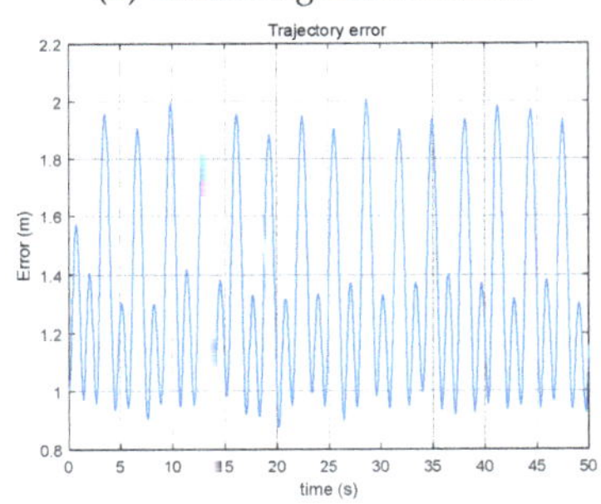

(**f**) Trajectory error simulation

Figure 14. Simulation of MPPI control with MLP on figure-eight trajectory (the blue line represents the results of MPPI control with MLP; the red line represents the desired result); RMSE = 1.1093 m.

5. Discussion

Although existing MPPI methods perform well in path planning and trajectory tracking, they typically require high computational resources and manual parameter adjustments to maintain control performance. In this study, we proposed an improved MPPI controller that incorporates an adaptive mechanism based on system errors, allowing a dynamic adjustment of the sampling number and prediction horizon, thereby improving the controller's robustness and efficiency in complex dynamic environments.

This paper used a multilayer perceptron (MLP) neural network to learn the dynamics of the quadcopter and combined it with the MPPI controller. Compared to other control methods based on traditional dynamic models, our approach can handle modeling inaccuracies and external disturbances by learning the system's nonlinear dynamic behavior. Compared to classical model-based approaches, this data-driven control method better adapts to real-world uncertainties.

Unlike other studies that use traditional Model Predictive Control (MPC) and robust control methods, our research continuously learns the system dynamics using a neural network and introduces an adaptive adjustment mechanism, enabling the system to maintain good performance even in the face of dynamic environmental changes and partial observability. This characteristic is particularly suitable for UAVs operating in unknown or dynamic environments, such as forests and urban canyons.

6. Conclusions

This paper presented an approach that integrated MPPI control and an MLP. The improved MPPI method could adjust the sampling quantity according to trajectory error to achieve a better trajectory. The MLP neural network could learn the dynamics of a quadrotor by its state parameter and model the quadcopter. The improved MPPI employed the model to guide and control the quadcopter along the desired trajectory. Three trajectories were employed to test the effectiveness of the proposed method. The results were compared with MPPI control without learning, MPPI control with two hidden layers, MPC with MLP, and improved MPPI control utilizing MLP. The experimental results showed that improved MPPI control with MLP had better performance than MPPI control without learning and MPPI control with two hidden layers. In the future, reinforcement learning methods could be integrated, allowing the neural network to automatically optimize control parameters based on flight variations, thereby improving adaptability and resilience to external disturbances. We could extend the improved MPPI method to not only perform single-objective trajectory tracking but also to execute multi-objective path planning and collision avoidance tasks in complex environments. Future work could incorporate graph search algorithms and global path optimization techniques to improve navigation efficiency and safety.

Author Contributions: Conceptualization, Y.L. and Q.Z.; methodology, Y.L.; software, Y.L.; validation, Y.L. and A.E.; formal analysis, Y.L.; investigation, Y.L.; resources, Y.L.; data curation, Y.L.; writing—original draft preparation, Y.L.; writing—review and editing, A.E.; visualization, Y.L.; supervision, Q.Z. All authors have read and agreed to the published version of the manuscript.

Funding: This research received no external funding.

Data Availability Statement: All data are available within the paper.

Conflicts of Interest: The authors declare no conflicts of interest.

References

1. Hoffmann, G.; Huang, H.; Waslander, S.; Tomlin, C. Quadrotor Helicopter Flight Dynamics and Control: Theory and experiment. In Proceedings of the AIAA Guidance, Navigation, and Control Conference, Hilton Head, SC, USA, 20–23 August 2007; Volume 2.
2. Brookner, E. *Tracking and Kalman Filtering Made Easy*; John Wiley and Sons, Inc.: New York, NY, USA, 1998.
3. Wan, E.A.; Van Der Merwe, R. The Unscented Kalman Filter for Nonlinear Estimation. In Proceedings of the IEEE Adaptive Systems for Signal Processing, Communications, and Control Symposium 2000, Lake Louise, AB, Canada, 4 October 2000; pp. 153–158.
4. Mellinger, D.; Michael, N.; Kumar, V. Trajectory generation and control for precise aggressive maneuvers with quadrotors. *Int. J. Robot. Res.* **2012**, *31*, 664–674. [CrossRef]
5. Gonzalez, D.; Perez, J.; Milanes, V.; Nashashibi, F. A review of motion planning techniques for automated vehicles. *IEEE Trans. Intell. Transp. Syst.* **2015**, *17*, 1135–1145. [CrossRef]
6. Mohta, K.; Watterson, M.; Mulgaonkar, Y.; Liu, S.; Qu, C.; Makineni, A.; Saulnier, K.; Sun, K.; Zhu, A.; Delmerico, J.; et al. Fast, autonomous flight in GPS-denied and cluttered environments. *J. Field Robot.* **2018**, *35*, 101–120. [CrossRef]
7. Baca, T.; Hert, D.; Loianno, G.; Saska, M.; Kumar, V. Model Predictive Trajectory Tracking and Collision Avoidance for Reliable outdoor Deployment of Unmanned Aerial Vehicles. In Proceedings of the IEEE/RSJ International Conference on Intelligent Robots and Systems, Madrid, Spain, 1–5 October 2018; pp. 6753–6760.

8. Ryll, M.; Ware, J.; Carter, J.; Roy, N. Efficient Trajectory Planning for High Speed Flight in Unknown Environments. In Proceedings of the International Conference on Robotics and Automation, Montreal, QC, Canada, 20–24 May 2019; pp. 732–738.
9. Theodorou, E.; Buchli, J.; Schaal, S. A Generalized Path Integral Approach to Reinforcement Learning. *J. Mach. Learn. Res.* **2010**, *11*, 3137–3181.
10. Kappen, H.J. Linear Theory for Control of Nonlinear Stochastic Systems. *Phys. Rev. Lett.* **2005**, *95*, 200201. [CrossRef] [PubMed]
11. Theodorou, E.A. Nonlinear stochastic control and information theoretic dualities: Connections interdependencies and thermodynamic interpretations. *Entropy* **2015**, *17*, 3352–3375. [CrossRef]
12. Karatzas, I.; Shreve, S. *Brownian Motion and Stochastic Calculus*, 2nd ed.; Graduate Texts in Mathematics; Springer: New York, NY, USA, 1991.
13. Friedman, A. *Stochastic Differential Equations and Applications*; Academic Press: New York, NY, USA, 1975.
14. Williams, G.; Drews, P.; Goldfain, B.; Rehg, J.M.; Theodorou, E.A. Aggressive Driving with Model Predictive Path Integral Control. In Proceedings of the 2016 IEEE International Conference on Robotics and Automation (ICRA), Stockholm, Sweden, 16–21 May 2016; IEEE: Piscataway, NJ, USA, 2016.
15. Romero, A.; Sun, S.; Foehn, P.; Scaramuzza, D. Model predictive contouring control for time-optimal quadrotor flight. *IEEE Trans. Robot.* **2022**, *38*, 3340–3356. [CrossRef]
16. Ruchika, N.R. Model predictive control: History and development. *Int. J. Eng. Trends Technol. IJETT* **2013**, *4*, 2600–2602.
17. Gómez, V.; Thijssen, S.; Symington, A.; Hailes, S.; Kappen, H.J. Real-Time Stochastic Optimal Control for Multi-Agent Quadrotor Systems. In Proceedings of the International Conference on Automated Planning and Scheduling (ICAPS), London, UK, 12–17 June 2016; pp. 468–476.
18. Mohamed, I.S.; Allibert, G.; Martinet, P. Model Predictive Path Integral Control Framework for Partially Observable navigation: A Quadrotor Case Study. In Proceedings of the 2020 16th International Conference on Control, Automation, Robotics and Vision (ICARCV), Shenzhen, China, 13–15 December 2020; IEEE: Piscataway, NJ, USA, 2020.
19. Williams, G.; Drews, P.; Goldfain, B.; Rehg, J.M.; Theodorou, E.A. Information-Theoretic Model Predictive Control: Theory and Applications to Autonomous Driving. *IEEE Trans. Robot.* **2018**, *34*, 1603–1622. [CrossRef]
20. Kusumoto, R.; Palmieri, L.; Spies, M.; Csiszar, A.; Arras, K.O. Informed Information Theoretic Model Predictive Control. In Proceedings of the International Conference on Robotics and Automation, Montreal, QC, Canada, 20–24 May 2019; pp. 2047–2053.
21. Nicol, C.; Macnab, C.; Ramirez-Serrano, A. Robust adaptive control of a quadrotor helicopter. *Mechatronics* **2011**, *216*, 927–938. [CrossRef]
22. Raffo, G.V.; Ortega, M.G.; Rubio, F.R. Sliding Mode Control of a Quadrotor Helicopter. In Proceedings of the 45th IEEE Conference on Decision and Control, San Diego, CA, USA, 13–15 December 2006; pp. 4957–4962.
23. Raffo, G.V.; Ortega, M.G.; Rubio, F.R. An integral predictive/nonlinear H∞ control structure for a quadrotor helicopter. *Automatica* **2010**, *46*, 29–39. [CrossRef]
24. Voos, I. Nonlinear Control of a Quadrotor Micro-UAV Using Feedback-Linearization. In Proceedings of the 2009 IEEE International Conference on Mechatronics, Malaga, Spain, 14–17 April 2009; pp. 1–6.
25. Zhaowei, M.; Tianjiang, H.; Lincheng, S.; Weiwei, K.; Boxin, Z.; Kaidi, Y. An Iterative Learning Controller for Quadrotor UAV Path Following at a Constant Altitude. In Proceedings of the 2015 34th Chinese Control Conference (CCC), Hangzhou, China, 28–30 July 2015; IEEE: Piscataway, NJ, USA, 2015; pp. 4406–4411.
26. Hehn, M.; D'Andrea, R. An Iterative Learning Scheme for High Performance, Periodic Quadrocopter Trajectories. In Proceedings of the European Control Conference (ECC), Zurich, Switzerland, 17–19 July 2013; IEEE: Piscataway, NJ, USA, 2013; pp. 1799–1804.
27. Bansal, S.; Akametalu, A.K.; Jiang, F.J.; Laine, F.; Tomlin, C.J. Learning Quadrotor Dynamics Using Neural Network for Flight Control. In Proceedings of the 2016 IEEE 55th Conference on Decision and Control (CDC), Las Vegas, NV, USA, 12–14 December 2016; IEEE: Piscataway, NJ, USA, 2016.
28. LeCun, Y.; Bengio, Y.; Hinton, G. Deep learning. *Nature* **2015**, *521*, 436–444. [CrossRef]
29. Kappen, H.J. Path integrals and symmetry breaking for optimal control theory. *J. Stat. Mech. Theory Exp.* **2005**, *2005*, P11011 [CrossRef]
30. Williams, G.; Aldrich, A.; Theodorou, E.A. Model predictive path integral control: From theory to parallel computation. *J. Guid. Control Dyn.* **2017**, *40*, 344–357. [CrossRef]
31. Ruggiero, F.; Cacace, J.; Sadeghian, H.; Lippiello, V. Passivity-based control of VToL UAVs with a momentum-based estimator of external wrench and unmodeled dynamics. *Robot. Auton. Syst.* **2015**, *72*, 139–151. [CrossRef]

drones

MDPI

Article

Distributed Control for Multi-Robot Interactive Swarming Using Voronoi Partioning†

Alexandre Eudes, Sylvain Bertrand, Julien Marzat * and Ioannis Sarras

DTIS, ONERA, Université Paris-Saclay, F-91123 Palaiseau, France; alexandre.eudes@onera.fr (A.E.); sylvain.bertrand@onera.fr (S.B.); ioannis.sarras@onera.fr (I.S.)

* Correspondence: julien.marzat@onera.fr

† This paper is an extended version of our paper: Bertrand, S.; Sarras, I.; Eudes, A.; Marzat, J. Voronoi-Based Geometric Distributed Fleet Control of a Multi-Robot System. In Proceedings of the 16th International Conference on Control, Automation, Robotics and Vision (ICARCV), Shenzhen, China, 13–15 December 2020; pp. 85–91.

Abstract: The problem of safe navigation of a human-multi-robot system is addressed in this paper. More precisely, we propose a novel distributed algorithm to control a swarm of unmanned ground robots interacting with human operators in presence of obstacles. Contrary to many existing algorithms that consider formation control, the proposed approach results in non-rigid motion for the swarm, which more easily enables interactions with human operators and navigation in cluttered environments. Each vehicle calculates distributively and dynamically its own safety zone in which it generates a reference point to be tracked. The algorithm relies on purely geometric reasoning through the use of Voronoi partitioning and collision cones, which allows to naturally account for inter-robot, human-robot and robot-obstacle interactions. Different interaction modes have been defined from this common basis to address the following practical problems: autonomous waypoint navigation, velocity-guided motion, and follow a localized operator. The effectiveness of the algorithm is illustrated by outdoor and indoor field experiments.

Keywords: multi-robot swarms; human-robot interactions; distributed control; voronoi partioning

1. Introduction

The deployment of swarms of unmanned vehicles for both civil and defense missions has radically increased in the last years. Recent progresses in vision or laser-based localization and mapping, along with the increase in embedded computational power, have led to the development of mobile and aerial robots of reduced dimensions allowing larger swarms of robotic vehicles to effectively undertake such missions under realistic environmental and communication conditions. Nevertheless, interaction with humans and obstacles or the practical limitations of inter-vehicle communication data links still pose serious challenges that need to be consistently addressed for on-field deployment of teams of autonomous robots [1,2]. This requires the synthesis of distributed control algorithms with increased capabilities in terms of autonomy, safety and resilience.

Several paradigms have been proposed for distributed multi-vehicle control [3–5] such as: leader-following, behavioral rules, virtual structure, artificial potential function, graph-rigidity. As indicated by its name, the leader-following approaches require to define one robot as the leader. In this setting, the leader has access to information such as the final destination or visibility to a target, which is unavailable to the other vehicles. However, the role selection of a particular robot as a leader is strongly related to the time-dependent mission and environment scenario as well as the swarm status. As such, if a leader change is necessary, particular additional rules need to be established in order to define the hierarchy alternations [6]. The second category of methods for cooperative control is inspired by initial works studying team behaviors in nature. These are based on behavioral rules that each agent should follow according to its local task and its interactions with the environment [7,8]. Such approaches are nevertheless usually very problem-dependent and are not easy to modify whenever an unexpected event occurs. The third category of methods hinges upon virtual geometrical structures in which the swarm of agents should remain globally. The control law must first ensure that the agents are located within the structure and then define a suitable structure evolution depending on the mission requirements [9,10]. A common approach is to design potential fields and navigation functions that are sometimes difficult to construct with good properties (differentiable, without multiple critical points).

Most of these methods require that the geometric formation of the swarm is quasi-explicitly defined through fixed, desired relative positions or distances to be attained [11–13], and they cannot incorporate naturally the interactions with external agents such as a human, be it a pedestrian that has to be avoided or an operator that has to be followed at a distance. With respect to a fleet formation, fixing inter-agent distances or positions restricts the relative motions of the agents, as well as of the global formation, and does not allow much flexibility and adaptability when dealing with uncertain, dynamic and cluttered environments. Such type of formations are usually called rigid [4]. An alternative approach has been derived in [14] where it is ensured that all the vehicles stay inside a region, with a minimum distance between neighbors, whose shape can be assigned by modifying the associated potential function. However, the gain selection for practical deployment of the approach is not easy.

A less rigid behavior, more suitable to interactions with a cluttered environment or human operators, can be achieved by partitioning the space for motion coordination. Voronoi diagrams have been widely considered as a natural way to define space partitioning. In the context of multi-agent swarming, it also a convenient way to define the topology of the interaction network between the agents [15], by considering the graph associated to the corresponding Delaunay triangulation [16]. For motion coordination of multi-agent systems, Voronoi diagrams have been mostly used for allocation and coverage tasks [17–19] and more recently for cooperative pursuit of a single target [20] or multiple targets [21] by multi-agent systems, as well as cooperative exploration using dynamic centroid-based area partition [22]. A simplified version of the swarm navigation problem compared to the one addressed in this paper has been addressed in [23–25]. These algorithms rely on user-defined navigation functions to compute the centroid of the Voronoi cell of each agent, which is used as the reference position to be tracked by the robots, and collision avoidance is handled directly by the space partioning. An improved version relying on geometric constraints has been proposed by the authors in [26], with a more intuitive and direct management of swarm navigation and collision avoidance behaviors between the vehicles. The work presented in this paper is an extension of the latter method, with the inclusion of obstacle avoidance and interaction with a human operator. Contrary to the majority of the literature on Voronoi tesselations for multi-vehicle applications that treat obstacle avoidance through the partitioning, the proposed solution prefers to exploit the appealing concept of collision cones that has been very successful, especially in mono-vehicle applications [27,28]. An alternative approach relying also on Voronoi partitions but with artificial potential fields for collision avoidance has been proposed in [29] and applied to a fleet of quadrotors in simulation and experiments.

We thus propose a new Voronoi-partioning swarm control algorithm which allows to define three different modes of interactions from a common basis. They allow operators to be included in a swarm of autonomous vehicles and guide the robots with their own velocity and/or position in a safe coordinated motion while evolving in unknown cluttered environments. Indeed, extending the earlier concept of swarm teleoperation by a human operator [30], more advanced interactions can be integrated in control algorithms for human–multi-robot swarming. Different types of interactions can be considered depending on information flows available between the robots and human operators (one-way/two-ways) and the nature of the interactions themselves (physical/non-physical), see e.g., [31] for a large overview.

The contributions of the paper can be summarized as follows:

- Proposition of a distributed control algorithm enabling non-rigid motion for human-multi-robot swarming in cluttered environments.
- Design of a purely geometric approach applied by each robot to define distributively a reference point to be tracked inside its Voronoi cell, accounting for other robots and obstacles (collision avoidance), as well as human operators (collision avoidance and other possible interactions, see below).
- Decoupling between the mechanisms of obstacle avoidance and collision avoidance. This allows to reduce the design complexity when accounting for obstacles, as opposed to navigation functions for example, and to render the gain tuning more straightforward.
- Possibility to handle different modes of interactions between human operator and robots of the swarm. These modes of interactions correspond to practical problems of interest which are autonomous waypoint navigation, velocity-guided motion and follow a localized operator.
- Implementation and real-world field experiments in indoor and outdoor environments with self-localized ground mobile robots and human operator, in presence of various obstacles.

The problem definition and the proposed swarm control method are described in Section 2, a corresponding system architecture is defined in Section 3 and field experiments with up to three mobile robots and a localized human operator are reported in Section 4. A Video of the experimental setup is available at https://tinyurl.com/OneraHumanRobotSwarm (accessed on 19 September 2023).

2. Swarm Control Method

2.1. Problem Definition

The problem studied is the guidance of a swarm of N Unmanned Ground Vehicles (UGVs) to a waypoint, denoted by $P^* \in \mathbb{R}^2$, and by extension to successive waypoints either on a given path or defined dynamically (see Section 2.3), in a cluttered environment with no prior map available. A typical applicative context is search-and-rescue or tactical missions, where human operators are assisted by a swarm of autonomous robots for transportation of critical resources, wounded persons or communication link maintenance. A fully autonomous behavior is expected from the swarm under safety constraints with respect to the presence of obstacles and humans, and allowing an automatic reconfiguration in case of vehicle loss(es). It is assumed that each vehicle is able to estimate its own position with respect to a common global fixed reference frame, where P^* is also defined, and to broadcast it to all other vehicles within a given range. The position of the ith agent (referred as Robot i thereafter) will be denoted by $p_i \in \mathbb{R}^2$ and the set of its neighbor robots is indexed by $\mathcal{N}^i = \{j \,|\, j = 1,\ldots,N, j \neq i, \|p_i - p_j\| \leq r_{com}\}$ where $r_{com} > 0$ is the communication range assumed to be constant. The number of neighbors of Robot i will be referred to as $N^i = Card\{\mathcal{N}^i\}$. Each robot is assumed to be able to localize a set of surrounding obstacles, which are modeled as disks with a radius incorporating the desired safety distance. Human operators are assumed to be equipped with equivalent localization devices, and are considered as additional vehicles with no control input computed. Three levels of interaction between the swarm and a human operator have been studied:

1. An *Autonomous* mode, in which the swarm has to follow autonomously a predefined path at a given nominal speed. Examples of tasks that can be performed with this mode are transfers of equipment or injured people between two locations. Other tasks could be the persistent surveillance of zones in order to detect abnormal events, by making the UGVs autonomously and repeatedly move along a surveillance path composed of predefined waypoints.
2. A *Velocity-Guided* mode, where the swarm follows a predefined path at the same speed as a human operator. In other words, the desired positions and orientations of

the robots with respect to the waypoints are the same as in the Autonomous mode, but the velocity to reach them is defined by the motion of a human operator.

3. A *Follow* mode, where the current waypoint to be tracked is defined with respect to the localized operator and also takes into account the positions of all the robots. Note that the human operator could be replaced by a tele-operated robot or a virtual point to obtain a platooning behavior, using the same underlying control algorithm.

All the modes share the common constraint that each robot should remain at a desired safety distance from any human operator, other robot and obstacle.

2.2. Algorithm Description

The main idea of the proposed distributed algorithm is that each vehicle computes online a Voronoi partition of the space involving other physical agents (other vehicles, human operators), and virtual (mirror) agents that are added to maintain the coherence of the swarm (see Figure 1). A reference position to be tracked by a lower-level controller is then computed by each robot inside its own Voronoi cell. A geometric approach has been preferred for this calculation, which is done by considering lines of sight between the vehicle, the waypoint (for attraction), the boundaries of possible obstacles (for avoidance), and other UGVs or human operators (for collision avoidance). The algorithm allows to obtain different behaviors and patterns (e.g., side-by-side, group, convoy-like) by only modifying the initial relative placement of the vehicles. Finally, the distributed nature of the algorithm also grants robustness to online modification (removing or adding) of the number of entities (robots, humans) in the swarm, while also addressing the mono-robot and 2-robot scenarios. If a robot suffers from a failure (e.g., loss of communication or mobility capabilities), the same distributed algorithm is applied without this robot and the swarm can carry on with the given mission. The main steps of the algorithm are the following.

Figure 1. Voronoi diagram for a swarm with four agents: without spacer segments (**left**) or with spacer segments (**right**) to enforce anti-collision between agents. Green dots are mirror agents, blue ones are robots' positions. The cell boundaries of each robot is in pink and spacers are blue segments.

Step 1: Voronoi partitioning

Each robot i computes a Voronoi partition accounting for the other real agents (robots, human operators) in the swarm and virtual mirror agents. The mirror agents are introduced as a means to guarantee the feasibility of the computation of the Voronoi partition, especially with one and two robots, and to adjust the size and bound of each robot's Voronoi cell.

The mechanism proposed in [23] generates mirror neighbors to ensure that the Voronoi cell of Robot i is bounded and help control the expansion of the swarm. As introduced by the authors in [26], let us first define the placement operator $\Psi : \mathbb{R}^2 \times \mathbb{R}^2 \times \mathbb{R} \to \mathbb{R}^2$ by

$$\Psi(p_i, p_j, d) = \begin{cases} p_i - d\dfrac{p_j - p_i}{\|p_j - p_i\|} & \text{if} \quad p_i \neq p_j \\ p_i & \text{otherwise} \end{cases} \tag{1}$$

If Robot i is not in the convex hull generated by the position of the $\mathcal{N}^i$ agents, then N^i mirror neighbors are defined with positions computed as

$$m^i_j = \Psi(p_i, p_j, d_{mir}), j \in \mathcal{N}^i \tag{2}$$

where $m^i_j \in \mathbb{R}^2$ and $d_{mir} > 0$ respectively denote the position of the mirror of neighbor j and the distance of placement of the mirror neighbor with respect to Robot i (see Figure 2). The set of all mirror neighbors for Robot i will be denoted by $\mathcal{M}^i = \left\{m^i_j, j \in \mathcal{N}^i\right\}$.

Now, Robot i computes its own Voronoi partition using the set of points $\{p_i\} \cup \{p_j, j \in \mathcal{N}^i\} \cup \mathcal{M}^i$. From this Voronoi partition, only the edges and vertices that correspond to the partition where Robot i belongs are kept. This Voronoi *cell* of Robot i is denoted by $\mathcal{C}^i = (\mathcal{V}^i, \mathcal{E}^i)$, where $\mathcal{V}^i$ and $\mathcal{E}^i$ are the sets of vertices and edges of the cell. This cell defines the space in which the reference position P^i_* to be tracked by Robot i is placed, as defined by the next steps of the approach.

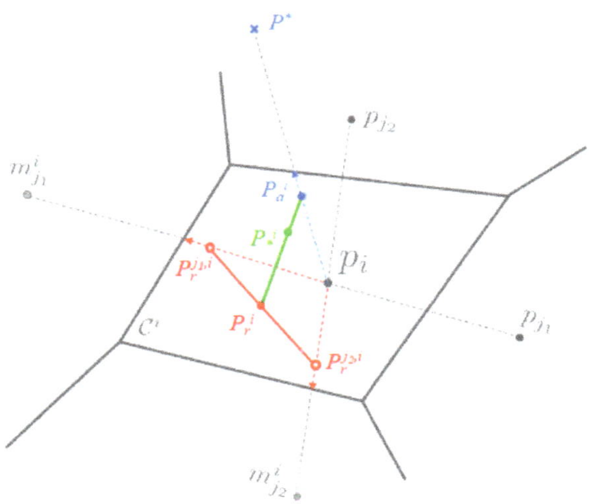

Figure 2. Illustration of computation by robot i of its reference position P^i_* to be tracked. Case with three robots (i, j_1, j_2) and two mirror agents $m^i_{j_1}$ and $m^i_{j_2}$ added to bound the Voronoi cell. Gray points are positions, symbols in blue are related to attraction to waypoint, in red to collision avoidance with other agents, and in green to reference position to be tracked by the robot (Updated from [26]).

An example of such a construction of the Voronoi cells for a swarm of $N = 4$ agents is illustrated on the left part of Figure 1, where real and mirror agents are respectively represented by blue and green dots. The Voronoi cell computed by each of the four agents is represented by magenta lines. In case of collision risk(s) with agent(s), the Voronoi cell is adapted as follows.

Compared to [26], a new feature is introduced in the construction of the cells, for each real agent located at a distance lower than a predefined threshold. In that case, a spacer segment is inserted between the robot and this agent to modify the construction of the Voronoi cell (see right part of Figure 1, with spacer segments in blue). More details on how spacers are built are given in Step 4 on collision avoidance with other agents.

In our implementation, the Voronoi partitions were computed using the Boost polygon library [32], which has been used in other multi-robot planning algorithms [33,34]. This library is based on the sweepline algorithm for Voronoi diagrams initially proposed in [35].

Step 2: Attraction to waypoint

This step is similar to the one proposed by the authors in [26]. An attraction point P_a^i is defined inside the Voronoi cell C^i of the vehicle i, on the segment directed along the line of sight between the vehicle and the waypoint P^*, and limited inside the Voronoi cell C^i (see Figure 2).

If the waypoint P^* to be reached is located inside the Voronoi cell C^i of Robot i, then the attraction point is simply defined as $P_a^i = P^*$. If not, P_a^i will be placed inside C^i along the line of sight between Robot i and the waypoint, as defined by the following procedure.

Let us denote by I_a^i the intersection point of the geometric segment $\overline{p_i\, P^*}$ with edges of C^i (see Figure 2).The attraction point for Robot i is finally defined as

$$P_a^i = \Psi(p_i, I_a^i, -d_a^i) \tag{3}$$

with the distance

$$d_a^i = \min(d_a^{max}, \lambda_a \left\| I_a^i - p_i \right\|) \tag{4}$$

and where $0 < \lambda_a < 1$ and $d_a^{max} > 0$ are two tuning parameters used to set the position of the attraction point P_a^i on the segment $\overline{p_i\, I_a^i}$ and to limit its distance to Robot i. Parameter λ_a enables to define a margin for the placement of the attraction point inside the Voronoi cell. A value of $\lambda_a = 1$ would correspond to an attraction point located on the edge of the Voronoi cell. A value $\lambda_a < 1$ is therefore preferred to possibly account for uncertainty (e.g., due localization) in the definition of the Voronoi cells and practically ensure with more robustness the belonging of the attraction point to the Voronoi cell. Parameter d_a^{max} is used to limit the distance of the attraction point with respect to the current robot position. This can be useful in cases of large Voronoi cells, e.g., when robots are moving far from each others, to avoid attraction points that would result for the low-level controller in large control input values for the robot.

Special cases $N^i = 0$ or $N^i = 1$: During the mission, the number of neighbors of a robot may change, temporarily or definitively, e.g., due to loss of communication links, loss of robots, etc. If at a given instant Robot i has zero or one neighbor, one additional step is performed before the standard algorithm. This step is described in Appendix A.

Step 3: Obstacle avoidance

The distances between the vehicle i and the detected obstacles are evaluated to identify obstacles in proximity which need to be checked for collision risk. A map of obstacles is built online thanks to on-board sensors of the UGV (see Section 3.2 for more details). For collision risk evaluation, a disk model of obstacles (including a safety distance) is considered.

Denote by $\{O_l^i, ro_l^i\}, l = 1, \ldots, N_o^i$ the set of N_o^i obstacles detected by Robot i and modeled by disks of centers O_l^i and radius ro_l^i. A first step then consists in computing a cone of "unsafe directions" $\mathcal{U}_l^i$ englobing and tangent to each obstacle l, with the robot's position p_i as vertex (see Figure 3).

A test is then realized to check whether the line of sight (p_i, P^*) between the robot and the waypoint belongs to at least one of these cones $\mathcal{U}_l^i$:

- If not, there is no collision risk with any of the obstacles, and direct straight motion to the waypoint is safe for the robot (as described in Figure 3). The attraction point P_a^i computed at Step 2 is still valid and the algorithm proceeds to the next step.

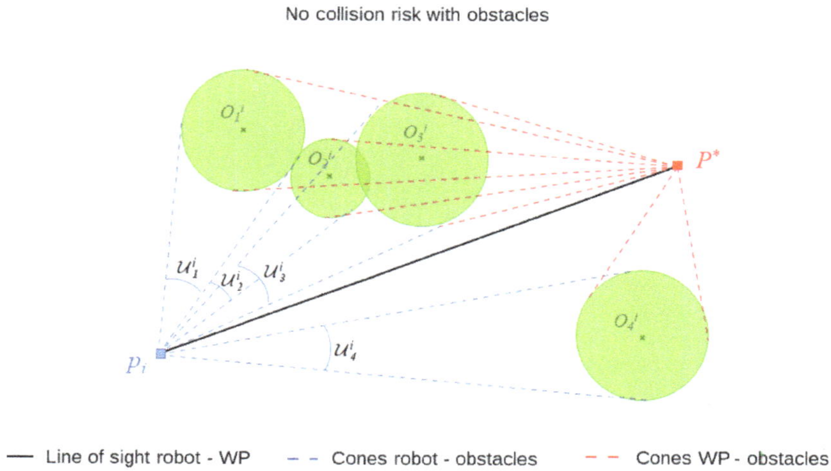

Figure 3. Collision cones and direct safe path towards waypoint in case of no collision risk.

- If there is at least one obstacle with collision risk, the cone of this obstacle is considered. It is enlarged step by step by considering adjacent and intersecting cones related to other obstacles, so as to obtain a larger cone $\bar{\mathcal{U}}^i$ containing a cluster of the obstacles with collision risk for robot i. An example is provided on the left part of Figure 4: the collision cone of obstacle O_2^i is merged with the intersecting collision cone of obstacle O_3^i, which is further merged with collision cone of obstacle O_1^i. This iterative procedure is stopped as there are no other intersection collision cones. The resulting cone is depicted by dashed blue lines. The same procedure is repeated to build another cone $\bar{\mathcal{U}}^*$, but this time by considering the waypoint P^* as vertex. The two intersection points T_1^{i*} and T_2^{i*} between these bounding cones are then computed. They correspond to two intermediate target points for the robot, each of them defining a possible obstacle-free path towards the waypoint. Some heuristics are used at this stage to select the shortest among the two available paths. The target point corresponding to the selected path is considered, instead of the waypoint, to compute a new attraction point P_a^i, in the same way as in Step 2. This new attraction point replaces the one computed in Step 2 and is used instead for the rest of the algorithm.

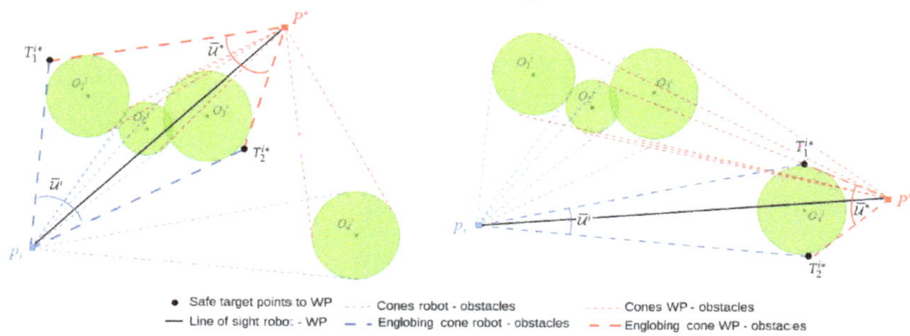

Figure 4. Two examples of collision cones and computation of a safe path towards a given waypoint in case of collision risk with obstacles.

Step 4: Collision avoidance with other agents

This step details the mechanism used to avoid collision between agents. For that purpose, two tools are used to ensure that agents stay at a safety distance from each other. For each other real agent j (either UGV or human operator) at collision risk (distance criterion), a spacer segment is introduced to reduce the Voronoi cell in the direction of the potential encounter. This spacer segment is used in a range of distance between agents of $\sigma_{col}d_{col}$ and d_{col}, where $d_{col} > 0$ defines a distance threshold representing a collision risk and where $\sigma_{col} \geq 1$ is a smoothing factor to account for some margin in the collision test. Figure 5 illustrates this mechanism. If for any reason, the distance between agents drops under the targeted collision distance $\sigma_{rep}d_{col}$, with $\sigma_{col} > \sigma_{rep} \geq 1$, an additional avoidance mechanism is used that builds a repulsion point. This repulsion point P_r^i is computed as the mean of all the agents individual repulsion point $P_r^{j,i}$ that are defined inside the Voronoi cell of the vehicle i, on the segment directed along the line of sight between agent j and agent i, and limited to the Voronoi cell C^i (see Figure 2). This second mechanism is usually not triggered, due to the use of spacer segments first, but could happen if a robot overshoots its Voronoi cell or for the special case of a localized object like a pedestrian where the Voronoi cell is not enforced. The segment and repulsion points could be used together to exhibit different repulsion behavior. The repulsion point has only a radial influence on the collision avoidance whereas the spacer segment will enforce more parallel trajectories by more aggressively limiting the cell (see Figures 1 and 5). Those mechanisms are compatible with each other as the segment will reduce the Voronoi cell and the repulsion point is defined in the cell itself.

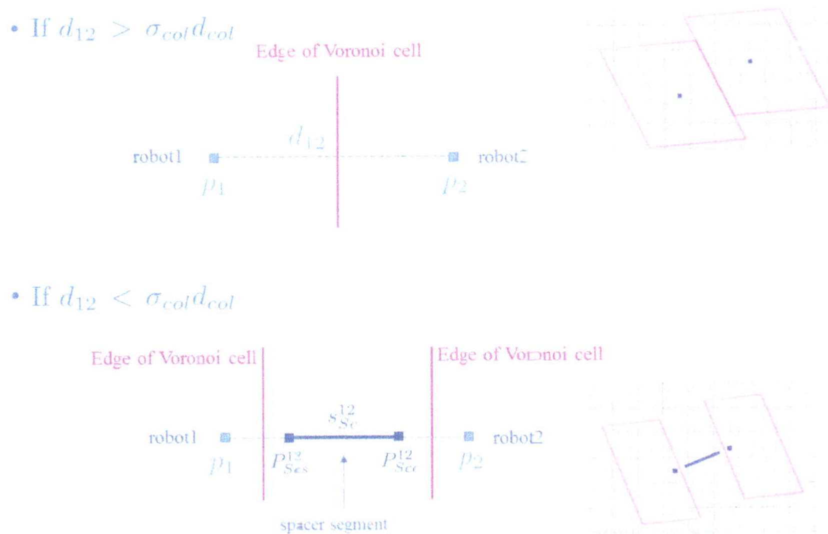

Figure 5. Construction of Voronoi cells: without spacer segments (**top part**) or with spacer segment (**bottom part**) in case of agent too close proximity.

More formally, let the set of robots with collision risk be $\mathcal{N}_{col}$ and the set of neighbor of Robots i that need to be considered in repulsion $\mathcal{N}_{rep}^i$ such that

$$\mathcal{N}_{col} = \left\{(i,j) \in (1,\ldots,N)^2 \,|\, j < i, d_{ij} \leq \sigma_{col}d_{col}\right\} \quad (5)$$

$$\mathcal{N}_{rep}^i = \left\{j \in \mathcal{N}^i \,|\, d_{ij} \leq \sigma_{rep}d_{col}\right\} \quad (6)$$

with $d_{ij} = ||p_j - p_i||$ and where d_{col} s.t. $d_{min} > d_{col} > 0$ is used to define a distance threshold representing a collision risk and where $\sigma_{col} > \sigma_{rep} \geq 1$ is used as a smoothing factor to account for some margin in the collision test. Let $N_{col} = Card\{\mathcal{N}_{col}\}$ be the number of agents with collision risk. If $N_{col} = 0$, the remaining part of Step 4 is skipped. The set of segments $\mathcal{S}_c = \left\{ S_c^{ij}, \forall (i,j) \in \mathcal{N}_{col} \right\}$ is added during the Voronoi partitioning at Step 1. Each segment S_c^{ij} starts at P_{Scs}^{ij}, ends at P_{Sce}^{ij} and is centered between p_i and p_j (see Figure 5):

$$P_{Scs}^{ij} = \Psi\left(p_i, p_j, \frac{1}{2}(d_{ij} - s_{Sc}^{ij})\right) \tag{7}$$

$$P_{Sce}^{ij} = \Psi\left(p_i, p_j, \frac{1}{2}(d_{ij} + s_{Sc}^{ij})\right) \tag{8}$$

with s_{Sc}^{ij} the size of the segment:

$$s_{Sc}^{ij} = \begin{cases} \dfrac{\sigma_{col} d_{col} - d_{ij}}{\sigma_{col} - 1} & \text{if } d_{ij} > d_{col} \\ d_{ij} - 2d_{vres} & \text{if } d_{ij} \leq d_{col} \end{cases} \tag{9}$$

where d_{vres} is the smallest distance such that $p_i \neq \Psi(p_i, p_j, d_{vres})$ in the Voronoi partitioning step. This is done for all robots pairs in collision and not only for neighbors of robot i, to ensure that the same cell edge is obtained by each robot in the distributed process.

When the repulsion is triggered for a neighbor $j \in \mathcal{N}_{col}^i$ of the robot i, a repulsion point $P_r^{j,i}$ is defined as follows. Let us consider the two intersection points of the geometrical line $(p_i p_j)$ with edges of the Voronoi cell $\mathcal{C}^i$ of Robot i. We denote by I_r^{ji} the intersection point such that the dot product $\overrightarrow{p_i p_j} . \overrightarrow{p_i I_{ji}^r}$ is negative, i.e., I_r^{ji} is located on the edge of $\mathcal{C}^i$ opposite to p_j with respect to p_i . The repulsion point for Robot i to avoid collision with Robot j is then defined by

$$P_r^{j,i} = \Psi\left(p_i, I_r^{ji}, -d_r^{ji}\right) \tag{10}$$

with the distance

$$d_r^{ji} = \min\left(d_r^{max}, \lambda_r \left\| I_r^{ji} - p_i \right\|\right). \tag{11}$$

The parameter λ_r, such that $0 < \lambda_r < 1$, is used to set the position of the repulsion point $P_r^{j,i}$ on the segment $\overline{p_i\, I_r^{ji}}$, and $d_r^{max} > 0$ to limit its distance to Robot i.

Following this procedure, one repulsion point is computed by Robot i for each robot with collision risk. A global repulsion point is then deduced for Robot i by

$$P_r^i = \frac{1}{N_{rep}^i} \sum_{j=1}^{N_{rep}^i} P_r^{j,i} \tag{12}$$

with $N_{rep}^i = Card\left\{\mathcal{N}_{rep}^i\right\}$ the number of robots to be considered for repulsion. Since all the $P_r^{j,i}$ are located inside the Voronoi cell $\mathcal{C}^i$, so does the global repulsion point P_r^i. Note that, by relation (12), P_i^r is computed as a mean of the $P_r^{j,i}$. A weighted mean could also be used for example to give more influence to repulsion points corresponding to the closest robots.

Step 5: Computation of reference

Similarly to [26], the reference position P_*^i that will be tracked by robot i is computed as a weighted mean of the attraction point P_a^i and the repulsion point P_r^i as

$$P_*^i = (1 - \beta) P_r^i + \beta P_a^i \tag{13}$$

where the weighting coefficient $0 \leq \beta \leq 1$ is adapted online depending on the minimum distance to other agents with collision risk. It enables to give more weight on repulsion if some UGVs are very close or more weight on attraction to the waypoint otherwise. If there are no collision risks between the agents ($\beta = 1$), this algorithm results in $P_*^i = P_a^i$, leading to pure attraction to the waypoint.

2.3. Waypoint and Velocity Management

Navigation to successive waypoints has been managed in the following way. For the *Autonomous* and *Velocity-Guided* modes, all vehicles dispose of the full list of waypoints assigned to the swarm from the mission path definition. In *Follow* mode, the waypoint is not predefined from a list but generated dynamically from the position of a target (which could be a localized human operator, a tele-operated robot, or a virtual point). An isosceles triangle of spacers is built with its height defined as the segment from the target towards the centroid of the swarm with a length of d_{follow} and a base of length equal to d_{follow_base} The waypoint is set at the intersection of the height and the base. An additional *little thumb* sub-mode has been specified, in which waypoints are recorded in a list to be tracked by the robots of the swarm, whenever the target is located at a distance greater than d_{little_thumb} from the robots (note that the distance used here is built in coherence with the mode of validation defined in the following paragraph). An illustration of these definitions is presented in Figure 6.

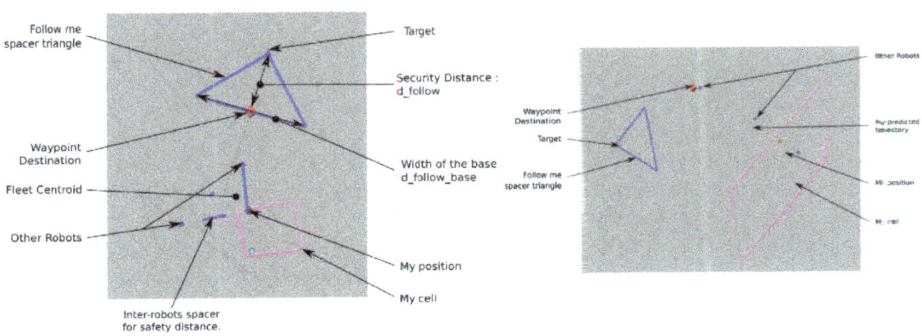

Figure 6. Left: *Follow* mode waypoint and triangle of spacers. **Right**: *little thumb* sub-mode.

Different validation strategies can be defined for the UGVs to determine that the current waypoint has been reached and that they should continue to the next waypoint. We have studied the following strategies:

1. *Selfish*: each robot has to validate its current waypoint (given a parameterized validation distance d_{val}), then it moves to the next waypoint in the list. This way, all the robots will cross each waypoint and stay close to the path. On the other hand, this does not impose any waiting behavior between the robots.
2. *First*: when a robot is the first to validate the current waypoint, all robots head to the next waypoint in the list by sharing its index. This strategy can be applied in large environments where deviation from the path can be allowed. There is also no waiting behavior around each waypoint in this case, however the UGVs always agree on and head towards the same current waypoint.
3. *WaitForAll*: in this strategy, the waypoint is validated only if each robot either gets closer to the waypoint than the validation distance d_{val} or if it is near the avoidance distance of another robot ($\sigma_{chain} d_{col}$) which validates one of these conditions. This is a more collective behavior, where all robots should wait for the others before heading to the next waypoint. This also creates a kind of validation chain between the UGVs, which is a useful feature for large swarms where all the UGVs cannot get closer to the waypoint than the validation distance because of the collision avoidance constraints.

Since the positions of the robots are shared within the swarm at all time instants, each robot is able to compute the validation conditions for itself and its neighbors in a distributed way. In the experiments described in this paper, the *WaitForAll* mode has been preferred to demonstrate group motion around obstacles.

In addition to the computation of the reference point to be tracked by the robot, a speed ratio is also produced by the algorithm. This speed ratio multiplies the speed value that has been chosen initially by the user, so as to provide the current reference speed to the low-level controller. In the *Autonomous* mode, the speed ratio is set to 0.5 in case of collision risk (slow motion in presence of obstacles) and to 1 otherwise (full-speed motion). In the *Velocity-Guided* mode, the speed ratio is set to copy the speed of the human operator, considered as command for the swarm, with a saturation at 1, which means the operator can go faster than the robot's maximal reference speed. In case of collision risk, the robot's speed ratio is saturated at 0.5, while still copying the operator's speed below this value. In the *Follow* mode, the speed ratio is set to 0 if the robot enters the triangle, otherwise it is set to 1. This speed modulation makes it possible for the target to turn around and come back towards the robots, which are forced to stop during the crossing.

2.4. Main Properties of the Algorithm

2.4.1. Safety Regions

Voronoi cells can be viewed as safety regions in the sense that if each robot performs a trajectory within its cell to reach its reference point to be tracked, collisions between the vehicles can be avoided. This safety consideration is enforced by the additional spacers introduced, which isolate the robot cells at a desired distance from each other.

Note that since the approach is distributed, each robot will compute its own Voronoi partition and cell. Non-overlapping of the cells can only be guaranteed in case of fully connected communication graphs (i.e., $N^i = N-1, \forall i$) and if this computation is done in a synchronized way, with all the robots disposing of information corresponding to the same situation of the swarm. In practice, as we do not want to enforce a synchronization mechanism, non-overlapping of the cells can be obtained if the vehicle dynamics are slower than the computation period of the Voronoi partition, as mentioned also in [23,24]. In addition, parameters λ_a and λ_r can be chosen to define margins with respect to the edges of the Voronoi cell in the placement of P_a^i, P_r^i and hence P_*^i, and to keep each robot and its reference to be tracked in a segregated partition of the space. In case of non fully connected communication graphs, other mechanisms must be looked at to provide non-overlapping guarantees for the Voronoi cells.

2.4.2. Flexibility and Pattern of the Swarm

Flexibility of the swarm can be adjusted by the parameters d_{mir} and d_{col} which set a compromise between attraction and repulsion between the robots. Choosing $d_{mir} \gg d_{col}$ adds more flexibility to the swarm. A swarm behavior close to a more rigid-formation can be obtained on the contrary for $d_{mir} \approx d_{col}$.

The pattern obtained for the swarm is not pre-specified but can be influenced by the initial positioning of the robots, making this feature an interesting one for practical applications. For instance a pattern close to a "platooning-like" formation can be obtained for an initial positioning of the robots close to a single line. This can be of interest for motion in narrow corridors. More regular patterns (triangle, square, etc.) can also be obtained during the motion by the same consideration. The formation shape will although be distorted in presence of obstacles between or near the waypoints. This trade-off makes it possible to carry out the autonomous navigation of a swarm of UGVs in large-scale environments with various levels of obstacle density, which cannot be achieved with control methods based on rigid formations or virtual structures.

2.4.3. Decentralized Algorithm for Robustness to Robot Failure and Communication Loss

In practice, the number of robots in the swarm and/or in the neighborhood of each robot may vary during the mission: loss or addition of robots, communication links temporarily/definitively unavailable, etc. Robustness with respect to these issues is ensured in practice by the proposed algorithm, being fully distributed and handling the limit cases with one or two robots. The experimental testing of this behavior was reported with a swarm of four mobile robots in [26] using a previous version of the algorithm which did not take obstacles into account.

3. System Architecture

3.1. Architecture

A global architecture to deploy the proposed algorithm in a swarm of autonomous robots is summarized in Figure 7. A global mission supervision module is available to the operator on a portable ground station to select a swarm mode, its parameters, and a reference path (for the *Autonomous* or *Velocity-Guided* mode) which are sent to all the UGVs. Once the mission has started, the algorithm ensures the self-organization of the swarm and the supervision layer is only used for monitoring progress. Each robot runs its own localization, mapping and control algorithms in a distributed scheme with limited exchanges of information.

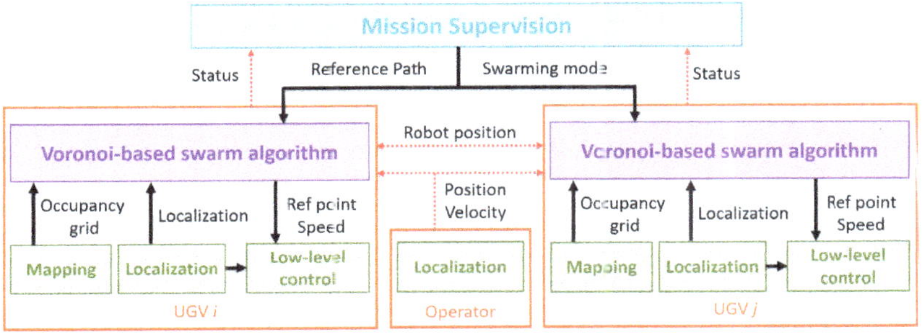

Figure 7. Architecture of the proposed multi-robot system.

The robots and the operator are equipped with localization sensors which provide their own global position, orientation and velocity with respect to a common reference frame (e.g., a WGS-84 local frame or a reference landmark). The mission path is also defined or converted with respect to this reference frame, such that all UGVs share the same list of waypoints. The position of each agent is the only information shared with the others during the mission to be able to execute the distributed swarming algorithm. The robots individually run a mapping process using their own embedded depth measurements to provide an occupancy grid centered on their current position, which is then processed by the proposed algorithm to estimate and keep track of the position of the closest obstacles while distinguishing them from the other robots (see next section for more details).

Based on each robot's own localization and mapping, and the shared localization of the other active agents of the swarm (UGVs and operator), the Voronoi-based swarm algorithm generates a local reference point and a speed modulation ratio on each robot which takes into account all the mission objectives and constraints. A low-level controller is then used to track this reference point and modulate the speed as requested. In our architecture, a proportional controller for a unicycle model derived from [36] was used to compute the steering inputs. The controller has been implemented so as to prefer trajectories close to straight lines in the direction of waypoints by correcting first large orientation errors at a lower speed. This is of a particular importance especially when dealing with nonholonomic vehicles, to obtain trajectories remaining inside the Voronoi cells.

The exchange of information between elements of the system has been kept as simple as possible, both in terms of the nature of the information and of the associated data flow to facilitate interoperability capacities as much as possible. This way, they do not depend on the choice of the technology composing the system, and the required communication bandwith is also quite limited. In a more prospective view, interoperability with aerial vehicles or manned wheeled vehicles could be facilitated and envisaged as long as the same type of communication interfaces can be handled. At the robot level, the modularity of the swarm module would also enable to modify or mix the types of vehicles, by only adapting the low-level control layer and distance parameters, without changing the complete architecture of the system.

3.2. Local Mapping from Embedded Depth Sensors

As mentioned in Step 3 of the algorithm description, obstacle avoidance is based on a local map of the environment. This local map is built from an occupancy grid in robot frame provided by a pre-processing mapping algorithm from raw depth data (either from a LiDAR or stereo cameras). It aims to remove suspicious obstacles by temporal filtering and extend obstacle memory when obstacles go out of sight. The local map is updated when a new occupancy grid is received. The track of each obstacle is kept in memory as: the position of the obstacle (in world frame), the number of times it has been observed in successive occupancy grids and at which time it occurred last. Four main steps are carried out in the mapping process, as described in Figure 8. The predicted trajectory mentioned in the following is obtained by repeating the main algorithm steps on a given time horizon.

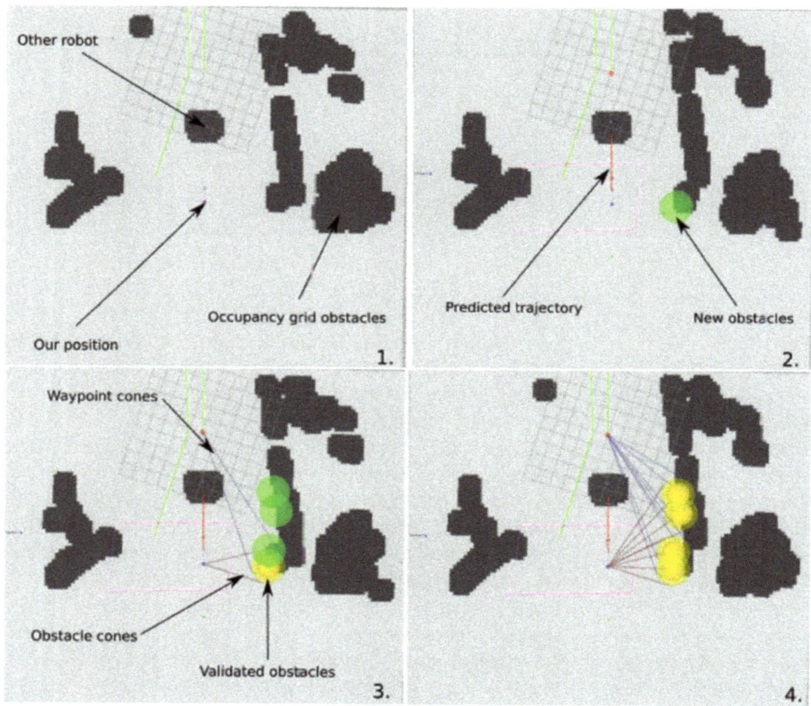

Figure 8. Local map building process. 1—the raw occupancy grid is in the background and consists of dark gray spots. 2—the closest obstacle to the robot is initialized in green. 3—Obstacles closest to the predicted trajectory are initialized and the first obstacle is validated (yellow, obstacle cones are visible). 4—All obstacles validated are in yellow, we can observe that another robot in front of the one building this map is not added to the obstacle list.

1. *Obstacle initialization*

   The local map does not keep track of all obstacles detected by the sensor but focuses only on the most threatening ones. Those are obstacles which are closest to the current position or to the future position along the predicted trajectory. When a possible new obstacle is found, we add it to the local map if it is not already tracked. The created obstacle is a disk of radius d_{des} (parameterized safety distance) centered on the grid point of a threatening obstacle, displayed in green in Figure 8 (sub-figures 2 and 3). All the new obstacles are created and tracked but not taken into account for avoidance immediately. This requires that the corresponding location has been seen as occupied a certain number of times in successive occupancy grids to filter out measurement artefacts (vegetation, dust, . . .).

2. *Obstacle life cycle*
When a new occupancy grid becomes available, each tracked obstacle is projected in the occupancy grid to check if it is still present. If this is the case, we increase the number of observations of this obstacle and reset the last time it has been seen. When an obstacle reaches a certain number of observations (set to 3 in our experiments), it is validated and considered for obstacle avoidance (in yellow in the map). In this algorithm, the radius of the obstacle is enlarged by an additional distance d_{safe} to take into account the size of the robot or the drift of localization (in light yellow on the map pictures). This number of observations is saturated at a given threshold (set to 100 in our experiments) to be able to remove obstacles that have been seen during a long period at a given location but have moved away afterwards.
3. *Obstacle removal*
After new obstacles are initialized, tracked and updated, obstacle removal is carried out. This process is based on the number of views and the last time an obstacle has been seen. When an obstacle is not seen anymore: after a certain amount of time passed from the last time it has been seen (set to 1 s in our experiments), the number of observations is decremented. When the number of observations reaches the minimum-view threshold, the obstacle is removed from the map. This process allows to keep track of an obstacle that has been seen for an extended period of time while allowing for a fast removal of an obstacle that has been seen just a short number of times. This process will also remove the obstacles that have been initialized but never seen again (validated).
4. *Occupancy grid pre-filtering*
Due to the multi-agent context, the input occupancy grid is pre-processed to remove views of the other agents (UGV or operator) such that they are not considered as obstacles, since they are moving and taken into account directly at Step 4 of the main algorithm. The global positions of the other robots and of the operator are projected in the local grid. All obstacle cells present in a given radius (chosen to be consistent with the robot dimensions and localization uncertainty) around these positions are then removed. This can be seen in Figure 8, where the spot in front of the robot currently building the map is at the location of another robot but is not considered as an obstacle.

4. Experimental Results

Outdoor and indoor experiments were carried out for the different modes of the swarming algorithm using the following robotic platforms (see Figures 9 and 10):

- 3 Robotnik Summit XL UGVs of mass 65 kg and base dimension 72 × 61 cm, equipped with a calibrated stereo-rig of IDS UI-3041LE cameras (baseline of 35 cm, running at 20 Hz) and an embedded Intel-NUC CPU.
- 1 operator Portable Localization Kit comprising an Intel RealSense d455 depth camera and a Intel-NUC CPU.
- A standard WiFi network connecting all the embedded computers and a ground station for mission supervision and visualization. Transmission of the positions between the robots was carried out using Ultra Wideband (UWB) DW1000 radio modules in a similar way as in [37].

Figure 9. *Velocity-Guided* outdoor experiment: two UGVs and localized operator.

Each agent (UGV or operator) computed on-board its localization from the stereo-vision data using the eVO visual odometry algorithm [38]. An additional initialization procedure was used to align the local frame of each individual visual odometry with the same global frame. In our experiments, the global frame was defined by a cube of AprilTags [39] and each robot computed its initial global position automatically by estimating its relative position and orientation with respect to this cube (Figure 10). This procedure allows to estimate a global position as long as the visual odometry presents a limited drift. In the presented experiments, the drift at the end of the trajectories was observed to be less than 0.5 m for trajectories of about 100 m. This can be considered accurate enough despite possible perturbations on visual odometry due to the presence of the other moving robots in the field of view, and this did not disturb the demonstration of control performance and was compatible with the desired inter-robot safety distances. For a larger-scale scenario, other visual localization methods should be considered like collaborative localization and distributed SLAM [40]. The UGVs used on-board the ELAS algorithm [41] to estimate the depth from the stereo images, which was then converted into a point cloud and projected in an occupancy grid of obstacles with a resolution of 25 cm as an input for the mapping process. The update rate of the swarming algorithm was set to 4 Hz, which was also the availability rate of the obstacle grid.

Algorithm tuning

The following parameters were used for all the experiments:

- Collisions: $d_{col} = 2.0$ m, $\sigma_{col} = 2$, $\sigma_{rep} = 1$
- Attraction: $d_a^{max} = 2.0$ m, $\lambda_a = 0.75$
- Repulsion: $d_r^{max} = 2.0$ m, $\lambda_r = 0.9$
- Mirrors: $d_{mir} = 3.0$ m
- Validation: $d_{val} = 2.0$ m, $\sigma_{chain} = 1.8$
- Obstacles: $d_{des} = 1.1$ m, $d_{safe} = 0.5$ m

Figure 10. Indoor experimental environment: UGVs and portable localization kit with reference cube.

4.1. Autonomous Mode

Figure 11 presents the results of a repeated experiment on the same reference path in *Autonomous* mode, where the number and starting positions of the robots varied. The experiment comprises 5 runs with a single robot, 6 runs with two robots, and 3 runs with three robots. The path is a loop of length 68.3 m, with two main obstacles along the way: a box (blue in Figure 10) placed in collision between two waypoints, and a large pillar (top left in Figure 10) which is not in direct collision but where there is not enough space for two robots to pass side by side (thus forcing them to change the shape of the formation or wait for the path to clear). It could be seen that with one robot, the achieved path is close to be the same for each run: some variability only occurs on the choice to go left or right around the obstacle box which is placed nearly symmetrically. With two robots, the first one keeps a trajectory close to the one-robot case but it could seen that the leading robot can change during the mission, around the pillar the second one needs to move closer to the obstacle to keep its safety distance with the first one. With three robots, the trajectories exhibit the same kind of pattern but with more diversity. The lengths of the traveled paths obtained during all the experiments are reported in Table 1, with associated statistics. It is interesting to note that experiments with more robots did not increase the path length as late robots are able to find shortcuts under the *WaitForAll* validation mode, since they are not required to reach exactly the waypoint but should only be at a validation distance from it. Another validation mode which would combine the *First* and *WaitForAll* behaviors could be designed to enforce the crossing of the waypoints by each robot if this is of interest for a given mission.

Table 1. *Autonomous* mode experiments: lengths of robot trajectories.

	Nb of Runs	Mean (m)	Std (m)	Max (m)	Min (m)
Waypoints		68.3			
1 Robot	5	67.29	2.35	71.33	65.16
2 Robots	6	69.19	2.86	73.28	64.84
3 Robots	3	65.05	4.37	74.10	58.29

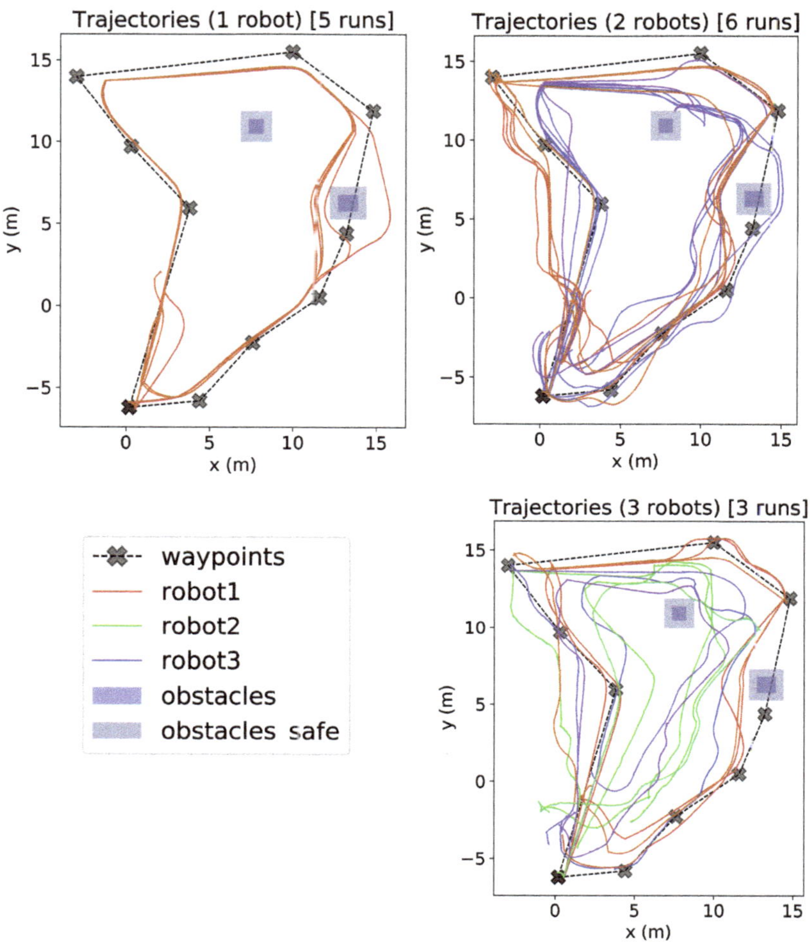

Figure 11. *Autonomous* mode: repeated experiment with 1, 2 and 3 robots on the same reference path.

If a human operator is present in this *Autonomous* mode, it is not considered as an obstacle but as an additional agent of the swarm in the Voronoi partition, while the UGVs follow the path autonomously. In this case, the repulsion mechanism described in Step 4 of the swarming algorithm applies. A dedicated test presented in Figure 12 illustrates the way the localized operator can influence the other robots, where the operator walks between two robots and forces them to make room: the red robot stops and the green one is forced to deviate from its trajectory.

Figure 12. *Autonomous* mode with Human Operator interaction. Each vignette shows a situation with time flowing from left to right on the top then the bottom rows. The Voronoi cell of each robot is represented in magenta, the trajectory of each robot is in red and green, the operator is in blue.

4.2. Velocity-Guided Mode

In this test, two UGVs follow a 8-shaped trajectory (Figure 13). It can be checked that both robots adapt their velocity to the one of the localized human operator, and that safety distances are respected at all times. In particular, it can be seen that the velocities of the robots and the operator are correctly superposed, except when the operator goes faster than the UGV nominal velocity (e.g., a little after time 80 s) or when an obstacle is along the way (at the end of the trajectory). These behaviors are fully consistent with the distributed algorithm and the imposed requirements.

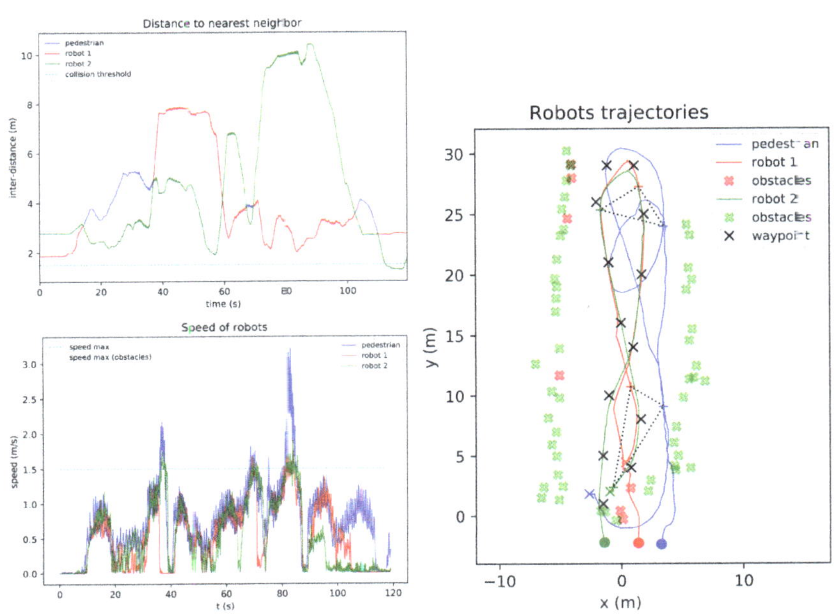

Figure 13. *Velocity-Guided* mode for two UGVs interacting with a localized human operator. **Right**: Trajectories followed by the agents. **Up left**: inter-distances between robots and with operator (safety is always ensured). **Bottom left**: superposition of robots and operator velocities.

4.3. Follow Mode

As described in Section 2.3, the behavior of the *Follow* mode is parameterized by:

- The minimal distance d_{follow} to which a robot could approach the target.
- The d_{follow_base} parameter, which has an influence on the shape of the robots formation behind the target. If the base length is short, the robots will be more in line, while if it is larger a triangular shape will emerge. A standard tuning was chosen equal to $2d_{col}$.

In the following experiments, $d_{follow} = 1.0$ m and $d_{follow\_base} = 2.0$ m. The *little thumb* sub-mode threshold was set to 8 m but never reached during the presented experiments, where the UGVs always followed closer the localized operator.

Figure 14 presents the trajectories of a localized pedestrian and one robot in *Follow* mode, while evolving in the same indoor setup from Figure 10 as the *Autonomous* mode tests, with a total path length of 237 m. In this experiment, the pedestrian forces the robot to make multiple loops around a pillar located around $x = 10$ m, $y = -10$ m and a box located at $x = 15$ m, $y = 5$ m to successfully demonstrate the obstacle avoidance capability in this mode. It can also be seen that the respective velocities are correctly synchronized in time, even if the human operator walks faster than the robot reference speed.

In Figure 15, the human operator traveled a path of 190 m with three robot followers, still in the same indoor environment. The graph on the left shows the relative distances between each agent and its nearest neighbor, which allows to verify that the UGVs and the human operator respect their safety distance from each other during the experiment (the distance values before time 80 s are related to a manual initialization phase before the start of the mission). Note that some of the obstacles detected by robot 2 are in fact the pedestrian, which is an artefact showing that the localization of this robot presented a drift (thus, of the order of magnitude of d_{col}). This is however a "fail-safe" case, since an obstacle is created at the perceived location and not erased in the mapping process, thus the avoidance is still achieved even if this might degrade the actual tracking of the operator. The localization of the robots could be improved by using inter-robot measurements (e.g., from UWB radio signals or vision processing), but this was out of the scope of this paper which is centered on the swarming control and operator interaction performances. The proposed swarm algorithm will remain applicable with relative positioning of all the agents, although the mapping process should be adapted accordingly.

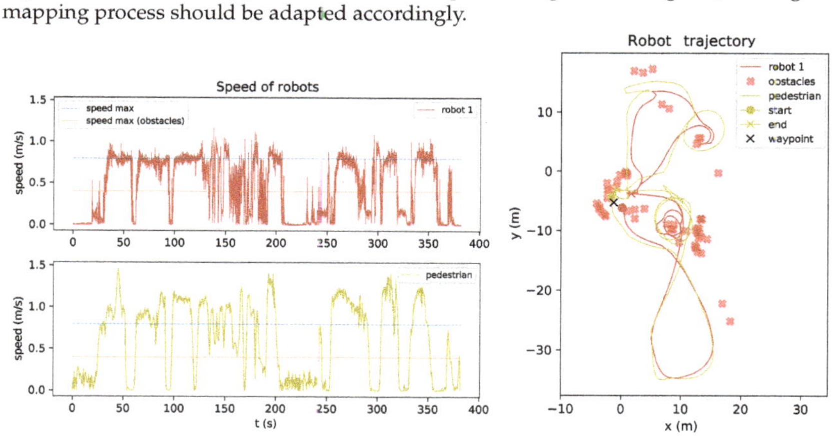

Figure 14. *Follow* mode with 1 robot: robot and pedestrian velocities (**left**) and trajectories (**right**).

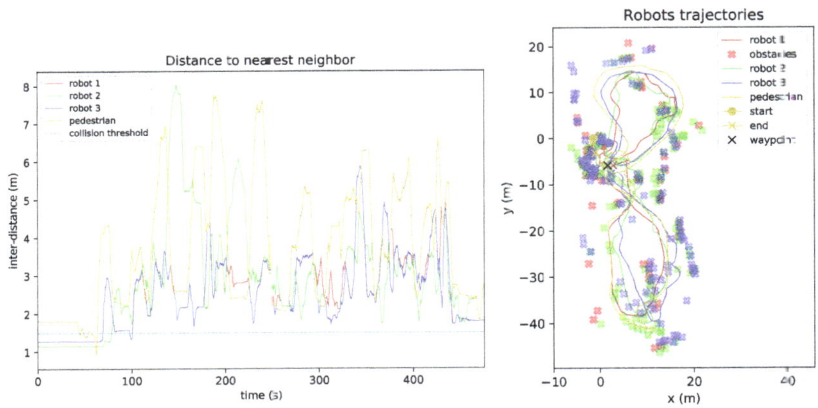

Figure 15. *Follow* mode with 3 robots: relative distances (**left**) and trajectories (**right**).

5. Conclusions and Perspectives

A new distributed control algorithm based on Voronoi partitioning and collision cones has been proposed to coordinate the navigation of a swarm of unmanned ground vehicles interacting with a localized human operator in unknown cluttered environments. Results from field experiments with mobile ground robots have been presented, illustrating a non-rigid swarm motion capability for several navigation modes, all including collision and obstacle avoidance. The behavior of the entire swarm can be easily reshaped by only modifying how the waypoint objectives and the reference speeds are defined, resulting in a range of possible interactions with the swarm for the operator. More elaborate interaction modes are foreseen for future work (e.g., formation split-and-merge or adaptation to multiple operators), as well as larger-scale experiments.

Author Contributions: All authors contributed equally. All authors have read and agreed to the published version of the manuscript.

Funding: This project has received funding from the European Commission through the European Union's Preparatory Action for Defence Research—PADR programme under grant agreement No. 883465—ARTUS.

Conflicts of Interest: The authors declare no conflict of interest.

Appendix A

This Appendix addresses the special cases corresponding to $N^i = 0$ or $N^i = 1$. Is has been proposed by the authors in [26] and is recalled here for completeness.

If at a given instant, Robot i has zero or one neighbor, the following additional step is performed before the standard algorithm.

- If $N^i = 0$, Robot i has no neighbors. In this case, the robot only executes Steps 2 & 3 for obstacle avoidance and does not compute the Voronoi partitioning and collision avoidance with neighbors. Finally, the robot will track the reference points $P^i_* = P^i_a$.
- If $N^i = 1$, Robot i has only one neighbor. In this case, the Voronoi cell obtained in Step 1 will be degenerated. To avoid this problem, if $N^i = 1$, two virtual robots are defined as follows and their indices v_1, v_2 are added to $\mathcal{N}^i$.

Let $(u_{ij}, u_{ij}^{\perp})$ denote the local reference frame attached to Robot i such that u_{ij} is the unit vector directed from Robot i to its neighbor Robot j, i.e., $u_{ij} = (p_j - p_i)/d_{ij}$ with $d_{ij} = \|p_j - p_i\|$ and $j \in \mathcal{N}^i$. The second unit vector $u_{ij}^{\perp}$ completes the orthonormal basis (see Figure A1). The positions of the two virtual robots are defined by:

$$V_{ij}^1 = p_i + \frac{d_{ij}}{2}u_{ij} + d_{mir}u_{ij}^{\perp} \tag{A1}$$

$$V_{ij}^2 = p_i + \frac{d_{ij}}{2}u_{ij} - d_{mir}u_{ij}^{\perp} \tag{A2}$$

They are artificially added to the neighborhood of Robot i, i.e., $\mathcal{N}^i \cup \{v_1, v_2\} \rightarrow \mathcal{N}^i$, and with $p_{v_1} = V_{ij}^1, p_{v_2} = V_{ij}^2$. Steps 1 to 5 of the algorithm are then executed by Robot i. By construction, these two virtual robots are located at a distance greater than d_{col} and will not be involved in collision risk.

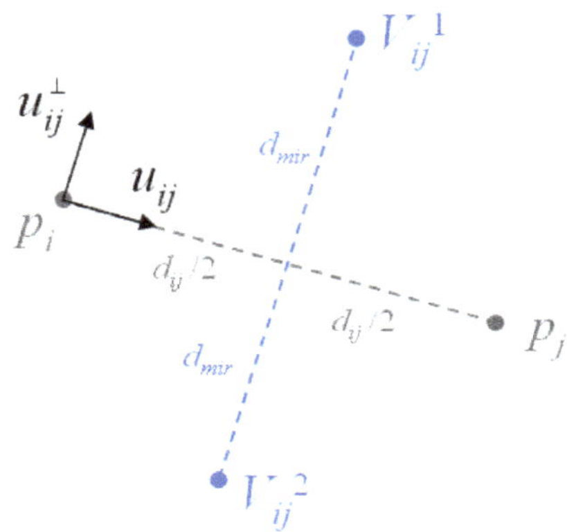

Figure A1. Positioning of virtual robots in case of $N_i = 1$. Reproduced from [26].

References

1. Moussa, M.; Beltrame, G. On the robustness of consensus-based behaviors for robot swarms. *Swarm Intell.* **2020**, *14*, 205–231. [CrossRef]
2. Adoni, W.Y.H.; Lorenz, S.; Fareedh, J.S.; Gloaguen, R.; Bussmann, M. Investigation of Autonomous Multi-UAV Systems for Target Detection in Distributed Environment: Current Developments and Open Challenges. *Drones* **2023**, *7*, 263. [CrossRef]
3. Murray, R.M. Recent research in cooperative control of multivehicle systems. *J. Dyn. Syst. Meas. Control* **2007**, *129*, 571–583. [CrossRef]
4. Mesbahi, M.; Egerstedt, M. *Graph Theoretic Methods in Multi-Agent Networks*; Princeton University Press: Princeton, NJ, USA, 2010.
5. Ren, W.; Beard, R.W. *Distributed Consensus in Multi-Vehicle Cooperative Control*; Springer: Berlin/Heidelberg, Germany, 2010. [CrossRef]
6. Canepa, D.; Potop-Butucaru, M.G. Stabilizing Flocking via Leader Election in Robot Networks. In *Stabilization, Safety, and Security of Distributed Systems*; Springer: Berlin/Heidelberg, Germany, 2007; pp. 52–66. [CrossRef]
7. Balch, T.; Arkin, R.C. Behavior-based formation control for multirobot teams. *IEEE Trans. Robot. Autom.* **1998**, *14*, 926–939. [CrossRef]
8. Lee, G.; Chwa, D. Decentralized behavior-based formation control of multiple robots considering obstacle avoidance. *Intell. Serv. Robot.* **2018**, *11*, 127–138. [CrossRef]
9. Zhou, D.; Wang, Z.; Schwager, M. Agile coordination and assistive collision avoidance for quadrotor swarms using virtual structures. *IEEE Trans. Robot.* **2018**, *34*, 916–923. [CrossRef]
10. Kahn, A.; Marzat, J.; Piet-Lahanier, H. Formation flying control via elliptical virtual structure. In Proceedings of the IEEE International Conference on Networking, Sensing and Control, Paris-Evry, France, 10–12 April 2013; pp. 158–163. [CrossRef]
11. Lafferriere, G.; Williams, A.; Caughman, J.; Veerman, J.J.P. Decentralized control of vehicle formations. *Syst. Control Lett.* **2005**, *54*, 899–910. [CrossRef]
12. Oh, K.K.; Park, M.C.; Ahn, H.S. A survey of multi-agent formation control. *Automatica* **2015**, *53*, 424–440. [CrossRef]

13. Fathian, K.; Rachinskii, D.I.; Spong, M.W.; Summers, T.H.; Gans, N.R. Distributed formation control via mixed barycentric coordinate and distance-based approach. In Proceedings of the American Control Conference, Philadelphia, PA, USA, 10–12 July 2019; pp. 51–58. [CrossRef]
14. Cheah, C.C.; Hou, S.P.; Slotine, J.J.E. Region-based shape control for a swarm of robots. *Automatica* **2009**, *45*, 2406–2411. [CrossRef]
15. Strandburg-Peshkin, A.; Twomey, C.R.; Bode, N.W.; Kao, A.B.; Katz, Y.; Ioannou, C.C.; Rosenthal, S.B.; Torney, C.J.; Wu, H.S.; Levin, S.A.; et al. Visual sensory networks and effective information transfer in animal groups. *Curr. Biol.* **2013**, *23*, R709–R711. [CrossRef]
16. Kolpas, A.; Busch, M.; Li, H.; Couzin, I.D.; Petzold, L.; Moehlis, J. How the Spatial Position of Individuals Affects Their Influence on Swarms: A Numerical Comparison of Two Popular Swarm Dynamics Models. *PLoS ONE* **2013**, *8*, e58525. [CrossRef] [PubMed]
17. Cortes, J.; Martinez, S.; Karatas, T.; Bullo, F. Coverage control for mobile sensing networks. *IEEE Trans. Robot. Autom.* **2004**, *20*, 243–255. [CrossRef]
18. Guruprasad, K.R.; Dasgupta, P. Distributed Voronoi partitioning for multi-robot systems with limited range sensors. In Proceedings of the IEEE/RSJ International Conference on Intelligent Robots and Systems, Vilamoura, Portugal, 7–12 October 2012; pp. 3546–3552. [CrossRef]
19. Hatleskog, J.; Olaru, S.; Hovd, M. Voronoi-based deployment of multi-agent systems. In Proceedings of the IEEE Conference on Decision and Control, Miami, FL, USA, 17–19 December 2018; pp. 5403–5408. [CrossRef]
20. Zhou, Z.; Zhang, W.; Ding, J.; Huang, H.; Stipanović, D.M.; Tomlin, C.J. Cooperative pursuit with Voronoi partitions. *Automatica* **2016**, *72*, 64–72. [CrossRef]
21. Kouzeghar, M.; Song, Y.; Meghjani, M.; Bouffanais, R. Multi-Target Pursuit by a Decentralized Heterogeneous UAV Swarm using Deep Multi-Agent Reinforcement Learning. In Proceedings of the IEEE International Conference on Robotics and Automation, London, UK, 29 May–2 June 2023.
22. Gui, J.; Yu, T.; Deng, B.; Zhu, X.; Yao, W. Decentralized Multi-UAV Cooperative Exploration Using Dynamic Centroid-Based Area Partition. *Drones* **2023**, *7*, 337. [CrossRef]
23. Lindhé, M.; Ogren, P.; Johansson, K.H. Flocking with obstacle avoidance: A new distributed coordination algorithm based on Voronoi partitions. In Proceedings of the IEEE International Conference on Robotics and Automation, Barcelona, Spain, 18–22 April 2005; pp. 1785–1790. [CrossRef]
24. Lindhé, M.; Johansson, K.H. A Formation Control Algorithm using Voronoi Regions. In *Taming Heterogeneity and Complexity of Embedded Control*; John Wiley & Sons, Ltd.: Hoboken, NJ, USA, 2013; Chapter 24; pp. 419–434. [CrossRef]
25. Jiang, Q. An improved algorithm for coordination control of multi-agent system based on r-limited Voronoi partitions. In Proceedings of the IEEE International Conference on Automation Science and Engineering, Shanghai, China, 7–10 October 2006; pp. 667–671. [CrossRef]
26. Bertrand, S.; Sarras, I.; Eudes, A.; Marzat, J. Voronoi-based Geometric Distributed Fleet Control of a Multi-Robot System. In Proceedings of the 16th International Conference on Control, Automation, Robotics and Vision (ICARCV), Shenzhen, China, 13–15 December 2020; pp. 85–91. [CrossRef]
27. Chakravarthy, A.; Ghose, D. Obstacle avoidance in a dynamic environment: A collision cone approach. *IEEE Trans. Syst. Man, Cybern. Part A Syst. Hmans* **1998**, *28*, 562–574 [CrossRef]
28. Sunkara, V.; Chakravarthy, A.; Ghose, D. Collision Avoidance of Arbitrarily Shaped Deforming Objects Using Collision Cones. *IEEE Robot. Autom. Lett.* **2019**, *4*, 2156–2163. [CrossRef]
29. Hu, J.; Wang, M.; Zhao, C.; Pan, Q.; Du, C. Formation control and collision avoidance for multi-UAV systems based on Voronoi partition. *Sci. China Technol. Sci.* **2020**, *63*, 65–72. [CrossRef]
30. Moniruzzaman, M.D.; Rassau, A.; Chai, D.; Islam, S.M.S. Teleoperation methods and enhancement techniques for mobile robots: A comprehensive survey. *Robot. Auton. Syst.* **2022**, *150*, 103973. [CrossRef]
31. Aggravi, M.; Sirignano, G.; Giordano, P.R.; Pacchierotti, C. Decentralized Control of a Heterogeneous Human–Robot Team for Exploration and Patrolling. *IEEE Trans. Autom. Sci. Eng.* **2021**, *19*, 3109–3125. [CrossRef]
32. Sydorchuk, A. The Boost Polygon Voronoi Extensions. 2013. Available online: https://www.boost.org/doc/libs/1_60_0/libs/polygon/doc/voronoi_main.htm (accessed on 3 July 2023).
33. Pereyra, E.; Araguás, G.; Kulich, M. Path planning for a formation of mobile robots with split and merge. In Proceedings of the 4th International Conference on Modelling and Simulation for Autonomous Systems, Rome, Italy, 24–26 October 2017; Springer: Berlin/Heidelberg, Germany, 2018; pp. 59–71. [CrossRef]
34. Salvado, J.; Mansouri, M.; Pecora, F. Combining multi-robot motion planning and goal allocation using roadmaps. In Proceedings of the IEEE International Conference on Robotics and Automation (ICRA), Xi'an, China, 30 May–5 June 2021; pp. 10016–10022. [CrossRef]
35. Fortune, S. A sweepline algorithm for Voronoi diagrams. In Proceedings of the Second Annual Symposium on Computational Geometry, Yorktown Heights, NY, USA, 2–4 June 1986; pp. 313–322.
36. Bak, M.; Poulsen, N.K.; Ravn, O. *Path Following Mobile Robot in the Presence of Velocity Constraints*; Technical Report, Informatics and Mathematical Modelling; Technical University of Denmark: Lyngby, Denmark, 2001; Available online: http://www2.compute.dtu.dk/pubdb/pubs/189-full.html (accessed on 19 September 2023).
37. Guo, K.; Li, X.; Xie, L. Ultra-wideband and Odometry-Based Cooperative Relative Localization with Application to Multi-UAV Formation Control. *IEEE Trans. Cybern.* **2020**, *50*, 2590–2603. [CrossRef]

38. Sanfourche, M.; Vittori, V.; Le Besnerais, G. eVO: A realtime embedded stereo odometry for MAV applications. In Proceedings of the IEEE/RSJ IROS, Tokyo, Japan, 3–7 November 2013; pp. 2107–2114. [CrossRef]
39. Olson, E. AprilTag: A robust and flexible visual fiducial system. In Proceedings of the IEEE international Conference on Robotics and Automation, Shanghai, China, 9–13 May 2011; pp. 3400–3407. [CrossRef]
40. Tong, P.; Yang, X.; Yang, Y.; Liu, W.; Wu, P. Multi-UAV Collaborative Absolute Vision Positioning and Navigation: A Survey and Discussion. *Drones* **2023**, *7*, 261. [CrossRef]
41. Geiger, A.; Roser, M.; Urtasun, R. Efficient large-scale stereo matching. In Proceedings of the 10th Asian Conference on Computer Vision (ACCV), Queenstown, New Zealand, 8–12 November 2010; pp. 25–38. [CrossRef]

MDPI AG
Grosspeteranlage 5
4052 Basel
Switzerland
Tel.: +41 61 683 77 34

Drones Editorial Office
E-mail: drones@mdpi.com
www.mdpi.com/journal/drones

www.ingramcontent.com/pod-product-compliance
Lightning Source LLC
Chambersburg PA
CBHW040800250926
51850CB00007BA/1046